PEDOLOGY
A Systematic Approach to
Soil Science

PEDOLOGY

A SYSTEMATIC APPROACH TO SOIL SCIENCE

EWART ADSIL <u>FITZPATRICK</u>

D.I.C.T.A. (Diploma of the Imperial College of Agriculture) Ph.D.

1971

OLIVER & BOYD

OLIVER & BOYD

Tweeddale Court, 14 High St., Edinburgh, EH1 1YL
(A Division of Longman Group Ltd.)

ISBN: Paperback 0 05 002335 7
Hardback 0 05 002336 5

Photoset in Malta by St Paul's Press Ltd

Printed in Great Britain by T. & A. Constable, Ltd, Edinburgh

PREFACE

Over the past few years the need for a text on pedology has become more and more urgent. This need is particularly strongly felt among undergraduate students who have to depend almost entirely on lecture notes, a few general references and the very dilute treatment found in some general books on soil science. In order to overcome this deficiency a number of ideas were considered one of which was to provide detailed lecture notes but it soon became apparent that a fuller and more comprehensive treatment was necessary, especially the need to use large amounts of illustrative material.

It was clear at the outset that it would be necessary to use single sets of terminology in order to be consistent throughout. This applies both to the soil and other phenomena, particularly climate – the desire being to produce an integrated whole. It is appreciated that some readers may not agree with the soil terminology which I have used but the intention has been to present the information in such a manner that except in a number of specific cases the text can be read and understood without the reader being committed to the jargon. Certainly the first five chapters are virtually free of jargon.

Some of the material contained in this book was either given specially or taken from the work of other authors and for this I am extremely grateful. It is a pleasure to acknowledge this valuable help below. Except where stated the photographs have been produced by the author who is also responsible for the line drawings, most of which are original though a small number have been adapted from the work of others. It was felt that a uniform level of artwork was important; in addition it was necessary to redraw the diagrams of others to incorporate new symbols.

ACKNOWLEDGEMENTS

I should like to thank those persons who have helped in the preparation of the book through their suggestions or by providing material in one form or another. To Dr J.W. Parsons, Dr J. Tinsley and particularly Mr C.P. Burnham, Dr M.N. Court and Mr K. Simpson my grateful thanks for reading and commenting on the material at various stages.

To the following also for their courtesy and permission to adapt or use these illustrations I am much obliged: Fig. 29, by permission from the *Annals* of the Association of American Geographers, Volume 40, 1950; John Bartholomew and Son Ltd for their equal area projection used in Figs; 80, 85, 88, 97, 98, 100, 109, 112, 115, 117 and 137; The British Museum for Figs. 8 and 12; Canadian Department of Agriculture for Plate IVd; E.G. Hallsworth for Figs. 45 and 46; J.D. McCraw for Plate Ib; K.H. Northcote for Figs. 61, 119 and 120 and Plates IIC and IVC; I.M. Scott for Plate Vd; and the United States Department of Agriculture for Fig. 53.

CONTENTS

CONTENTS

*Most of the classes are discussed under the following sub-headings:
    Derivation of name. Approximate equivalent names.
    General characteristics:
        Morphology. Analytical data. Genesis.
    Principal variations in the properties of the class:
        Parent material. Climate. Vegetation. Topography. Age. Pedounit.
    Distribution.
    Utilisation.

TABLES

INTRODUCTION

HISTORICAL

The development of soil studies naturally falls into two stages; the first stretches back many centuries and references to agricultural practices are found frequently in early religious literature. Extensive discussions about soils occur in the writings of the Greeks, particularly Plato who witnessed the erosion of the hills around Athens and warned of its dangers. In Roman times there was a considerable literature about agriculture, but it was merely a collection of facts about the systems in use up to that time and rather speculative interpretations about the results but surprisingly some of the conclusions have been sustained subsequently.

The second stage is relatively recent having developed only during the last two centuries; but it has a sound foundation being based on experimentation and the application of the scientific method. This stage started with the work of Théodore de Saussure (1804) who founded the quantitative experimental method and applied it to the oxygen and carbon dioxide relationships in plants. He was able to show that carbon dioxide is absorbed and oxygen released by plants. This work ushered in a completely new approach to the problem of plant nutrition and plant physiology to which the outstanding contributors included Liebig and more especially Boussingault. At first sight these studies of de Saussure on plants might appear to be unconnected with soil science but in the early days there was considerable controversy about the origin of the material in plants and the contributions made by the soil, therefore his discovery was a fundamental contribution to the study of soil-plant relationships.

In the nineteenth century Dokuchaev (1883) produced his classical work on the chernozems of Russia and recognised for the first time that soils are made up of several layers. In addition, he and his students conducted surveys and discovered that the nature and properties of soils vary with their environmental factors, particularly climate and vegetation. These contributions initiated a major revolution in soil science for, as indicated above most of the previous work was devoted to soil fertility and plant nutrition. Thus quite suddenly soils became objects for academic study and the investigation of soils as a natural phenomenon emerged as a separate scientific discipline known as 'Pedology' with its own set of principles, concepts and methodology. These discoveries in Russia probably took place as a result of the enormous size of the country and its span over many different types of climate and vegetation. At the same time in Germany, Müller, (1887), and others were making detailed descriptions and investigations of podzols in a manner similar to those being conducted in Russia. Principally because of language difficulties, knowledge of the Russian work spread very slowly to the rest of the world but wherever it reached, the impact was great and acceptance relatively quick, with the result that now there is quite a body of data about most of the major soils of the world.

Ideally, pedology is the study of soils as naturally occurring phenomena, taking into account their composition, distribution and method of formation. However such studies cannot be separated entirely from other ramifications of soil science, including some fairly detailed aspects of soil chemistry, soil physics, soil microbiology and land utilisation.

Unfortunately pedology is still somewhat fragmented and lacking unification, largely because of its relationship with so many other disciplines as is shown in Fig. 1. Most sciences contribute to the study of soils, whereas the fundamental sciences make a unilateral contribution, the others enter into reciprocal relationships, the greatest and most complete being between pedology and the applied sciences of agriculture and forestry. In most disciplines unification comes through a common purpose or a generally accepted system of classification. Both are lacking in pedology, particularly the latter, but it has engaged the attention of most pedologists and in this book a new approach is attempted which it is hoped will go some way towards meeting these deficiencies. Also because of the association of pedology with so many sciences the terminology is large and diverse, which requires readers to have a wide knowledge of many disciplines or otherwise numerous definitions must be given in the text. Neither of these alternatives is realistic or satisfactory consequently a glossary of terms is given which commences on p. 284.

PURPOSE OF THE BOOK

At present there are a number of text books covering various aspects of soil science; some include chapters on pedology, but only a few are devoted entirely to this subject. There are many reasons for this situation. Firstly the relevant information required to produce such a text has accumulated only during the last two or three decades following the conclusion of a significant number of purely pedological investigations conducted in many different parts of the world. The second reason concerns the general lack of agreement among pedologists about methods of nomenclature and classification of soils. This is probably the foremost factor that has caused many potential authors to refrain from making the effort. A further reason is the subject's youth and its rapid growth, creating the possibility that any text-book could quickly become out of date, irrespective of the care and effort spent over its preparation.

This book has been prepared as an aid to teaching. It is produced primarily for undergraduate students taking pedology as a part of their general training in soil science, including those in agriculture, forestry, geography and soil science itself. Postgraduate students having advanced training in soil science, but specialising in a branch other than pedology, should also find it useful. It must be emphasised that this is not a treatise for the specialist pedologist. The intention is to present a modest level of information about the fundamental concepts in pedology, the factors of soil formation and the characteristic soils of the world, their relationships to each other as well as to their factors of formation. Readers in specific parts of the world may find many deficiencies about their local area. It is not proposed to give detailed information about every possible pedological situation. Any reader requiring specific local knowledge is referred to the regional soil survey reports and memoirs.

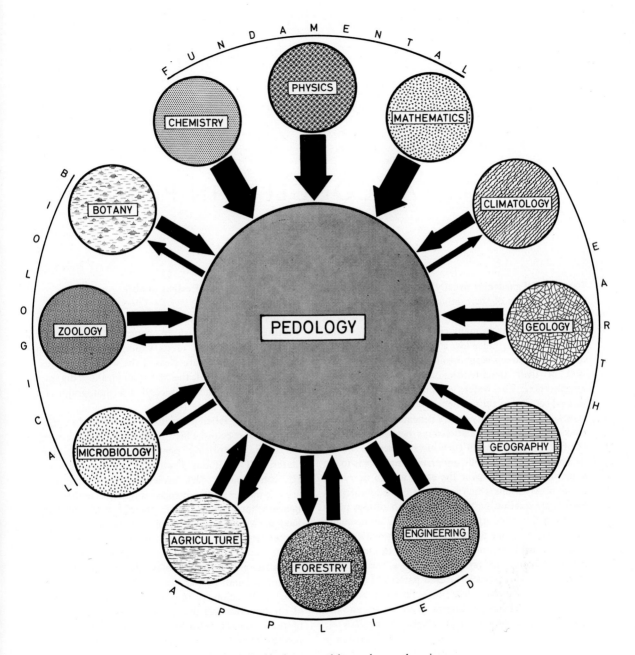

FIG. 1.   The relationships between pedology and some other sciences.

1

FUNDAMENTAL CONCEPTS IN PEDOLOGY

THE SOIL PROFILE

Farmers and gardeners usually regard soil as the upper few centimetres of the earth's crust which are either cultivated or permeated by plant roots. This concept, however adequate in many cases, is at the same time limited and does not allow a full appreciation of the difference between various soils, nor a full comprehension of their potentialities and the problems of amelioration. The wider, comprehensive and fundamental concept of the pedologist recognises the surface layers, as well as many others beneath and includes a consideration of the relationships between soils and the factors and forces of their environment.

The first step towards understanding soils is to dig a pit into the surface of the earth and to carry out visual observations which are supplemented sometimes by simple qualitative tests. The depth of the pit is determined by the nature of the soil itself, but normally this varies from one to three metres below which is the relatively unaltered material generally termed *parent material*. When the pit is dug and a vertical face exposed, a characteristic layered pattern is revealed, in which the different colours of the layers may be seen. Each individual layer is known as an *horizon* and the set of layers in a single pit is called a *soil profile*.

In some cases the contrast between horizons is dramatic and self-evident, while in others it is very subtle. An insight into these profile characteristics is gained by considering the two examples given below. In order to facilitate the discussion of these two profiles it is necessary to introduce a limited number of names and symbols for horizons and parent materials. The full lists are presented and discussed in Chapters 2 and 5.

A Podzol profile

This soil has been chosen as the first example because it displays sharp contrasts between the horizons and because of its widespread distribution in the humid temperate areas of the world. Two illustrations are provided, the first is an idealised diagram given in Fig. 2 and the second is a photograph of a profile shown in Plate IIId.

The freshly fallen plant litter, Lt, forms the uppermost horizon below which is the dark brown fermenton, Fm, comprising partially decomposed plant material. This grades downwards into the humifon, Hf, composed of dark brown or black amorphous organic matter in an advanced stage of decomposition. Beneath the organic layers there is the dark grey modon, Mo, which is a mixture of black organic matter and light coloured mineral grains, mainly quartz and felspars. This horizon is underlain by the very pale, grey or white zolon, Zo, composed mainly of quartz and felspar; in many cases dominantly quartz. The zolon is formed by weathering (hydrolysis) of the minerals and the removal by percolating water of the colouring substances, which are mainly compounds of iron. The weathering process is aided considerably by the acid decomposition products released from the organic matter at the surface. Much of the material removed from the zolon accumulates immediately below to form the husesquon, Hs, of very dark brown colour. Next, there is the pale cream acid drift, AST, showing little sign of alteration, and identical to the material in which the upper modon, zolon and husesquon have developed.

In the diagram, acid rock, AK, is shown underlying the entire profile, since it is quite common for podzols to develop in superficial deposits that overlie solid rock. These superficial deposits include alluvium, glacial drift and wind blown material.

A Krasnozem profile

This is an example of a soil which has developed under humid tropical conditions through progressive *in situ* weathering of consolidated acid rock. It also illustrates those soils which exhibit distinct, but subtle changes from one horizon to the other. A photograph of a similar profile is given on Plate IIIa, while Fig. 3 is an idealised diagram showing the profile with a gradation through weathered rock into fresh rock. At the surface there is a thin loose litter, Lt, of leaves and twigs containing numerous earthworm casts. Beneath the litter is the greyish-red tannon, Tn, which is a mixture of organic and mineral material brought about mainly by earthworms and termites. The tannon grades fairly sharply into the bright red krasnon, Ks, which has a high content of clay and varying amounts of iron concretions. This horizon is formed by the complete hydrolysis of the primary

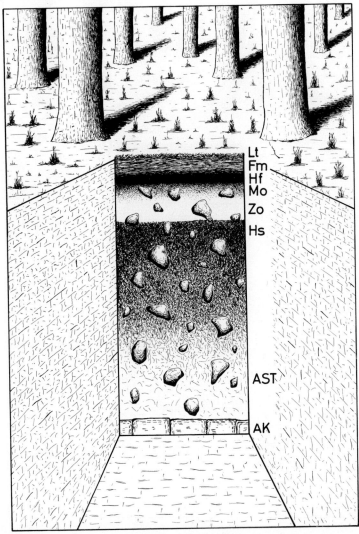

FIG. 2. A podzol profile.

silicates followed by the oxidation, hydration and reorganisation of the decomposition products. The krasnon, in turn, grades into the easily recognisable chemically weathered acid rock, AKw, with its high content of unweathered primary silicates. This grades with depth into the solid rock, AK, through a zone containing rounded unweathered areas of rock known as *core stones* which get larger with depth. Finally at the base of the profile there are the initial stages of weathering along the faces of the joint blocks.

These two examples serve to illustrate the wide variations that exist between soils. They also demon-strate that soil formation takes place through the interaction of certain specific factors and therefore that soils are largely the products of their environment. There are five factors of soil formation viz. parent material, climate, organisms, topography and time. Briefly, the parent material or mineral matter is altered and differentiated through physical, chemical and biological processes into horizons. Alteration takes place mainly through the interplay of moisture and temperature which are the two main components of climate. Some organisms contribute organic matter to soils while others aid organic matter decomposition

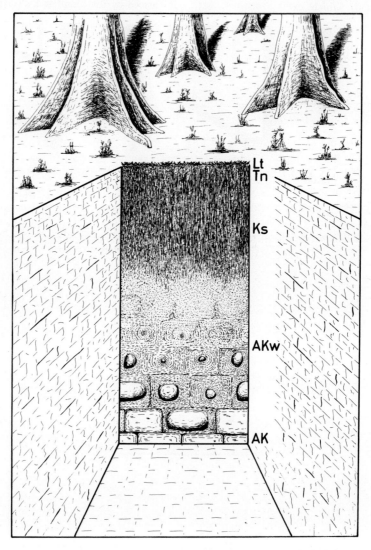

Fig. 3.   A krasnozem profile.

and development of horizons, which occur very gradually and occupy a span of time. Topography is not illustrated above, but is fully considered in Chapter 2.

SOIL – A THREE DIMENSIONAL CONTINUUM

The profile is simply a two dimensional section through soil which actually extends laterally in all directions over the surface of the earth forming a three dimensional continuum. This concept is of considerable fundamental importance since false ideas can develop about soils when profiles only are examined. Repeated observations have shown that the constituent horizons in a soil do not remain uniform throughout their lateral extent but exhibit changes in their degree of development as well as in their intrinsic properties.

Figs. 4 and 5 illustrate the three-dimensional character of a soil which shows various types of change in the thickness of the horizons. In these two diagrams the horizons at the surface are fairly uniform in thickness, but immediately below they have wavy boundaries. In the middle and lower parts of the soil the horizons are very irregular and change in an apparently haphazard manner. This overall pattern of change is very common, occurring in many different types of soils throughout the world. Also the tops and the bottoms of the horizons may have different patterns of change. When conducting a long traverse, one or more horizons gradually change laterally into other horizons. These changes are often shown by variations in colour which are accompanied by variations in their physical and chemical properties which may be of greater importance than the changes in colour. Horizons with characteristic and distinctive properties are known as *standard* horizons, but as horizons change from one to the other there is a stage or *intergrade* having properties of both horizons. The distance through which individual horizons change is seldom the same; whereas one or more may change completely over a given distance others continue unaltered and have greater spatial distribution. The dissimilar rate of change in horizons is a further important soil characteristic.

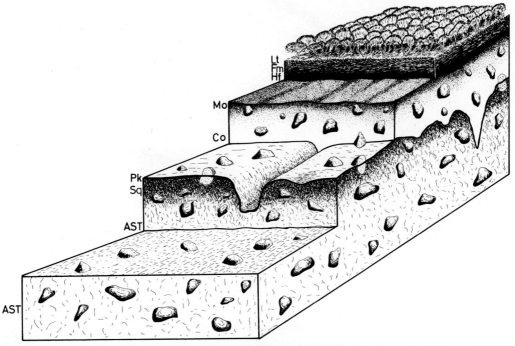

FIG. 4. Soil – a three dimensional continuum.

An attempt is made in Fig. 6 to demonstrate how horizons intergrade. Each of the six symbols in the diagram represents a different standard horizon. The uniform lowest horizon represented by solid hexagons is the relatively unaltered underlying parent material. Starting at Stage III the diagram shows a sequence of three horizons, two of which change laterally into other horizons, thus forming new sequences – each indicated by a Roman numeral. To the right of Stage III the solid circles change into open circles while the open squares change into solid squares. In both cases the rate of change is the same. As a result, Stage IV is composed of two intergrade horizons and the underlying material, but at Stage V the intergrades have changed into two new standard horizons forming a new sequence. To the left of Stage III the solid circles remain constant through Stages II and I but the open squares change into solid triangles. Consequently Stage II has a standard horizon overlying an intergrade which in turn overlies the relatively unaltered material, but Stage I again has two standard horizons. It should be noted that one of the standard horizons – the solid circle – continues through three stages and immediately overlies three different horizons.

In addition to the intergrades caused by lateral variations, many are formed by vertical changes from one horizon to the other. Such a situation is shown at Stage VI where there is an intergrade horizon of

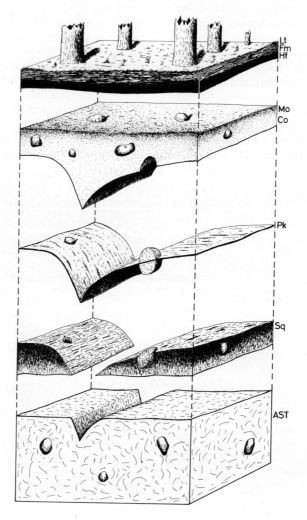

Fig. 5.　Exploded soil showing variations in the thickness of horizons.

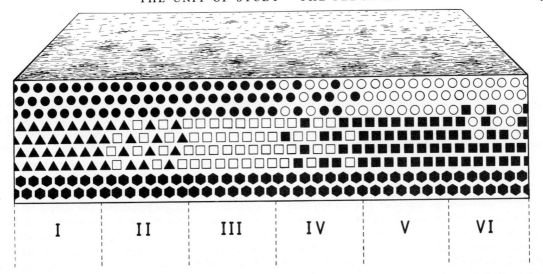

FIG. 6.   Intergrading of horizons.

solid squares and open circles situated between the open circles above and solid squares beneath. The continuous spatial variation of soils and their constituent horizons is not a unique feature of pedology, but it is of such immense importance that it is regarded as a fundamental and distinctive characteristic of soils.

SOIL – A CONTINUUM IN TIME

The soil surface of the earth is not static; for example, many soils are constantly losing ions in drainage water or through uptake by plants. Also the temperature at any one point has both a certain diurnal and annual cycle. The result of these and other continuous and progressive changes is a complex, dynamic system forming a continuum with three tangible dimensions as well as changing with time.

These progressive changes in time may cause profound alteration in the character of the horizons which steadily change and intergrade from one to the other. Thus standard and intergrade horizons can develop through time as well as laterally in space. It follows therefore that soils have four dimensions and form a space-time continuum.

THE UNIT OF STUDY – THE PEDOUNIT

Phenomena such as soils that change continuously in space and time pose a number of practical and theoretical problems with regard to their study, designation and classification.

Although the inspection of a profile is usually the first step towards categorising a soil, profiles in themselves are not the units of complete study since they have only two dimensions, – breadth and depth. Thus profiles can be studied only by visual examination, whereas, it is the soil continuum that must be investigated.

The research carried out on soils aimed at their quantitative characterisation is extremely varied and is determined primarily by the requirements of individual workers. The assessment of a tangible property such as the total amount of iron, is usually conducted in the laboratory on samples collected at points in the continuum. On the other hand, the measurements of moisture or temperature fluctuations are more satisfactorily carried out by inserting suitable probes into the soil and taking continuous or intermittent readings. The data obtained by these techniques yield information about the intrinsic properties of the continuum at any one point and also allow inferences to be made about the rate and type of change from point to point.

The unit of study in pedology, or the *pedounit* can now be defined as follows: *The pedounit is a selected column of soil containing sufficient material in each horizon for adequate laboratory characterisation.*

The concept of the pedounit is similar to the concept of the pedon (Simonson and Gardiner 1960) but the term pedounit seems more appropriate. Further the -on ending is reserved here for the names of horizons.

Fig. 7 illustrates the pedounit and various other units of study. The pedounit occupies the volume of soil extending below ABCD and the soil profile is the exposed section ADEF. A monolith is shown extending below G and there are two cores H and I. Ideally it is desirable to have cores with flat faces, but it is more convenient to collect circular columns. The study of cores or monoliths is a useful alternative method of studying soils, having the advantage that it allows investigations to be conducted in the relative comfort of the laboratory. However this method has certain disadvantages since the study is conducted away from the natural environment of the soil. Wherever possible,

it is imperative to study soils in their normal surroundings, which play a major role in their formation. The *purely* pedological characterisation of the soil is normally performed in the laboratory on bulk or undisturbed samples obtained from the pedounit. Duplicate or triplicate samples are collected from each horizon in a vertical sequence which ideally forms a column. The size of the individual horizon sample should always be minimal because many small samples are statistically more valid than a few large ones. However their precise amount is determined by the nature of the soil itself and by the requirements of the quantitative analytical techniques used to assess the various soil properties. The sample can be as small as a few grams, but if information is required about the proportion of stones, several kilograms may be needed. However a smaller sample may have to suffice if the horizon is very thin. A rectangular column with surface dimensions of about 100 × 30 cm and extending down to include the relatively unaltered underlying material is the usual size of the pedounit being large enough to give triplicate samples. This is not to be regarded as a fixed amount since the volume of soil examined will vary with the nature of the soil.

The selection of sample points should be based on some system of randomisation but it is normally made

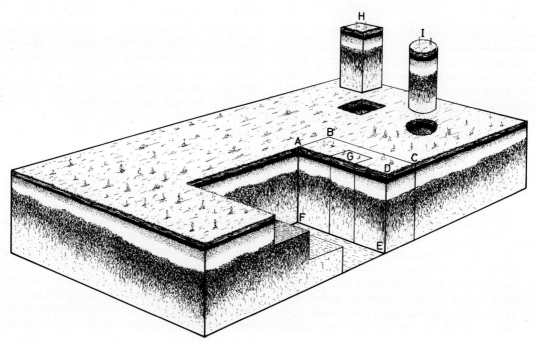

FIG. 7. The pedounit.

subjectively, being chosen at places to represent areas of apparent uniformity as well as to determine the type and rate of change within the continuum. The number of points examined is increased as the rate of change increases.

DEFINITION OF SOIL

After considering the foregoing fundamental concepts, it should be possible to offer a definition of soil. However, this is not an easy task since a number of divergent attitudes towards soil are adopted by different workers. Some hold the extreme view that soils are substrates for plant growth while others regard soils as natural phenomena justifying academic study, without any consideration of the applicability and practicability of the results that are obtained. It is virtually impossible to resolve such widely dissimilar views and to formulate a single simple definition. Perhaps the attempt made below should be regarded as a short description rather than as a definition: *Soil is the space-time continuum forming the upper part of the earth's crust.*

This definition like most others may have many deficiencies. It includes as soil, the sands of the deserts and the bare rock surfaces of mountainous areas, thereby excluding the presence of plants and a stable surface as essential for soil. If an allowance is made for these latter two features then many stone walls with mossy surfaces would be included since they are stable and are covered by vegetation. Ultimately it may be necessary to use a nonsense definition such as *Soil is anything so called by a competent authority.*

2

FACTORS OF SOIL FORMATION

Dokuchaev firmly established that soils develop as a result of the interplay of the five factors, parent material, climate, organisms, topography and time; the first four are the tangible factors interacting through time to create a number of spcific processes leading to horizon differentiation and soil formation. Some workers, particularly Jenny (1941), have tried to demonstrate, quite unconvincingly, that these factors are independent variables, i.e. each of them can change and vary from place to place without the influence of any of the others. Only time can be regarded as an independent variable, the other four depend to a greater or lesser extent upon each other, upon the soil itself or upon some other factor. For example, it is now generally accepted that vegetation is a function of climate which is itself a function of wind currents, latitude, proximity to water and elevation. In addition, some parent materials such as glacial deposits, result from geological processes which are greatly influenced by climate but some may form as a result of tectonic phenomena and are therefore independent of the other factors of soil formation.

Many attempts have been made to show that some factors are more important than others and therefore play a major role in soil formation. Such efforts are a little unrealistic since each factor is absolutely essential and none can be considered more important than any other, but locally one factor may exert a particularly strong influence.

Below are presented the more important characteristics of these factors as they are related to soil formation; this is followed in the next two chapters by discussions of the principal processes and properties of the soil system. Very little is said about how each factor influences soil processes and properties. As far as possible the discussions are confined to factual statements about the factors and their influence generally.

PARENT MATERIAL

Jenny (1941) defines parent material as 'the initial state of the soil system'. The precision of this definition cannot be questioned but most attempts to determine the initial stages of soils are fraught with difficulties, for in a number of cases the original character of the material has been changed so drastically by a long period of pedogenesis that it is possible only to speculate about the full composition of its pristine state. Sometimes, where soil formation has proceeded for a short time, the nature of the original material can be ascertained fairly accurately, but even in these cases some deductions may be necessary, particularly about soluble substances which are easily lost or redistributed in the soil system.

Usually the relatively unaltered underlying material is similar to the material in which the overlying horizons have developed, but this is not always true, especially in stratified sediments and folded metamorphic rocks where one stratum may have an entirely different chemical and mineralogical composition and structure from the material below or when there are thin superficial deposits overlying rock. In spite of these difficulties, which are concerned primarily with the evaluation of the contribution of the parent material to the soil, it is possible to state the composition of parent materials without referring them to any particular soil or set of pedogenetic processes.

Parent materials are made up of mineral material or organic matter or a mixture of both. The organic matter is usually composed predominantly of unconsolidated, dead and decaying plant remains while the mineral material which is the most widespread type of parent material contains a large number of different rock-forming minerals and can be in either a consolidated or unconsolidated state. The consolidated mineral material includes rocks like granite, basalt and conglomerate and the unconsolidated material comprises a wide range of superficial deposits of which glacial drift and loess are two important representatives. It may be of interest to know the origin of the parent material be it igneous rock or glacial drift, but the chemical and mineralogical composition are more important properties; these are responsible largely for the course of soil formation, and the resulting chemical and physical composition of the soil including the secondary products of weathering. Parent materials also contribute to soil formation through their permeability and specific surface area.

Structure of minerals

A full discussion of the rock-forming minerals is beyond the scope of this book, nevertheless it is necessary to outline the structure of the main types in order to understand the reasons for their behaviour when subjected to various processes. Also it is of paramount importance to understand the composition and structure of the sheet silicates which include the all-important clay minerals. Minerals can be divided into non-silicates and silicates.

Non-silicates

As seen from Table 3, (page 22), this group contains oxides, hydroxides, sulphates, chlorides, carbonates and phosphates. All have relatively simple structures but they vary widely in their solubility and resistance to decomposition.

Silicates

Generally these minerals have very complex structures in which the fundamental unit is the silicon-oxygen tetrahedron. This is composed of a central silicon ion surrounded by four closely-packed and equally-spaced oxygen ions. The whole forms a pyramidal structure the base of which is composed of three of the oxygen ions with the fourth forming the apex (Figs. 8 and 9). The four positive charges of Si^{++++} are balanced by four negative charges from the four oxygen ions O^{--}, one from each ion, thus each tetrahedron has four negative charges. The tetrahedra themselves are linked together in a number of different ways forming a variety of distinctive and characteristic patterns which form the basis of the classification of these minerals. Furthermore, the type of linkage determines the crystal structure as well at its resistance to weathering. A significant variation in the tetrahedral structure is the substitution of Al for Si. This is known as *isomorphous replacement* and when it occurs

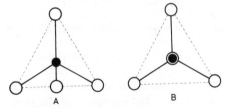

FIG. 9. Diagrammatic representation of a silicon-oxygen tetrahedron A: vertical; B: plan.

FIG. 8. Models of silicon-oxygen tetrahedron (*left*) complete model; (*right*) model with apical oxygen ion removed to show the smaller central silicon ion.

it produces an imbalance in the charges within the structure which is satisfied by the introduction of cations such as Na, K, Ca and Mg. Silicates are divided broadly into framework silicates, chain silicates, ortho- and ring silicates and sheet silicates.

Framework silicates

These minerals are composed of tetrahedra linked through their corners into a continuous three dimensional structure. The simplest member of this class is quartz which is a colourless mineral composed entirely of silicon-oxygen tetrahedra, so that the resulting structure has twice as many oxygen ions as silicon ions and the formula can be written $(SiO_2)_n$. Such minerals with a continuous structure are extremely hard and resistant to weathering. The other main group of framework silicates is the colourless or white felspars which also have a three dimensional arrangement of tetrahedra, but in this case there is a considerable amount of isomorphous replacement and therefore they contain a high proportion of basic cations. In many felspars about one quarter of the silicon ions are replaced by aluminium ions and when the extra charges are balanced by sodium the new formula becomes $NaAlSi_3O_8$.

There are three principal types of felspars, known respectively as sodium, potassium and calcium felspars which form the ternary system, $NaAlSi_3O_8 - KAlSi_3O_8 - CaAl_2Si_2O_8$. The members of the first series, Na to K, are known as alkali felspars and those between Na and Ca are the plagioclase felspars. Generally the alkali felspars contain up to 10% of calcium 'molecule', and in a similar way the plagioclase felspars contain up to 10% of the potassium 'molecule.'

Chain silicates

There are two main divisions within this group — the pyroxenes and the amphiboles. The former, of which enstatite and hypersthene are good examples, are composed of tetrahedra linked to each other by sharing two of the three basal corners to form continuous chains (Fig. 10). These chains have various dispositions with respect to each other and are linked laterally by various cations such as Ca, Mg, Fe, Na and Al, thus forming a variety of structures.

The amphiboles exemplified by green hornblende, have chains that are double the width of those of the pyroxenes and can be regarded as a single band of tetrahedra arranged in a hexagonal pattern. The bands have various dispositions with respect to each other and like the pyroxenes are linked by Ca, Mg, Fe, Na

and Al ions (Fig. 11).

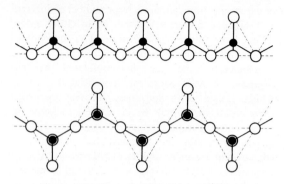

FIG. 10. Diagrammatic representation of a pyroxene chain, (a) vertical view and (b) plan view. The points of linkage are through the unsatisfied negative charges on the oxygen ions.

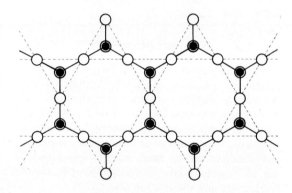

FIG. 11. Diagrammatic representation of an amphibole chain, plan view. The vertical view is identical to that of the pyroxene chain. The points of linkage are through the unsatisfied negative charges on the oxygen ions.

Ortho- and ring silicates

Probably the greatest variations in the structure of primary minerals occur within this group which includes the olivines, zircon, titanite and garnet. It is most convenient to start with the olivines because of their relatively simple structure with separate silicon-oxygen tetrahedra arranged in sheets and linked by Mg and/or Fe ions. Olivines also differ from many of the other members of this group by being more common and forming a frequent constituent of many basic and ultrabasic igneous rocks. The other important mineral in which the linkages are provided by a cation is zircon in which each zirconium ion is surrounded by eight oxygen ions resulting in a very strong structure. Contrasted with these two are garnet, sillimanite and kyanite which have various complex linkages provided by aluminium octahedra as described on page 16. Thus the structure of the members of this group varies from relatively simple to highly complex, the latter being found in the majority.

Sheet silicates

This group includes minerals such as muscovite, biotite, pyrophyllite, talc and the secondary clay minerals. It is better to consider these two groups separately but they have a number of features in common and both can be regarded as being composed of various combinations of three basic sheets namely the silicon tetrahedral sheet, the gibbsite sheet and the brucite sheet.

Silicon tetrahedral sheet. This is composed of silicon-oxygen tetrahedra linked together in a hexagonal arrangement with the three basal oxygen ions of each tetrahedron in the same plane and all the apical oxygen ions in a second plane. Thus the silicon sheet is a hexagonal planar pattern of silicon-oxygen tetrahedra (Figs. 12 and 13).

Gibbsite sheet. This sheet has as its basic unit, aluminium-hydroxyl octahedra in which each aluminium ion is surrounded by six closely packed hydroxyl groups (Fig. 14) in such a way that there are two planes of hydroxyl ions with a third plane containing

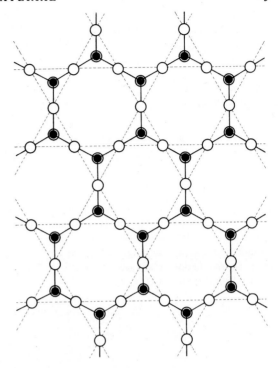

FIG. 13. Diagrammatic representation of a silicon-oxygen tetrahedral sheet.

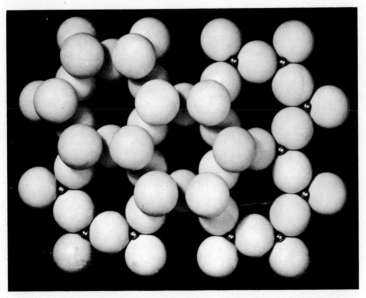

FIG. 12. Model of silicon-oxygen tetrahedral sheet with some of the oxygen ions removed to show the smaller silicon ions.

aluminium ions sandwiched between the two hydroxyl plains. In order that all the valencies of the structure be satisfied only two out of every three positions in the aluminium sheet are occupied by aluminium ions forming what is known as a dioctahedral structure.

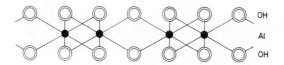

FIG. 14.   Diagrammatic representation of a gibbsite sheet.

Brucite sheet.   This has a similar structure to the gibbsite sheet but the aluminium is replaced by magnesium and because magnesium is divalent all the sites in the middle plane are occupied, forming a trioctahedral structure (Fig. 15).

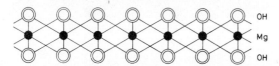

FIG. 15.   Diagrammatic representation of a brucite sheet.

Pyrophyllite and talc.   Although these minerals are infrequent and of little importance in soil studies it is convenient to start with them since the structure of the primary and clay micas, montmorillonite and vermiculite can be thought of as being derived from their structure by relatively simple substitutions. Pyrophyllite is composed of one gibbsite sheet lying between two silica sheets and is known as a 2:1 type mineral. In this structure each silica sheet has its apical oxygen ions facing towards the other and replacing hydroxyl ions on either side of the gibbsite sheet. Therefore the top and bottom surfaces of each composite sheet are made up of oxygen ions in the silica sheets (Fig. 16). Talc has a similar structure but the gibbsite layer is replaced by a brucite sheet. Thus pyrophyllite is a dioctahedral mineral and talc is the trioctahedral equivalent.

Muscovite. The structure of this mineral can be derived from the dioctahedral pyrophyllite structure by substituting one quarter of the silicon ions by aluminium ions in the tetrahedral layers. This causes an imbalance in the charges that is satisfied by potassium which bonds the composite sheets together (Fig. 17).

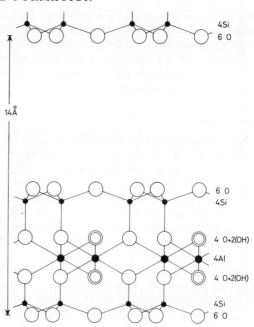

FIG. 16.   Diagrammatic representation of the structure of pyrophyllite.

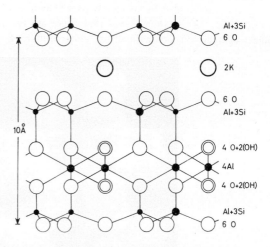

FIG. 17.   Diagrammatic representation of the structure of muscovite.

Biotite. The structure of this mineral can be derived from the trioctahedral talc structure by about one third substitution of iron — Fe^{++} for magnesium in the brucite layer. In a similar manner to muscovite the tetrahedral layers have negative charges that are satisfied by potassium which bonds the composite sheets together. There is also a small amount of substitution of aluminium in both the octahedral and tetrahedral layers.

Clay minerals

There are seven types of clay minerals important in soils, namely: kaolinite, halloysite, montmorillonite, mica, vermiculite, chlorite and allophane, the first six are crystalline and composed of silica, gibbsite and brucite sheets in various combinations. Allophane is not crystalline but it is convenient to consider it here. Clay minerals can be presented in terms of their structure, properties and methods of formation but before starting these discussions it must be emphasised that a full comprehension of their structure can be achieved only by a careful study of models.

Kaolinite group or kandites

This is the simplest type of clay mineral, being composed of one gibbsite sheet and one silicon tetrahedral sheet in which each apical oxygen of the silicon sheet replaces one hydroxyl group of the gibbsite sheet and forms what is known as a 1:1 type of structure. Kaolinite has a well developed pseudo-hexagonal crystal structure in which the individual crystals range from $0.2\ \mu - 2\ \mu$ and form by growth about all the axes. Development along the c axis takes place by a build up of the paired sheets in such a way that the oxygen ions of the basal plane of the silicon sheet are aligned opposite to hydroxyl groups of the gibbsite sheet, to which they become firmly attached by hydrogen bonding to produce a rigid structure, which cannot expand (Fig. 18).

The members of this group include nacrite, dickite, and kaolinite, the last of which is by far the most common.

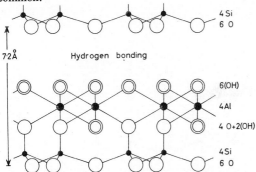

Fig. 18. Diagrammatic representation of the structure of kaolinite.

Halloysite

This is similar to kaolinite with the addition of a layer of water between the paired sheet. The water is firmly attached by hydrogen bonding resulting in a rigid structure. However, dehydration can take place very easily causing collapse of the structure forming a mineral similar to kaolinite. In a number of situations halloysite is regarded as the precursor of kaolinite.

Montmorillonite or smectite group

These minerals have a 2:1 type of structure and can be derived from the pyrophyllite structure by substituting one sixth of the aluminium ions by magnesium, which causes an inbalance of charge within the structure that is usually satisfied by basic cations (Fig. 19). Some substitution of iron can also take place. In these minerals there is weak bonding between successive composite layers. This imparts one of the most important properties to these minerals namely their ability to expand and contract in response to the addition and loss of moisture.

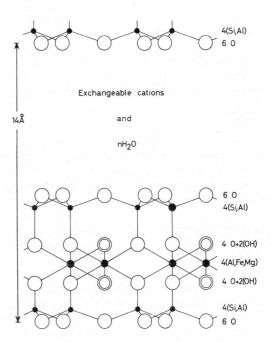

Fig. 19. Diagrammatic representation of the structure of montmorillonite.

Mica (Illite)

This group can be regarded as clay size particles of muscovite and biotite. Those similar to muscovite have the pyrophyllite structure and are of the dioctahedral variety. The trioctahedral variety is similar to biotite, which in turn is similar to talc. Both of these minerals are hydrated but neither is capable of expanding due to the additional linkages supplied by potassium but these linkages are fewer than in primary micas.

Vermiculite

This mineral exists in both dioctahedral and trioctahedral forms and can be regarded as hydrated micas from which potassium has been removed and replaced by calcium and magnesium and as a result it is capable of expanding because the linkages which are broken when potassium is removed are not reformed by calcium and magnesium (Fig 20).

Chlorite

This mineral is a 2:2 or 2:1:1 type of layered silicate in which a brucite-like sheet – $Mg(OH)_2$ is sandwiched between two mica layers and replaces the potassium in the mica structure (Fig. 21). The chemical composition of chlorite is very variable since Mg and Fe and other cations replace aluminium to various extents.

Allophane

This is a poorly categorised substance which is sometimes regarded as a clay mineral and at other times considered among the hydroxides. Some workers regard it as a mixture of silica gel and aluminium hydroxide while others consider it to be a hydrous aluminium silicate. Because of the difficulties in determining definite values for many of its properties it is probable that it is a variable mixture of silica gel and aluminium hydroxide. The molecular ratio of silica to alumina is known to vary from 100:112 to 100:180 but the precise composition depends upon the nature of the decomposing materials as well as upon the environment and age, since it appears that it slowly changes to halloysite and then to kaolinite. Allophane was at one time thought to be associated mainly with soils derived from volcanic ash but it now appears to be much more widespread.

Mixed layers

Clay minerals scarcely ever occur in a pure form

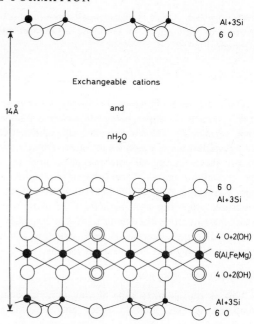

FIG. 20. Diagrammatic representation of the structure of trioctahedral vermiculite.

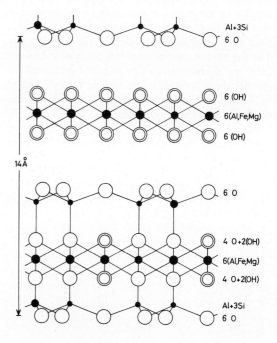

FIG. 21. Diagrammatic representation of the structure of trioctahedral chlorite.

in soils, often a single particle is composed of inter-stratified layers of two or more different 2:1 clay minerals. Also two or more may be intimately mixed or they may be enmeshed in large quantities of oxides hence the many difficulties encountered in their identification.

Properties of clay minerals. The two most important properties of clay minerals are their cation exchange capacity, i.e. their ability to absorb basic cations and secondly, their capacity to absorb water. Because of variations in their structure, clay minerals do not have a fixed exchange capacity but a range which is influenced by a number of factors of which the most important is pH as mentioned on page 92. Exchange reactions result from:

1. Unsatisfied bonds at the edges of crystals
2. Increased number of charges due to isomorphous replacement.

Minerals such as kaolinite with their rigid structure and absence of isomorphous replacement have their exchange sites confined to the broken edges of crystals where the charges of the relatively few broken bonds are satisfied by hydrogen or a basic cation, therefore the exchange capacity of kaolinite is very low. On the other hand montmorillonite and vermiculite have a relatively large amount of isomorphous replacement of silicon by aluminium giving a large number of exchange sites and a high exchange capacity. The indeterminate structure of allophane gives a very variable exchange capacity which is influenced very strongly by relatively small variations in the techniques used for its assessment. In addition to the above properties the micas are capable of adsorbing large amounts of potassium which is relatively unavailable to plants but it may be released slowly.

Flocculation and dispersion are two more properties of clays. Flocculation is the process whereby the individual particles of clay are coagulated to form floccular aggregates. Although this is clearly seen in the laboratory when a small amount of calcium hydroxide is added to a suspension of clay, a similar reaction takes place in the soil. The degree and permanence of flocculation depends upon the nature of the ions present; calcium and hydrogen are very effective in this role. At the other extreme is the state of dispersion in which the individual particles are kept separate one from the other. This is accomplished by potassium and more particularly by sodium. Thus, depending upon the nature of the cations present in the soil, it may either be in a floccular or aggregated state or in a dispersed and often massive condition.

A further distinctive property of clay minerals is that between the composite layers there is a characteristic spacing which is utilised in X-ray studies as a differentiating criterion. Set out in Table 1 below is the range in the exchange capacities and spacings for the commonly occurring clay minerals.

Table 1

CATION EXCHANGE CAPACITIES AND SPACINGS OF CLAY MINERALS

Clay mineral	C.E.C. me% clay, pH7	Spacing Å
Kaolinite	3–15	7.2
Halloysite (fully hydrated)	40–50	10.25
Mica	10–40	10
Montmorillonite	80–150	14
Vermiculite	100–150	14
Chlorite	10–40	14

Å = Angström unit = 10^{-8} cm

Fuller discussions about the structure and properties of minerals can be found in Deer *et al.* (1966), Jackson (1964) and Grim (1968).

Chemical and mineralogical composition of parent materials

Set out in Table 2 is a list of a few of the common types of rocks, their mineralogy and chemical composition. This is followed in Table 3 by a list of the important and widespread minerals found in parent materials, their formulae and approximate frequency. In this context it seems more appropriate to give the empirical formulae for most minerals rather than the unit cell formulae which can be found in Deer *et al.* (1966). In addition the chemical composition of the principal minerals is given in Table 4. These three sets of data give the significant properties of the wide range of mineral materials forming parent materials, of which quartz, felspars and the sheet silicates are the principal constituents. Also shown are the relatively few elements that are found in abundance in parent materials. As Clark (1924) has pointed out the upper 10 km of the surface of the earth is made up predominantly of O, Si, Al, Fe, Ca, Na, K, Mg, Ti, P, Mn, S, Cl and C, but only the first eight exceed 1%. Oxygen occupies over 90% by volume but usually occurs in combination with other elements, particularly with silicon the second most frequent element, forming quartz and occurring within the more complex silicate structures. Thus the majority of the resulting soils are composed predominantly of silica, either as quartz or in a combined form. The third most frequent element is aluminium which is found mainly in felspars and in the sheet silicates, but which occurs also in varying proportions in the other silicates. Iron is also of fairly widespread distribution and it should be noted that it is mainly in the ferrous state occurring in large amounts in relatively few minerals of which augite, biotite, hornblende, hypersthene and some olivines are the principal contributors. Sodium and potassium are found chiefly in the felspars, but significant amounts of potassium also occur in the micas. Calcium and magnesium have a wide distribution among the silicates as well as being the principal cations in many non-silicates such as calcite and dolomite which are the dominant minerals in limestones. Phosphorus is very restricted in its distribution occurring only in significant proportions in apatite which is fairly widespread in small amounts in parent materials but the total content of this mineral in soils seldom exceeds 0.2%. The microelements needed by plants occur in a number of minerals. Some have relatively high proportions of one or more elements. For example, tourmaline which is usually of low frequency in soils contains a large amount of boron but as this mineral is very resistant to weathering it is doubtful whether the boron is easily available to plants.

Most of the silicates are primary minerals and originate in igneous or metamorphic rocks, but the clay minerals are formed within soils or are inherited from parent materials such as lacustrine deposits and shale. Also included in Table 3 are a large number of minerals such as zircon, magnetite, kyanite, rutile and titanite that are usually referred to as *accessory minerals*. They are present only in very small amounts in soils but are important for their contribution of

Table 2

CHEMICAL COMPOSITION OF SOME COMMON ROCKS

ROCK TYPE	COMPOSITION (%)											
	SiO_2	Al_2O_3	Fe_2O_3	FeO	MgO	CaO	Na_2O	K_2O	TiO_2	P_2O_5	MnO	H_2O
BASALT Labradorite, augite and pseudomorphs in chlorite and goethite after olivine; and opaque minerals. (Guppy and Sabine, 1956)	53	18	5	2	4	7	4	2	1	<1	<1	2
DIORITE Hornblende, biotite, titanite, oligoclase, quartz and opaque minerals. (Guppy and Sabine, 1956)	58	17	3	4	3	5	4	3	2	1	<1	1
GRANITE Fine-grained grey granite, composed of quartz, microcline, oligoclase, biotite, muscovite, accessory minerals, apatite and opaque minerals. (Guppy and Sabine, 1956)	76	13	<1	1	<1	<1	3	5	<1	<1	<1	1
GARNETIFEROUS MICA-SCHIST Muscovite, biotite, chlorite, garnet, quartz and oligoclase, iron ore and apatite. (Guppy and Sabine, 1956)	58	19	1	7	2	2	2	4	1	<1	<1	2
LIMESTONE Calcite. (Clark, 1924)	1	<·1	Nil	·1	<1	55[+]	<·1	<1	Nil	<·1	Nil	<·1
SLATE Quartz, chlorite, muscovite, apatite, tourmaline, albite and opaque minerals. (Guppy and Sabine, 1956)	57	20	10*		3	1	3	3	1	<1	Nil	Nil

*Total Fe calculated as Fe_2O_3; [+]$CO_2 = 43\%$

microelements for plant nutrition and in detailed studies of soil formation as mentioned on page 57.

An extremely important characteristic of parent materials is their wide variability in composition within short distances. Some consolidated rocks, but more especially sediments can show a three or four fold variation in one or more constituents within a few cm, this is more common among the accessory minerals such as zircon but can be very important since these are often used as index minerals in weathering studies.

Surface area of parent materials

The specific surface area of the constituent particles in the parent material determines the amount of interaction that is possible with the environment, parti-

Table 3
SOIL FORMING MINERALS

CRYSTAL STRUCTURE	MINERAL	FORMULA	FREQUENCY
ORTHO- AND RING SILICATES	Andalusite	$Al_2O_3 \cdot SiO_2$	Rare
	+ Kyanite	$Al_2O_3 \cdot SiO_2$	Rare
	Sillimanite	$Al_2O_3 \cdot SiO_2$	Rare
	Epidote group		
	Zoisite	$4CaO \cdot 3Al_2O_3 \cdot 6SiO_2 \cdot H_2O$	Rare
	Clinozoisite	$4CaO \cdot 3Al_2O_3 \cdot 6SiO_2 \cdot H_2O$	Rare
	Epidote	$4CaO \cdot 3(Al,Fe)_2O_3 \cdot 6SiO_2 \cdot H_2O$	Rare – frequent
	Garnet group		
	+ Almandite	$Fe_3Al_2Si_3O_{12}$	Rare – frequent
	Olivine group		
	Olivine	$2(Mg,Fe)O \cdot SiO_2$	Absent – common
	+ Titanite	$CaO \cdot SiO_2 \cdot TiO_2$	Rare*
	+ Tourmaline	$Na_2O \cdot 8FeO \cdot 8Al_2O_3 \cdot 4B_2O_3 \cdot 16SiO_2 \cdot 5H_2O$	Rare
	+ Zircon	$ZrO_2 \cdot SiO_2$	Rare*
CHAIN SILICATES	*Amphibole group*		
	Hornblende	$Ca_3Na_2(Mg,Fe)_8(Al,Fe)_4Si_{14}O_{44}(OH)_4$	Rare – frequent
	Tremolite – Ferroactinolite	$2CaO \cdot 5(Mg,Fe)O \cdot 8SiO_2 \cdot H_2O$	Rare
	Pyroxene group		
	Enstatite	$MgO \cdot SiO_2$	Absent – common
	Hypersthene	$(Mg,Fe)O \cdot SiO_2$	Absent – common
	Diopside	$CaO \cdot MgO \cdot 2SiO_2$	Absent – rare
	Augite	$CaO \cdot 2(Mg,Fe)O \cdot (Al,Fe)_2O_3 \cdot 3SiO_2$	Absent – common
SHEET SILICATES	*Serpentine group*	$Mg_3Si_2O_5(OH)_4$	Absent – rare
	Mica group		
	Muscovite (dioctahedral)	$K_2Al_2Si_6Al_4O_{20}(OH)_4$	Rare – common
	Biotite (trioctahedral)	$K_2Al_2Si_6(Fe^{2+},Mg)_6O_{20}(OH)_4$	Rare – common
	Clay minerals		
	Kaolinite	$Si_4Al_4O_{10}(OH)_8$	Absent – dominant
	Mica	$K_1Al_4(Si_7Al_1)O_{20}(OH)_4 nH_2O$	Absent – dominant
	Montmorillonite	$Ca_{0.4}(Al_{0.3}Si_{7.7})Al_{2.6}(Fe^{3+}_{0.9}Mg_{0.5})O_{20}(OH)_4 nH_2O$	Absent – dominant
	Vermiculite (trioctahedral)	$Mg_{0.55}(Al_{2.3}Si_{5.7})(Al_{0.5}Fe^{3+}_{0.7}Mg_{4.8})O_{20}(OH)_4 nH_2O$	Absent – dominant
	Chlorite (trioctahedral)	$AlMg_5(OH)_{12}(Al_2Si_6)AlMg_5O_{20}(OH)_4$	Absent – rare

*usually present   + accessory minerals

cularly with water. Consolidated rocks have an extremely small surface area when compared with alluvial sands which in turn have a smaller surface area than lacustrine clays, thus the surface area increases as the particle size decreases. Variations in surface area and particle size distribution have a profound effect on the speed of soil formation, for it is found that a soil such as a podzol will develop in sediments much quicker than from consolidated rock of the same mineralogical composition.

Permeability of parent materials

The permeability of the parent material influences the rate of moisture movement which in turn influences the speed of soil formation. The most permeable materials and therefore those that allow free movement of moisture usually have a high content of sand, but as the particle size decreases the general tendency is for the material to become more impermeable so that clays allow only a very slow rate of moisture movement.

This can cause water to accumulate at the surface or within the body of the soil resulting in a number of important phenomena due to waterlogging, however, some sandy materials can be quite impermeable when compacted. The permeability of rocks varies with their structure; those with well developed jointing are very permeable whereas others without cracks and fissures are relatively impermeable.

Classification of parent materials

Recently, Whiteside (1953), and Brewer (1964) have suggested methods for classifying parent materials. In the system of Whiteside a somewhat general statement is made about the mineralogy and state of the material while that of Brewer is really an attempt to classify the potential of the material rather than a statement about the material itself. Thus neither of these proposals makes a clear and precise statement about parent materials. Presented below is a classification based upon the significant and easily recognis-

Table 3 continued

CRYSTAL STRUCTURE	MINERAL	FORMULA	FREQUENCY
FRAMEWORK SILICATES	*Felspar group* Alkali felspar Plagioclase Quartz	$(Na,K)_2O \cdot Al_2O_3 \cdot 6SiO_2$ $Na_2O \cdot Al_2O_3 \cdot 6SiO_2 \cdot - CaO \cdot Al_2O_3 \cdot 2SiO_2$ SiO_2	Rare − common Rare − common Frequent − dominant
NON-SILICATES	*Oxides* + Hematite + Ilmenite + Rutile + Anatase + Brookite + Magnetite	Fe_2O_3 $FeO \cdot TiO_2$ TiO_2 TiO_2 TiO_2 Fe_3O_4	Absent − frequent Absent − rare* Rare* Absent − rare* Rare Rare*
	Hydroxides Gibbsite Boehmite Goethite Lepidocrocite	$Al(OH)_3$ $\gamma AlO(OH)$ $\alpha FeO\text{-}OH$ $\gamma FeO\text{-}OH$	Absent − dominant Absent − dominant Rare − common Absent − rare
	Sulphates Gypsum Jarosite	$CaSO_4 . 2H_2O$ $KFe_3(OH)_6(SO_4)_2$	Absent − common Absent − rare
	Chlorides Halite	$NaCl$	Absent − common
	Carbonates Calcite Dolomite	$CaCO_3$ $(Ca,Mg)CO_3$	Absent − common Absent − common
	Phosphates + Apatite	$Ca_4(CaF \text{ or } Cl)(PO_4)_3$	Rare*

*usually present   + accessory minerals

able intrinsic characteristics employing a system of letter symbols for designating the parent material in the field and in written descriptions. Broadly, parent materials can be divided initially into nine classes according to their chemical and mineralogical composition; there are five classes established on the proportions of ferromagnesian minerals, two classes on the amount of carbonate, the eighth is established on the content of salts and the ninth on the amount of organic matter. The presence of ferromagnesian minerals rather than the amount of silica as employed by geologists is used to create the first five classes since it is the content of these minerals that is most often the important factor determining the chemical properties of the parent material. Set out in Fig. 22 are the nine classes, their percentage content of ferromagnesian minerals and letter symbols. There are a few exceptions to this rather simple rule, particularly rocks having a high content of easily hydrolysable minerals such as anorthite which are regarded as basic. Many volcanic rocks, because of their poor crystallinity do not fit easily into this scheme and have to be classified on the basis of their content of silica.

Each of these classes can exist in a number of forms and may be divided into consolidated and unconsolidated materials.

Consolidated materials

As shown in Fig. 22, these are subdivided into four types on the basis of the size distribution of the mineral grains as follows:

Non-crystalline: composed of a continuous phase without crystals; such as volcanic glass.
Fine grained: material containing minerals that are just visible with the unaided eye but there may be a few larger crystals or phenochrysts.
Medium grained: these rocks are composed mainly of minerals up to 5 mm in diameter.
Coarse grained: rocks with minerals greater than 5 mm.

Unconsolidated materials.

There are two principal subdivisions of these materials in order to accommodate particle sizes greater than, as well as less than, 2 mm, these are shown also in Fig. 22, bottom right.

Particles < 2 mm

These four classes are defined in terms of the percentage distribution of sand, silt and clay as shown in Fig. 54.

Table 4
CHEMICAL COMPOSITION OF SOME COMMON MINERALS

MINERAL TYPE	COMPOSITION (%)													
	SiO_2	Al_2O_3	Fe_2O_3	FeO	MgO	CaO	Na_2O	K_2O	H_2O	TiO_2	P_2O_5	MnO	F	CO_2
Apatite	—	—	< 1	—	< 1	56	—	—	< 1	—	42	< 1	4	—
Augite	47	3	1	20	7	17	1	< 1	< 1	1	—	1	—	—
Biotite	36	18	1	22	7	< 1	< 1	9	4	2	—	< 1	< 1	—
Calcite	—	—	—	—	< 1	56	—	—	—	—	—	—	—	44
Dolomite	—	—	—	< 1	21	31	—	—	—	—	—	—	—	47
Hornblende	45	11	3	13	10	12	1	1	2	1	—	< 1	—	—
Hypersthene	50	1	2	28	16	1	< 1	< 1	—	—	—	1	—	—
Muscovite	45	37	< 1	< 1	< 1	—	1	10	5	< 1	—	< 1	1	—
Olivine														
Fayalite	35	< 1	< 1	61	3	1	—	—	—	1	—	3	—	—
Forsterite	41	1	< 1	4	54	—	—	—	< 1	< 1	—	< 1	—	—
Serpentine (chrysotile)	43	1	1	1	41	< 1	< 1	—	12	tr	< 1	< 1	—	—
Alkali felspar														
Orthoclase	65	20	< 1	tr	tr	1	5	8	1	—	—	—	—	—
Plagioclase														
Labradorite	53	30	1	—	—	12	4	< 1	< 1	tr	—	—	—	—
Clay minerals														
Kaolinite	44	36	1	< 1	< 1	< 1	< 1	1	16	1	< 1	—	—	—
Montmorillonite	51	20	1	—	3	2	< 1	—	23	—	—	—	—	—

tr = trace

Particles > 2 mm

These are defined as follows:

Gravelly: The material contains more than 25% of material 2–10 mm in size.

Stony: The material contains 10–30% of material 1–10 cm in diameter.

Very stony: The material contains > 30% of material 1–10 cm in diameter.

Bouldery: The material contains > 20% material > 10 cm in diameter.

The principal rock types, their symbols and mineralogy are given in Table 5 and below is the list of the predominant types of superficial deposits:

Alluvial deposits	Marine clays
Dune sands	Pedi-sediments
Glacial drifts	Raised beaches
Lacustrine clays	Solifluction deposits
Loesses	Volcanic dust

It is not possible to specify these deposits as accurately as the individual rocks because of much wider variations in their properties. However a few examples are given in Table 6 to show how the system operates and definitions are given in the glossary (p. 284).

It is shown (Tables 5 and 6) that the same symbol is used to designate material of several origins thus an attempt is made to produce an objective categorisation of parent material and not to classify them on interpretations of their methods of formation since it is often difficult to be sure about their origin. For example it is not always easy to differentiate between alluvium and certain forms of glacial drift, also there is considerable controversy about the origin of loess.

All of these types of parent materials can undergo chemical and/or physical weathering whether originally consolidated or not. Then they are variously regarded as part of the soil or as parent material and can be designated by adding the suffixes *w* and *p* to the symbol

Type	COMPOSITION % ferromagnesian minerals	SiO$_2$ %	Symbol
Ultrabasic	> 90	—	U
Basic	40–90	45–55	B
Intermediate	20–40	55–65	I
Acid	5–20	65–85	A
Extremely acid	< 5	—	E

Type	COMPOSITION	Symbol
Carbonate	1–30% (Ca + Mg)CO$_3$	C
Limestone	> 30% (Ca + Mg)CO$_3$	L
Saline	> 2% soluble salts	H
Organic	> 30% organic remains	O

CONSOLIDATED   UNCONSOLIDATED

Mineral size	Symbol
Non-crystalline	N
Fine grained	F
Medium grained	M
Coarse grained	K

Particles < 2 mm	
Name	Symbol
Sand	S
Loam	L
Silt	Z
Clay	C

Particles > 2 mm	
Name	Symbol
Gravelly	G
Stony	T
Very stony	V
Bouldery	P

FIG. 22. Classification of parent materials.

of the unweathered state to indicate chemical and physical weathering respectively and are defined as follows: Chemically weathered material: $> 90\%$ of the material is disaggregated into separate minerals or form gravel size units; $< 20\%$ of the minerals are pseudomorphs composed of clay and hydroxides.

Table 5

MAJOR ROCK TYPES, THEIR SYMBOLS AND MINERALOGY

ROCK TYPE	SYMBOL	MINERALOGY
Ultrabasic		
Peridotite	UK	Olivine, enstatite, augite, hornblende, biotite, plagioclase
Serpentine	UM	Antigorite, chrysotile
Basic		
Amphibolite	BM	Hornblende, chlorite, epidote, plagioclase, garnet
	BK	Hornblende, chlorite, epidote, plagioclase, garnet
Basalt	BF	Augite, hypersthene, olivine, plagioclase, magnetite
Gabbro	BM	Augite, hornblende, biotite, enstatite, hypersthene, olivine, apatite, plagioclase
	BK	Augite, hornblende, biotite, enstatite, hypersthene, olivine, apatite, plagioclase
Intermediate		
Andesite	IF	Hornblende, biotite, augite, hypersthene, plagioclase, quartz
Diorite	IM	Hornblende, biotite, augite, plagioclase, quartz, apatite
	IK	Hornblende, biotite, augite, plagioclase, quartz, apatite
Syenite	IM	Orthoclase, plagioclase, biotite, apatite
	IK	Orthoclase, plagioclase, biotite, apatite
Acid		
Conglomerate (one type)	AK	Orthoclase, plagioclase, biotite, muscovite, quartz
Granite	AM	Orthoclase, plagioclase, muscovite, quartz, apatite, biotite
	AK	Orthoclase, plagioclase, quartz, biotite, muscovite, apatite
Gneiss	AM	Orthoclase, plagioclase, muscovite, quartz, apatite, biotite
	AK	Orthoclase, plagioclase, quartz, biotite, muscovite, apatite
Obsidian	AN	Non-crystalline $> 65\%$ SiO_2
Sandstone	AM	Quartz, felspars, biotite, muscovite
Schist	AM	Quartz, biotite, muscovite, felspars
	AF	Quartz, biotite, muscovite, felspars
Slate	AM	Quartz, chlorite, muscovite
	AF	Quartz, chlorite, muscovite
Extremely Acid		
Quartzite	EM	Quartz, muscovite, felspars
	EK	Quartz, muscovite, felspars
Sandstone	EM	Quartz, muscovite, felspars
Slate	EM	Quartz, chlorite, muscovite
	EF	Quartz, chlorite, muscovite
Carbonate		
Sandstone	CK	Quartz, felspar, mica, calcite, dolomite
	CM	Quartz, felspar, mica, calcite, dolomite
	CF	Quartz, felspar, mica, calcite, dolomite
Limestone		
Dolomite	LF	Dolomite, calcite, fossils
	LM	Dolomite, calcite, fossils
Limestone	LF	Calcite, dolomite, fossils
	LM	Calcite, dolomite, fossils
Marble	LM	Calcite, dolomite, epidomite, zoisite, diopside
Organic		
Coal	ON	Plant remains

Physically weathered material: disintegrated in its present position or moved < 1 m without any change in composition. The separates > 2 mm have at least one sharp angular edge if they are spherical or at least three if they are other shapes. Adjacent particles often fit together showing that they were at one time part of the same unit.

Distribution of parent materials

Superficial deposits including glacial drift, alluvium and aeolian deposits are the predominant parent materials having resulted from the widespread climatic changes and surface disturbances of the Pleistocene period (see page 48 *et seq.*). It is in tropical and subtropical countries that there are large contiguous areas where soils are developed from the *in situ* decomposition of rocks most of which are acid in composition being mainly granite, gneiss, schist, slate, sandstone and shale. Outside these areas it is mainly in mountain ranges that soils have developed from the underlying rock but these are usually shallow. Intermediate and basic rocks occur in many places and occasionally

Table 6
SOME SUPERFICIAL DEPOSITS AND THEIR SYMBOLS

SYMBOL	TYPE OF DEPOSIT
CZ	Calcareous loess without stones Calcareous alluvium without stones
AC	Lacustrine clay Marine clay
ASV	Acid sandy solifluction deposit with many stones Acid sandy alluvium with many stones Acid sandy raised beach with many stones Acid sandy glacial drift with many stones

become important. Limestone and other calcareous materials also are of restricted distribution but there are a few places where they are widespread, the area bordering the Mediterranean Sea is probably the best example. Where limestone, basic and intermediate rocks are present they often lead to the formation of soils which contrast with the more common ones formed in acid materials.

CLIMATE

Climate is the principal factor governing the rate and type of soil formation as well as being the main agent determining the distribution of vegetation and the type of geomorphological processes, therefore it forms the basis of many classifications of natural phenomena including soils.

The climate of a place is a description of the prevailing atmospheric conditions and for simplicity it is defined in terms of the averages of its components, the two most important being temperature and precipitation. The data on which the averages are based are usually accumulated for at least 35 years so as to take account of the differences – sometimes wide – in annual rainfall and temperature patterns. Although averages are most commonly used, diurnal and annual patterns and extremes are not ignored, since they give character and sometimes are important factors. The occurrence of occasional high winds can determine the development policy of an area. However, it must be stressed that the atmospheric climatic data do not always give a true picture of the soil climate. For example, the amount of water in the soil may vary considerably within a distance of a few metres from

permanently saturated to dry and quite freely drained; whereas there is virtually no difference between the amount of total precipitation of the two sites, one site may be in a depression where moisture can accumulate and the other on an adjacent slightly elevated situation. Regularly these differences in the moisture regime at the two sites, lead to the development of different soils and contrasting plant communities.

Data on soil climate are meagre because of the difficulties encountered in making the necessary measurements actually within the soil itself, and frequently the lack of adequate equipment; thus many inferences are made based either on occasional observations without the support of adequate measurements or upon extrapolations using atmospheric climatic data. In spite of these difficulties and deficiencies it is possible to make many general statements about soil climate and its effect using the available atmospheric and soil climatic data, coupled with reasoning based on chemical and physical principles.

Soil climate has the same two major components as atmospheric climate, namely temperature and moisture.

Temperature

Atmospheric and soil temperature variations are the most important manifestations of the solar energy reaching the surface of the earth. The atmosphere transmits most of the visible and heat rays through clean dry air but absorbs nearly all of the short wave radiation, and of the radiation reaching the soil surface part is absorbed and converted into heat while the remainder is reflected back. A proportion of the heat produced is maintained in the soil but some is lost to the atmosphere by convection of hot air from the soil and by back radiation. Also some may be used for evaporation of moisture into the atmosphere and is an additional loss. The utilisation of the greater part of the long wave solar radiation is given diagrammatically in Fig. 23.

The main effect of temperature on soils is to influence the rate of reactions, for every 10°C rise in temperature the speed of a chemical reaction increases by a factor of two or three. The principal reaction in soil to which this applies is the hydrolysis of primary silicates. The rate of biological breakdown of organic matter and the amount of moisture evaporating from the soil are also increased by a rise in temperature.

Development of vegetation can also be affected by soil temperature, in cool climates plants become active at 5°C and their rate of maturation is approximately doubled for each 10°C rise in temperature up to a maximum of about 20°C. The amounts of radiation reaching the surface and the soil temperatures are determined by:

1. Diurnal and seasonal variations
2. Latitude
3. Aspect
4. Altitude
5. Cloudiness
6. Humidity
7. Prevailing winds
8. Vegetational cover
9. Snow cover
10. Soil properties:
    colour
    moisture content
    conductivity and porosity

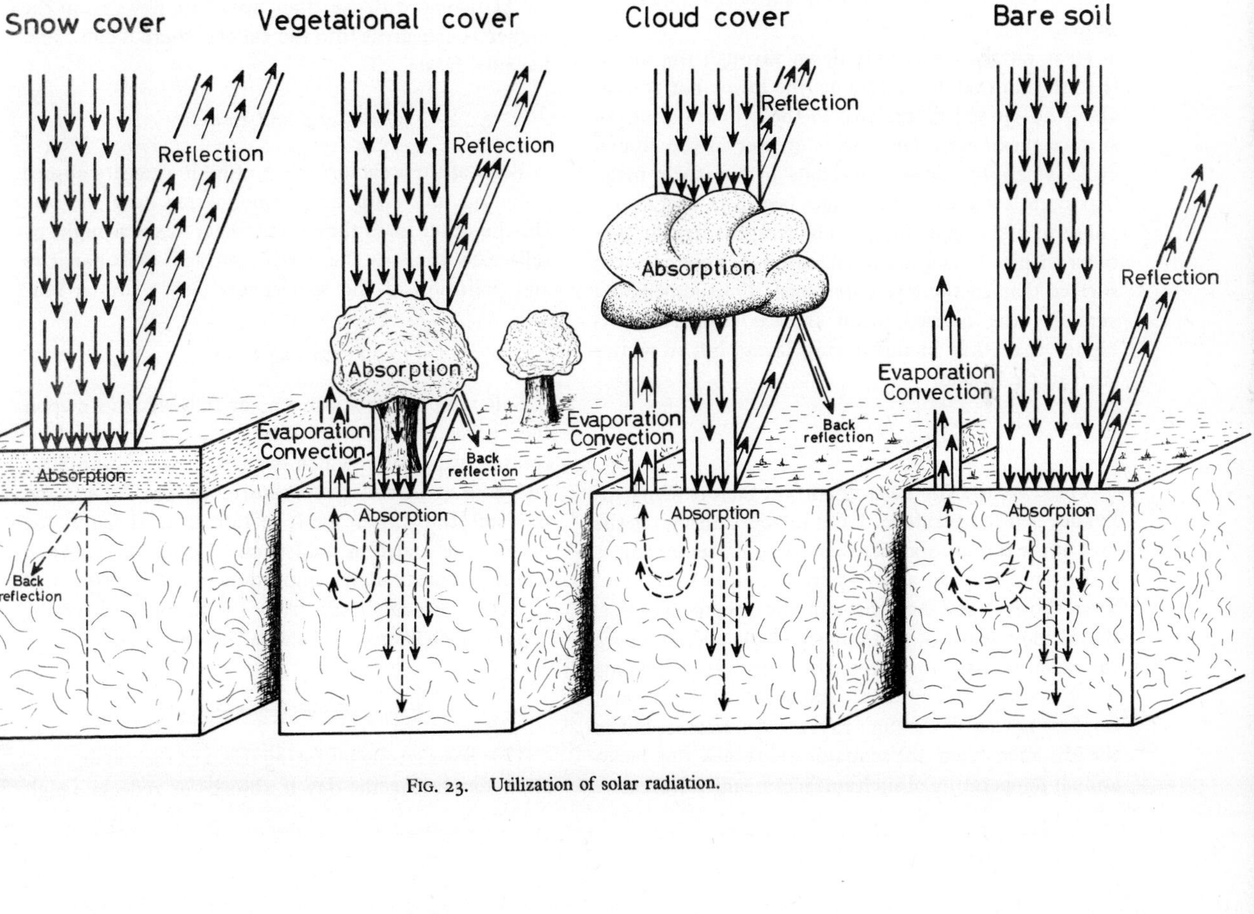

FIG. 23.   Utilization of solar radiation.

Diurnal and seasonal variations

Soils have well marked diurnal and annual temperature cycles. During the diurnal cycle in tropical and subtropical areas it is normal for heat to move downwards in the soil during the day from the surface warmed by incoming radiation and upwards during the night as the surface cools. This takes place also during the summer period in the middle and polar latitudes but in these areas, during the winter, the atmosphere is generally cooler than the soil and the incoming radiation is not sufficient to heat the soil which steadily cools and eventually may freeze from the surface downwards. In tropical areas the total incoming radiation varies little from one season to another but in the middle latitudes there is ten times more radiation in summer than in winter. This difference increases towards the poles where there are long periods during the winter without radiation. During the annual cycle in countries with contrasting seasonal climates the soil becomes warm during the summer and cool during the winter but the rhythm of these cycles, particularly the daily cycle, is regularly interrupted by wind and rain and occasionally by snow.

Heat moves very slowly down through the soil so that the diurnal maximum in the lower horizons at about 20–30 cm occurs two to three hours after the surface maximum. This lag is greater in the annual cycle when the lower horizons attain their mean maximum even after the surface begins to cool in response to a seasonal change. The temperature fluctuations within the soil between seasons is greater at the surface than in the lower horizons. Thus, during the summer, the diurnal mean surface temperature is higher than that in underlying layers but in winter the reverse is true.

Latitude and aspect

These are the two most important factors which influence soil temperature. Land surfaces normal to the rays of the sun are warmer than those at smaller angles and it is in tropical areas where the sun's rays travel their shortest distance through the atmosphere that this is most marked, particularly in the deserts where the radiation is not intercepted by water vapour, clouds or vegetation. To the north and south of the tropics, surfaces normal to the sun's rays are also warmer but as the distance from the equator increases the mean annual temperature of such surfaces steadily decreases.

This is because of the greater distance that the sun's rays have to travel through the atmosphere and also enhanced by the curvature of the earth. Differences in warmth between surfaces are particularly important in the middle latitudes where sloping land surfaces facing towards the equator usually carry a plant community indicative of a warmer and drier climate than land facing towards the poles. Therefore slopes facing the equator may be ready for spring cultivation days and even weeks before land facing towards the poles.

Altitude

A characteristic of the relationship between climate and altitude is the decrease of about 1°C for each 150 m rise in elevation but this can vary within fairly wide limits. With increasing altitude there is initially a steady increase in precipitation, which reaches a maximum and then decreases again. In very mountainous areas these two factors combine to produce a vertical, climatic, vegetational and soil zonation which often terminates in glacial phenomena. An additional characteristic feature of mountainous areas is that cold air being more dense than warm air flows from the higher cooler areas into the valleys where it can cause harmful frosts.

Cloudiness and humidity

Both of these factors when of high intensity absorb radiation and reduce the amount reaching the soil. On the other hand they reduce heat losses at night by reflection back to the earth's surface. Dust particles and pollution of the atmosphere play a similar role.

Prevailing winds

Soil temperatures can rise or fall depending upon the character of the prevailing wind, a dry wind increases evaporation which in turn causes a loss of heat and a drop in soil temperature. On the other hand the soil can either gain or lose heat through the presence of warm or cold winds. Ocean currents and winds are particularly important in transferring heat to extra-tropical areas from the tropics where the incoming solar energy exceeds the outgoing.

Vegetational cover

This has a strong buffering effect on the soil temperature for during the day it absorbs or reflects a high proportion of the radiant energy which would otherwise reach the soil surface and at night it reflects to the surface some of the heat radiating from the warm soil to the cooler atmosphere. Thus soils beneath vegetation are cooler during the day and warmer during the night than adjacent bare sites. The cooling effect of vegetation upon the soil is utilised in tropical agriculture where plants such as coffee and cocoa which are very susceptible to high light intensities and high soil temperatures are interplanted with shade-trees. The buffering effect of vegetation is also important in the middle latitudes where it prevents rapid heat losses during the winter and reduces the penetration of frost into the ground.

Snow cover

A cover of snow protects the surface from prevailing winds and reduces heat losses thereby preventing rapid temperature fluctuations. Sometimes an early shower of snow can reduce the penetration of frost during the winter months.

Soil colour

Dark coloured soils absorb the greatest amount of heat and may have temperatures a few degrees higher than neighbouring soils of lighter colour; as a consequence they lose more moisture by evaporation but they gain more by condensation at night. Because dark objects emit more heat than light coloured objects they become cooler at night hence the greater amount of condensation. Black soil absorbs about 92% of the solar energy whereas snow absorbs only 10–15%; the remainder of the radiation is reflected back into the atmosphere.

Soil moisture content

The moisture content of the soil can have a marked influence upon the absolute soil temperature as well as upon the rate of temperature change. Dry soils have a specific heat of about 0·20 cal/g whereas that of a saturated soil is about 0·85 cal/g, this means that moist soils warm up more slowly because of the considerably greater amount of heat required to raise the temperature. Some heat is also used up in evaporation and lost to the atmosphere, further preventing a rise in temperature. Therefore wet soils never attain the same maximum temperature as adjacent dry ones. Similarly wet soils cool much more slowly since they require to lose more heat to effect a change in temperature.

Soil conductivity and porosity

The subsurface horizons receive most of their heat by conduction from the surface. The rate of heat transfer depends upon a number of factors of which porosity and moisture are the two most important variables more especially the latter. Therefore it is necessary to consider the movement of heat in soils in the dry and the wet state. When the moisture content is low the conduction of heat is very slow because the conductivity of air in the pores space is very low. It is only at the points of contact between mineral grains that heat can be conducted rapidly from one to the other, thus the higher the porosity the lower the conductivity. Very often porosity increases with decrease in particle size consequently soils containing a high percentage of fine materials tend to heat up slowly; dry peat with its high content of air heats up even more slowly.

In a moist soil, water forms bridges between the particles thus increasing the areas of contact between them. Since water transfers heat both by conduction and convection better than air, heat will flow more rapidly through the bridges and therefore through the soil as a whole.

A limited amount of heat transfer takes place when warm water flows from the warmer upper part of the soil into the lower horizons.

Moisture

The moisture in soils includes all forms of water that enter the soil system and is derived mainly from precipitation as rain and snow or may be supplied by the lateral movement of water over the surface or within the body of the soil itself. Underground water may also contribute to the moisture in the soil. In a number of cases flooding by rivers contributes the greater part of the soil moisture, perhaps the best examples are to be found in the Nile valley.

The moisture entering soils contains appreciable amounts of dissolved CO_2 thus it is probably more correct to think of it as a dilute weak acid solution which is much more reactive than pure water.

Precipitation varies considerably from place to place but the total volume is no guide to the actual amount entering the soil. This is determined by:

1. Intensity of precipitation
2. Vegetational cover
3. Infiltration capacity
4. Permeability
5. Slope
6. Original moisture content of the soil
7. Moisture losses.

Intensity of precipitation

In areas of bare soil it is precipitation of moderate intensity that is most effective in entering the soil particularly when it is porous. Light showers of rain that hardly enter the soil are quickly lost by evaporation and transpiration. Heavy showers may fall at a rate greater than the capacity of the soil to absorb the moisture, therefore it accumulates on the surface or runs off, thereby creating an erosion hazard. Further, the impact of large rain drops on the soil during a heavy shower may cause puddling and sealing of the surface which prevents the entry of moisture and also creates an erosion hazard. The impact of the rain drops causes material to be splashed upward and outwards. On slopes, some of the material will fall below the original position thereby bringing about a certain amount of erosion down the slope.

Vegetational cover

This has a profound effect on the amount of precipitation reaching the soil surface. In many cases light showers are held on the foliage and returned to the atmosphere by evaporation. Only heavy showers provide sufficient moisture to saturate the foliage and produce an excess which falls on to the soil surface. As the density of the vegetation increases the amount of moisture held on the foliage increases. This attains its maximum in tropical rain forests which, with their layered structure can intercept and retain large volumes of rainfall.

Infiltration capacity

The capacity of the soil to absorb moisture varies considerably; soils with a well developed structure or coarse texture allow free entry of moisture whereas dispersed clays are virtually impermeable. Variations in the capacity of the soil to absorb moisture are particularly important in areas of high rainfall where low infiltration rates cause water to collect at the surface creating a serious erosion hazard. In areas of low rainfall, maximum infiltration is of prime importance so that the plants can utilize fully the annual precipitation.

Permeability and slope

The amount of moisture capable of percolating completely through the soil is dependent largely upon the permeability of the middle and lower horizons. When either of these is impermeable the upper layers can quickly become saturated with water, resulting in lateral movement through the soil and run-off over the surface. On sloping sites, erosion is the normal consequence while on flat sites temporary flooding takes place. Fig. 24 is a generalised attempt to show the moisture cycle under humid conditions and Fig. 25 shows the fate of moisture falling on the surface under a number of different climatic conditions.

Original moisture content of the soil

This also influences the entry of precipitation, obviously no further additions can take place if the soil is already saturated. Such conditions occur at the lower ends of slopes and in depressions where run-off usually accumulates and the water-table comes to the surface. On these sites precipitation accumulates with the formation of temporary or permanent ponds.

Moisture losses

The moisture entering the soil system is lost through percolation, evaporation and transpiration. Even in very wet and permanently saturated sites there is usually continuous percolation for seldom is the moisture stagnant. Losses by transpiration and evaporation are influenced by the nature of the vegetation and a number of atmospheric conditions including insolation, the character of the prevailing winds, humidity and cloudiness. Humidity is possibly the most important factor determining the rate of moisture loss by transpiration and evaporation. In areas of particularly high humidity, losses are reduced to a mimimum, sometimes resulting in water-logging of the surface horizons as in certain maritime areas of Western Europe and at high elevation in the wet tropics.

Moisture movement

The differentiation of horizons is determined very

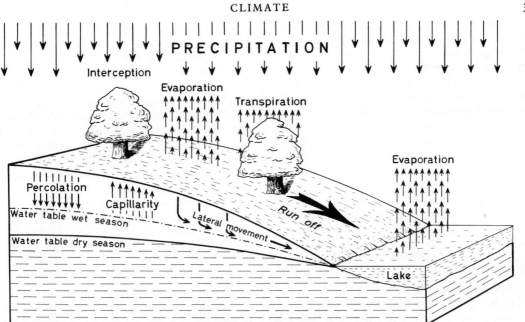

FIG. 24. The moisture cycle under humid conditions.

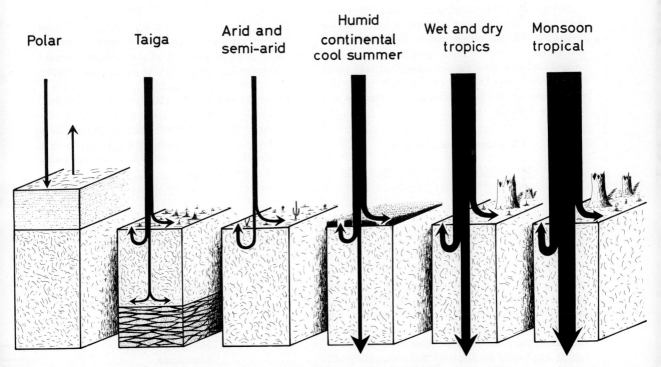

FIG. 25. The fate of moisture falling on the surface under a number of different climatic conditions.

largely by the movement of moisture; therefore this process is of paramount importance. In fact the soil solution might be regarded as the main 'conveyor belt' in soils whereby ions and small particles are translocated from one place to the other. The rate of movement is determined mainly by the volume of water entering the soil and the size of the pore spaces. When a large amount of moisture is present in a soil with large pore spaces it moves in response to gravity by flowing vertically downwards. With decreasing moisture content the moisture flows at first over the surface of the aggregates but this will cease when there is insufficient moisture to overcome surface tension. In soils with closely fitting particles and narrow pore spaces moisture cannot move in response to gravity because of restrictions induced by frictional forces. In can move only by surface flow and capillarity from one particle to the other. This accounts for the slow movement of moisture in compacted or fine textured soils.

Upward movement by capillarity takes place in response to surface drying or where there is ground water. Thus the situation is that, when rain is falling or snow melting, moisture moves downwards by gravity and surface flow, but during dry periods moisture moves upwards by capillarity.

Movement in the vapour phase is also important. Since water vapour moves to areas of lower vapour pressure it will move from wet to dry areas; it will also move to areas of lower temperature thus the normal tendency is for vapour transfer to take place downwards during the day and upwards during the night. However the full significance of this has not been demonstrated.

Classification of climates

Since climate plays the major role in determining the nature of the soil, vegetation and landscape, it would seem appropriate to include the classification of Critchfield (1966) given in Table 7. This is one of the many systems currently in use and seems to be useful in pedology.

A number of workers have introduced climatic indices based on atmospheric climatic data. Most are of little value, since atmospheric climatic data often have little pedological significance. For example, the precipitation:evaporation ratio, (P/E index) of Transeau (1905) shows the amount of percolation relative to evaporation so that as the value increases above unity, the amount of moisture available for percolation through the soil increases. This may be true for easily permeable soils on flat sites, with complete through drainage but it is of little value for soils with compact impermeable layers or those at the lower end of slopes and in depressions that receive a considerably larger volume of moisture by run-off.

Recently, Penman (1956) has produced a fairly reliable formula for calculating evapotranspiration from various surfaces using standard meteorological data. However the main value of this lies outside the field of pedology.

Climatic changes

One of the more fascinating aspects of climate is its constant and continuous change with time as revealed by an examination of the geological column which shows that the climate at any one point on the surface of the earth has changed many times, being at one time hot and dry, at other times hot and humid, or cold and wet. During the last two and a half million years some of the most dramatic changes have taken place and in many cases these fluctuations have led to the change from one set of soil-forming process to another (see page 47).

ORGANISMS

The organisms influencing the development of soils range from microscopic bacteria to large mammals including man. In fact, nearly every organism that lives on the surface of the earth or in the soil affects the development of soils in one way or another. The variation is so wide that any grouping in order of their taxonomy really has little meaning from the pedological standpoint. An arbitrary and somewhat crude classification of the more important soil organisms is as follows:

Higher plants
Vertebrates
    Mammals
    Other vertebrates
Microorganisms
Mesofauna
Man

Higher plants

Under natural conditions higher plants are normally found arranged in communities containing many different species, but occasionally there may be a very small number present, particularly in forest communities which may be dominated by a single tree species. When afforestation is practised it is customary to plant only one or sometimes two different tree species in a given area but there may be many species in the ground flora. Under advanced agricultural conditions there is almost invariably a single species planted but occasionally a mixture of two or three different species may be grown together. Set out below is a list of the common naturally occurring plant communities (Eyre 1963, Riley & Young 1966):

Boreal forests	Sand dune communities
Coniferous forests	Savannas
Deciduous forests	Savanna woodlands
Desert communities	Sclerophyllous woodlands
Grassland communities	Temperate rain forests
Heaths	Thorn woodlands
Hydroseres	Tropical rain forests
Mangrove swamps	Tundra and mountain communities

The higher plants influence the soil in many ways. By extending their roots into the soil they act as binders and so prevent erosion from taking place; grasses are effective in this role and it is suggested that the development and spread of flowering plants especially during the Tertiary period (see page 48) is one reason for the formation and preservation of many of the very deep soils from that period. Roots also bind together small groups of particles and help to develop a crumb or granular structure. This is also a characteristic feature of some grasses which are regularly planted purely for their improvement of the soil structure. As the large roots of trees grow and expand they bring about a certain amount of redistribution and compaction of the mineral soil. They also act as agents of physical weathering by opening cracks in stones and rocks. When plants die their root systems contribute organic matter which upon decay leave a network of pore space through which water and air can circulate more freely causing some soil processes to be accelerated. One of the best examples is seen in many poorly drained soils where the mineral soil that surrounds a decomposed root becomes oxidised to give yellow or reddish-brown colours. If this process is intense pipes of soil material can develop around roots through the cementing action of ferric hydroxide that gradually accumulates. A somewhat similar situation is found in arid and semi-arid areas where the freer movement of water down the old root channels leads to a greater deposition of calcium carbonate on their sides, eventually forming pipes.

The amount of organic matter contributed by plant roots varies widely; whereas trees contribute a very large amount when they die, the total averaged over the life span of a single tree really amounts to a very small annual addition. On the other hand the contribution of annuals or biennials may be quite considerable.

Plant roots exude various substances on which many microorganisms thrive so that the soil in immediate proximity to the root or *rhizosphere* is an area of prodigious microbiological activity; frequently the concentration of iron immediately next to the live root is reduced imparting a bleached appearance to the soil.

The addition of litter to the surface of the soil is probably the greatest contribution of the higher plants except in certain agricultural practices when it may be removed. The total amount of organic matter added by the different plant communities is variable but it is no guide to the amount in the soil which depends more upon the rate and type of breakdown than on the amount added. Table 8 gives the weight of litter supplied by some plant communities to the surface. These data demonstrate that tropical vegetation contributes the greatest amount of organic matter but their associated soils usually contain the least amounts since biological activity, particularly that of microorganisms and termites is much greater under warmer conditions. Thus, although the organic matter production is greater so also is its decomposition by the soil fauna and flora. On the other hand under cool humid conditions the supply of organic matter is small but the decomposition rate is slower so that the normal tendency is for organic matter to accumulate at the surface especially in moist habitats.

By intercepting rain, higher plants shelter the soil from the impact of rain drops which puddle the surface, reducing the permeability and creating an erosion hazard. Higher plants tend also to maintain more equable conditions by shading the soil from the direct rays of the sun thereby reducing losses by evaporation. It follows, therefore that the soil beneath dense vegetation neither becomes excessively wet when the rainfall is high nor very dry during periods of low rainfall.

Plants extract nutrients from the body of the soil and

Table 8

ORGANIC MATTER PRODUCTION BY VEGETATION

TYPE OF VEGETATION	ANNUAL PRODUCTION Metric tons per hectare (Air dry)	LOCATION	AUTHOR
Alpine meadow	< 0·5—0·9	Poland	Swederski (1931)
Short-grass prairie	1·6	Colorado U.S.A.	Clements and Weaver (1924)
Mixed tall-grass prairie	5·0	Colorado U.S.A.	Clements and Weaver (1924)
Quercus rubor	3·7	Netherlands	Witkamp and van der Drift (1961)
Pinus sylvestris	2·5	Norway (Eidseberg)	Bonnevie-Svendsen and Gjems (1957)
Fagus sylvatica	3·5—5·0	Germany	Danckelmann (1887)
Tropical forest	25·0	Java	Hardon (1936)

under natural conditions return most of them to the surface in their litter which decomposes and releases the nutrients, rendering them available for reabsorption by the plants. This cyclic process is fundamental and under natural conditions there is a delicate balance between the demands of the plant community, the quantity of nutrients in the cycle and the amount released by decomposition of the soil minerals. In tropical areas there is a number of high forest communities sustained solely by this process, the soils being completely devoid of minerals that yield plant nutrients. This paradoxical situation of virgin high forest on an inherently infertile soil leads to disastrous consequences when the forest is used as an indicator of high fertility and agriculture is attempted. An interesting example of the differing capacity of two plant communities to recycle nutrients is reported from Wisconsin (Jackson and Sherman 1953). Under a hardwood community the surface soil is maintained at pH 6·8 because the roots of the trees are able to penetrate to the relatively unaltered underlying calcareous material and extract calcium which is returned to the surface in the litter to maintain the high pH value. In an adjacent situation there is a tall grass community where the surface soil is at pH 5·5 because the calcareous material is well below the reach of the grass roots. This is also an excellent example of the variability in the composition of the litter supplied by different species to the soil surface.

Vertebrates

Mammals

Most mammals roam freely over the surface of the ground but there is a relatively small number that inhabit the soil and contribute directly towards its formation, they either live on other soil organisms or forage on the surface eating plant material. Included in this group are rabbits, moles, susliks and prairie-dogs, which burrow deeply into the soil and cause considerable mixing, often bringing subsoil to the surface and leaving a burrow down which topsoil can fall and accumulate within the subsoil. Perhaps the most conspicuous and classical examples of this phenomenon are the krotovinas found in many chernozems and kastanozems especially those of Europe (see plate IIa) where blind mole rats are principally responsible.

Moles are of common occurrence in soils beneath temperate deciduous forests or in cultivated fields, normally they do not burrow below the upper horizons so that there is not the same dramatic manifestation of their activity as with the blind mole rats. Their presence in grass fields is clearly demonstrated however by the small mounds of soil or 'mole hills' that they form (Fig. 26). Moles burrow in search of food particularly earthworms; consequently a high frequency of mole hills usually indicates a high earthworm population.

Other burrowing organisms are the African mole rats, Australian marsupial moles, chipmunks, gophers, ground squirrels, marmots, mice, mountain beavers, snow shoe rabbits, shrews, voles and woodchucks. Generally this group plays a minor role in soil formation but locally some of them can be important. The behaviour of beavers is somewhat unique in that they build dams which can cause water to accumulate, resulting in the development of very marshy conditions and waterlogging of the soil particularly when the dams are abandoned or poorly maintained.

In certain areas, where there is uncontrolled grazing and mammals such as goats are allowed to multiply freely they devour most of the vegetation leaving the soil surface bare for erosion to take place both by wind and water. This is conspicuous in many countries bordering the Mediterranean Sea where much top-soil has been lost through continuous overgrazing and progressive erosion. Even domesticated animals may be deleterious to the soil which can become compacted at the surface due to overgrazing by large animals such as cows and sheep, thereby reducing the permeability and increasing run-off which can be an erosion hazard.

Other vertebrates

These include various birds, snakes, lizards and tortoises that make their nest in the soil; however their importance is minimal.

Microorganisms

This group of organisms includes a large number of bacteria, fungi, actinomycetes, algae and protozoa, all of which enter into umpteen poorly understood processes in the soil. An interesting feature is the wide distribution of many species occurring in tropical as well as in polar areas.

The predominant microorganisms are the bacteria and fungi. Bacteria are the smallest and most numerous of the free-living microorganisms in the soil where they number several million per gram with a live weight variation of between 1000 and 6000 kg per ha in the top 15 cm. This weight is slightly less than that of the fungi but greater than that of all the other microorganisms combined (Clark 1967).

Unique among the microorganisms are the algae which are unicellular organisms containing chlorophyll and therefore are capable of photosynthesis. However, their requirement of light for photosynthesis makes it imperative that they occur predominantly at

FIG. 26. Mole hills in a field of cultivated grass in Britain. Each mole hill is about 30 cm in diameter and 15 cm high.

or near to the surface of the soil. This is probably not an ideal habitat rendering them vulnerable to rapid fluctuations in moisture and temperature.

Protozoa are also found in large numbers, mainly in the organic horizons. Their preferred habitat is in moist and wet situations but they can survive dry periods by encysting. The rhizopods and flagellates are the principal forms but also present are a few ciliates and altogether may number 10^3 to 10^5/g with a biomass of 5 to 20 g/m^2.

The distribution of microorganisms in the soil is determined largely by the presence of a food supply. Generally they need a substrate with adequate and balanced nutrition and occur in the greatest numbers in the surface organic matter, on faeces or in the rhizosphere. Most members have an optimum temperature range of 25–30°C but the thermophilic bacteria which are important in the desert require much higher temperatures. In addition microorganisms require an aerobic environment but some important bacteria exist in both aerobic and anaerobic conditions. Bacteria and actinomycetes thrive in soils at pH 7 or slightly higher but are not as frequent as fungi in acid soils, while in alkaline soils some actinomycetes find favourable conditions for their growth. As a result of their restricted mobility microorganisms can only travel short distances or are immobile as in the case of the fungi and actinomycetes which can penetrate and exploit new areas only through the growth of their mycelia. In the absence of food all microorganisms enter a resting stage until the arrival of a fresh supply. Thus the general picture of microorganisms in soils is one of the organisms consisting mainly of resting stages in a mosaic of microhabitats bursting into activity when some event brings a fresh food supply to their immediate environment (Warcup 1967).

Microorganisms are divided into two groups; the heterotrophs and autotrophs. The former obtain their food and energy from complex organic compounds while the latter derive their body carbon solely from the carbon dioxide of the atmosphere.

Heterotrophic microorganisms

These organisms are further subdivided according to their activity in the soil into the following four categories:

1. Saprophytes   3. Parasites
2. Symbionts   4. Predators

Saprophytes

This is the largest group of microorganisms that derive their energy and body carbon from complex carbon compounds and consequently play a major role in the transformation of the vast quantities of dead organic material, both animal and vegetable, that reach the soil. They are also of immense importance through their participation in nitrification, humification and other processes in the soil system. Included in this group are most of the fungi and bacteria and probably all of the actinomycetes, but the proportions of the various types vary rather dramatically from place to place depending upon environmental factors such as pH, temperature, nutrient level and moisture content.

Symbionts

There are not many members of this group of organisms but a few are of fundamental importance to some of the higher plants, forming relationships such as the composite fungus-root organ known as *mycorrhiza* which is found on many arborescent angiosperms and conifers especially the *Pinaccae*. Mycorrhizae are of two types, ectotrophic and endotrophic. The former develop a complete sheath of fungal tissue which encloses the terminal rootlets of the root system while the latter, which are more widespread, ramify through the cortex of the root. Their function is similar to that of root-hairs, aiding in the uptake of nutrients and water, for mycorrhiza are more competitive than plant roots particularly when substances are relatively unavailable.

Another type of symbiotic relationship exists between bacteria and the root systems of many plants particularly with members of the *Leguminoseae*. In this case bacteria (*Rhizobium* sp.) form little colonies or nodules beneath the epidermis of the root where they fix atmospheric nitrogen which then passes into the conducting system of the plant and serves as a nutrient.

Parasites

Among the microorganisms there are a large number that are parasitic on all forms of life be it plant or animal, but only a few live in the soil; however they can be of considerable economic importance when they attack crop plants. Important members of this group include *Poria hypobrunnea* of tea, *Fomes lignosis* of rubber and *Ophiobulus graminis* (Take all) of wheat and barley.

Predators

The protozoa are really the only predaceous microorganisms of significance in the soil. Apart from the relatively small amount of fine particles of organic matter that they consume their main food supply is composed of other microorganisms particularly bacteria. They do not enter into any reaction with the soil therefore their main function appears to be the control of the bacterial population.

Autotrophic microorganisms

Since the autotrophic organisms convert the carbon of carbon dioxide into body tissue they require to have an external source of energy which is obtained either from sunlight or by the oxidation of a substrate. When sunlight is the source, the organisms contain chlorophyll and are photosynthetic, when it is oxidation the organisms are chemosynthetic and perform inorganic oxidations.

The higher plants, most algae and some bacteria are photosynthetic autotrophic organisms and are the primary suppliers of organic matter to soils. Although the higher plants are usually the main contributors some algae may be important particularly in arid and semiarid areas. The photosynthetic behaviour of the algae is similar to that of the higher plants, but the photosynthetic bacteria can only oxidise substrates such as hydrogen sulphide to sulphur or sulphate or relatively simple organic compounds.

Some algae are capable of fixing atmospheric nitrogen and therefore supplement the nitrogen content of the soil. This is of considerable importance in rice soils where the amount of N fixed by algae is often sufficient to dispense with the application of nitrogenous fertilisers and has led to algal innoculations in some places in India and Japan.

The chemosynthetic organisms are bacteria that derive their energy from a wide variety of oxidation reactions. Probably the most important of these in the soil is the oxidation of hydrogen sulphide to sulphur and then sulphur to sulphate, ammonia to nitrites, nitrites to nitrates and ferrous iron to ferric iron. Autotrophic bacteria are unable to fix atmospheric nitrogen, nor can they easily utilise amino acids therefore they have to obtain their nitrogen supply from a mineral source such as ammonium or nitrate ions.

Mesofauna

This group of organisms comprises the large number of varied species that are in the main discernible with the naked eye and ingest organic matter. It includes earthworms, enchytraeid worms, nematodes, mites, springtails, millipedes, centipedes, a few crustacea, some gastropods and many insects, particularly termites. Like microorganisms, their distribution is determined almost entirely by their food supply and therefore they are concentrated in the top 2 to 5 cm of the soil; only a few, such as earthworms and termites, penetrate below 10 to 20 cm. Generally the mesofauna require a well aerated soil since they require atmospheric oxygen and thus cannot live in waterlogged or wet puddled soils, they also have a narrow temperature range with optimum conditions about 25–30°C. While most types prefer conditions around neutrality some are tolerant of acid conditions, particularly mites and springtails.

A simple classification according to their activity is as follows.

1. Organisms ingesting organic and mineral material: earthworms, termites, millipedes, enchytraeid worms, dipterous and coleopterous larvae.
2. Organisms ingesting organic material: earthworms, millipedes, enchytraeid worms, mites, springtails, snails, slugs, ants, termites, centipedes.
3. Organisms transporting material: earthworms, millipedes, enchytraeid worms, ants, termites.
4. Organisms improving aeration and structure: earthworms, millipedes, ants, termites.
5. Predators: nematodes, centipedes, mites, snails, slugs.
6. Parasites: nematodes, arthropods.

Organisms ingesting organic and mineral material

Earthworms are probably the best example of this class of organisms since they consume a greater volume of material than the others, however millipedes and enchytraeid worms may be important locally.

The passage together of mineral and organic material through the alimentary system of organisms is a unique process, during which some of the organic material is decomposed to provide energy and body tissue for the organism, and simultaneously the remainder is formed into a homogeneous blend with the mineral material. The resulting mixture is rich in ammonia from the urine of the earthworm and provides a suitable habitat for microorganisms causing them to proliferate rapidly. An interesting feature about the passage of material through the earthworm is that the microorganisms are in no way affected but continue their activity completely uninterrupted, indeed it would appear that the alimentary system of earthworms does not have a specific microflora, it is the same as that of the surrounding soil. A further unique behaviour of the earthworm is the secretion of calcium carbonate by the calciferous gland which gives the cast a slightly higher pH value than the surrounding soil.

Organisms ingesting organic material

This is the largest group of the mesofauna, of which the surface feeding earthworms and millipedes are the most conspicuous. However the most numerous and widespread are the arthropods including mites whose chief activity is the decomposition of the organic matter. A few arthropods are parasitic on plants or are vectors of plant and animal diseases and therefore are of considerable economic importance. The mites are most common in the accumulation of surface organic matter in acid soils of cool humid areas such as podzols, from which earthworms are normally absent. They exceed several thousand per sq m, and consume mainly the easily digestible plant remains. In the case of dead roots and stems they avoid the outer lignified tissues but burrow into the softer central part where their presence and progress is usually shown by the occurrence of small ovoid faecal pellets, easily seen in thin sections under the microscope (Fig. 104). Mites also consume large amounts of fungal mycelium, fragments of which are found in their faecal pellets. Among the insects that ingest organic matter are the ants and termites. The latter are found mainly in hot climates where they consume vast quantities of dead organic matter with such efficiency that they keep the soil surface almost bare of litter and decomposing organic material.

Organisms transporting material

Within many soils large amounts of material are transported from one place to the other by earthworms, enchytraeid worms and millipedes. Certain species of earthworms also bring material to the surface to form their characteristic casts, the shape of which varies with species. Some produce a roughly conical cast whereas others produce tower shaped forms. It has been estimated that under optimal conditions in Australia where earthworms have a biomass of 80 g/m² it takes about 60 years for a volume of top soil equal to that of the top 15 cm of soil to pass through the alimentary system of the earthworm population, but this figure can only be a very rough approximation (Barley 1959).

Ants and termites also transport considerable amounts of material from one place to the other, particularly the latter which bring large amounts of material to the surface to build their termitaria (Fig. 27). Termites are selective, moving only material less than 1 mm in diameter therefore they can alter the particle size distribution of soil in which they operate. It has been suggested that up to 2 m of fine textured top soil in East Africa and elsewhere have been formed by their activity (Webster 1965).

Organisms improving aeration and structure

When organisms burrow into the soil they produce a network of passages which facilitate the movement of water and so improve aeration. In addition, organisms that ingest both mineral and organic material produce faecal pellets which are invariably more stable than other soil peds and therefore represent an improvement in structure.

These processes can be very important in soils that tend to lose their structure easily and become compacted. In certain soils most of the aggregates in the upper horizons are fragments or entire faecel pellets of earthworms, hence their better aeration and high fertility status.

Predators

Among the mesofauna in the soil, the main predators are the nematodes, which are characterised by their very restricted feeding habits (Neilsen 1967). Their food seems invariably to be protoplasm, obtained by devouring entire organisms such as bacteria or by piercing the cell walls of fungi or plant cells with their stylets. There are over 10,000 species of nematodes of which about 2,000 inhabit the soil, the remainder occur in sea or fresh water or are parasites in plants, animals or man. The density of the population in the soil varies considerably from place to place but is generally 5—10 million per m².

A second group of predaceous organisms are centipedes which feed on a variety of smaller arthropods, worms, snails and slugs. Other predators include some mites, slugs and insect larvae.

Parasites

In addition to being the main group of predators, nematodes are also the main group of soil inhabiting parasites among the mesofauna but in this role they play a negligible part in soil formation.

Man

The activities of man are too many and too diverse to be considered adequately, without lengthy treatment. His influence has been so widespread that in many countries it is difficult to find soils that have escaped his influence of one sort or another. Generally his main role is the cultivation of soils for the production of food or tree crops but in carrying out these operations his practises are often very poor causing

FIG. 27. A termitarium in Western Nigeria.

impoverishment of the soil and erosion. Brief mention is made in Chapter 6 of the different ways the various soils can be utilised for agriculture and forestry, some indication being given of the main ameliorative methods and precautions that have to be adopted.

TOPOGRAPHY

Topography refers to the outline of the earth's surface and is synonymous with relief. It includes the dramatic mountain ranges and flat featureless plains both of which give the impression of considerable stability and permanence and that seem to be timeless. However this is not the case, for it is known from many investigations that all land surfaces even those in areas of very hard rocks such as granites are constantly changing through weathering and erosion. Further it is well established that ultimately all mountainous areas will be worn down to flat or undulating surfaces but this process is very slow, taking many millions of years in the case of mountains such as the Himalayas and the Andes. However there are a few topographic features such as sand dunes and volcanoes that can change or develop quite rapidly. Thus topography is not static, but forms a dynamic system, the study of which is known as *geomorphology*.

The nature of the topography can influence soils in many ways, for example, the thickness of the pedounit is often determined by the nature of the relief. On flat or gently sloping sites there is always the tendency for material to remain in place and for the pedounit to be thick but as the angle of slope increases so does the erosion hazard, resulting in thin soils on strongly sloping ground. Where there is a thick cover of vegetation, soil will accumulate on fairly steep slopes.

It follows that soils on steep mountain slopes are shallow and often stony, containing many primary minerals, but those of the less steeply sloping old land surfaces contain a high percentage of clay and resistant minerals. In addition it has been pointed out already that relief determines aspect and strongly influences the moisture regime of many soils. Also, in areas where the difference in elevation between the highest and the lowest point is great, then climatic changes are introduced. These differences in elevation, aspect, slope, moisture and soil characteristics lead to the formation of a number of interesting soil sequences some of which are discussed in Chapter 8.

Topographic features fall into three main categories, those produced by tectonic processes, those formed by erosion and those formed by deposition. Set out below in Table 9 are the principal topographic features and their processes of formation: definitions of the common phenomena are given in the glossary.

Tectonic processes

Initially all major relief features are produced by tectonic processes whether they are uplift, subsidence, differential lateral movement, or vulcanism; subsequently, the surfaces are acted upon by water, ice, wind, frost and mass movement which are the principal agencies of erosion and deposition.

Running water

This is the main agency of erosion and deposition but the exact processes and stages through which the landscape must pass are not well established. This is a subject surrounded by much controversy, however there are two principal schools of thought; those of Davis (1954) and Penck (1953). The former regards the land surface as going through a series of stages during which the surface gradually becomes subdued and eventually forms a peneplain. In contrast to this Penck suggests that after the initial incision and down cutting by running water, there is parallel retreat of the slopes to form pediments and pediplains. These two ideas are shown diagrammatically in Fig. 28. Although there is evidence for both theories, the study of very old land surfaces in Africa and elsewhere gives stronger support to the ideas of Penck. These contrasting theories apply more specifically to humid environments. Since the earth's surface has a variety of climates it is reasonable to assume that each set of climatic conditions will produce its own set of topographic features and this has led to the concept of morphogenetic regions (Peltier 1950). Each morphogenetic region has its own range of temperature and precipitation which together produce specific processes forming characteristic topographic features (Fig. 29). Surprisingly, when most land surfaces are examined in some detail they display well developed pediments or remnants of pediments which appear to have formed under humid conditions. This is probably the result of the prolonged period of erosion and deposition that existed during the Tertiary period when warm and humid conditions prevailed over most of the earth's surface. These conditions have left an indelible imprint and produced landscapes similar to those suggested by Penck but now, superimposed upon these Tertiary landscapes are the influences of many of the other morphogenetic processes. However they have changed only slightly the main outline of the landscapes, but are responsible for a large number of diverse and small topographic features such as terraces, moraines, dunes and solifluction deposits.

Disregarding the theories of landscape formation, the material removed by erosion is normally transported in rivers and later deposited to form alluvial deposits or taken to the sea to build deltas. In arid areas where rivers are few or absent the material produced by weathering either remains *in situ* or is carried only a short distance, usually to the bottom of the

slopes where it accumulates to form alluvial cones or bahadas.

In areas of limestone where solution is the principal process of weathering a unique type of landscape known as karst is developed. This is characterised by stony surfaces and the occurrence of dolines which are depressions formed by the gradual slumping of material down solution channels (Fig. 30).

Moving ice

Erosion by ice, or glacial erosion, is a relatively local process at present but was more widespread in the past when it was superimposed upon land forms that developed mainly under the influence of water during the Tertiary period. Glaciers can be powerful eroding agents particularly in very mountainous areas where their most spectacular effects are the formation of U-shaped valleys and the removal of pre-existing soils and weathered rocks. The material carried in glaciers (Fig. 31) is later deposited on land or in the sea. When

Table 9

PRINCIPAL TOPOGRAPHIC FEATURES AND THEIR METHODS OF FORMATION

PROCESSES		TOPOGRAPHIC FEATURES
TECTONIC	Uplift	Faults, folded mountains, plateaux, raised beaches, table lands
	Subsidence	Rift valleys (graben)
	Vulcanism	Ash and agglomerates, calderas, lava flows, volcanoes
RUNNING WATER	Erosion	Bad lands, cuestas, dolines, gorges, inselberge, karst, mesas and butes, pediments, peneplains, tors, valleys
	Deposition	Alluvial fans, alluvial plains, alluvial terraces, bahadas, deltas, lévées
MOVING ICE	Erosion	Aretes, cirques, crag-and-tail phenomena, hanging valleys, roches moutonnées U-shaped valleys
	Deposition	Drumlins, eskers, moraines, till (boulder clay)
FROST	Erosion	Gullies, jagged rock surfaces, tors
	Deposition	Talus cones, solifluction terraces
WIND	Erosion	Desert pavements or hamadas, dreikanters (ventifacts)
	Deposition	Dunes, loesses
MASS MOVEMENT	Erosion and Deposition	Landslides, soil-creep, solifluction terraces

on the land the configuration of the surface is usually altered by characteristic depositional features, such as drumlins, eskers, moraines and till (Fig. 32).

Wind

Today, vigorous wind action is confined largely to arid and semi-arid areas, and to coastal positions, where the formation of sand dunes is a most characteristic and conspicuous process. However during the Pleistocene period, wind activity was widespread in areas peripheral to glaciers. From these areas winds picked up much fine material from the deposits in front of melting glaciers and transported it long distances to form loess deposits such as those found in Europe and North America (Fig. 33). This process is still active in certain glaciated areas. In contrast, some very thick loess deposits such as those of northern China have been derived from material blown off the Gobi desert. Loess usually forms a thick featureless blanket

which entirely obliterates the pre-existing topographic features.

Frost

Repeated freezing and thawing of wet unconsolidated material on slopes causes a considerable amount of movement. This process is known as solifluction and is probably the principal geomorphological process taking place in polar areas leading to the accumulation of great thicknesses of material on the lower part of the slopes. Often a number of topographic features such as terraces, result from this process (Fig. 42). Since the action of frost is intimately associated with soil formation a fuller account is given in the next chapter (see page 63 et seq.).

Mass movement

Mass movement, without the aid of frost is often

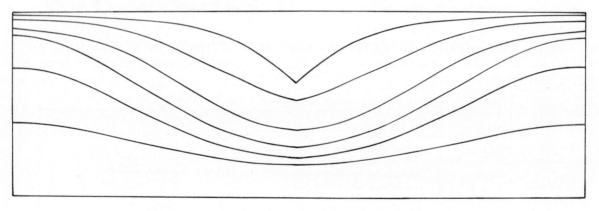

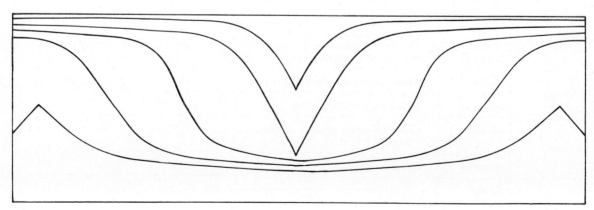

FIG. 28. Contrasted sequences of valley profiles from youth to old age. *Above*: according to Davis
Below: according to Penck.

ignored as a process of landscape formation but Sharpe (1938) has attempted to demonstrate that although the amount of soil creep and the frequency of landslides per annum may be small, the total number measured over many thousands of years may bring about considerable alteration to topography. In tropical and subtropical areas the high frequency of stone lines and the arching of material down the slope is a clear indication that mass movement is taking place.

The result of the various denudation and depositional processes discussed above is to produce a wide range of geomorphological features at various stages of development. Sometimes development may be arrested or accelerated by a change of climate or by renewed tectonic activity. For example many of the land forms in the Australian desert were produced by running water under humid conditions but development has virtually stopped since the climate became dry. Similarly, many old land surfaces with well developed pediments are found at high elevations in East Africa because of renewed tectonic activity and uplift during the Tertiary period. A further result of geomorphological processes is that large stretches of land are now covered with glacial drift, loess or alluvium which

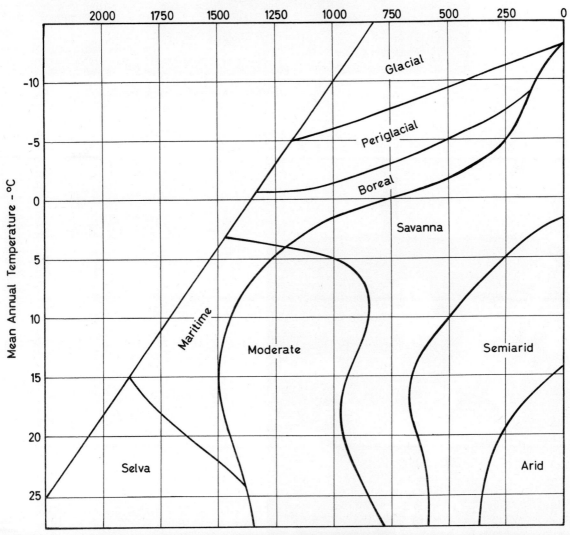

FIG. 29.   Diagram of climatic boundaries of morphogenetic regions (adapted from Peltier 1950).

FIG. 30. Sink holes in karst area of western Yugoslavia.

FIG. 31.
A melting glacier in Svalbard containing numerous layers of rock detritus some of which may be deposited, eventually to form parent material.

are new parent materials, often quite unrelated to the underlying material. So that at present the major elements of the landscape in most places reflect Tertiary erosion and sculpturing, but superimposed upon them are the various influences of the Pleistocene and Holocene periods.

FIG. 32. Stratified glacial deposits in Eastern Scotland. The two lower strata are unsorted compact till with angular stones and boulders. The upper stratum is a glacio-fluvial deposit hence the higher concentration of stones and boulders many of which are rounded or subrounded.

FIG. 33. A loess deposit in central Poland. Note the characteristic prismatic structure in the upper part of the section and the dark coloured buried soil near to the base.

TIME

Soil formation is a very slow process requiring thousands and even millions of years, and since this is much greater than the life span of any individual human being, it is impossible to make categorical statements about the various stages in the development of soils. A further complication is introduced by the periodic changes of climate and vegetation which often deflect the paths of soil formation in one direction or another. Therefore, all that is said about time as a factor in soil development is in part speculation and in part deductions based mainly on geological, geomorphological and palynological evidence. In a number of cases radiocarbon dates have been used to establish the age of a soil.

Not all soils have been developing for the same length of time. Most started their development at various points during the last hundred million years. In Table 10 are set out the principal events of the last 75 million years as related to soil formation. The divisions in the Table are not fixed date boundaries for soils, indeed many soils that started their development in one period have continued to develop in a succeeding period or periods.

Some horizons differentiate before others, especially those at the surface which may take only a few decades to form in unconsolidated deposits. Middle horizons differentiate more slowly, particularly when a considerable amount of translocation of material or weathering is necessary; some taking 4000–5000 years to develop. Some other horizons require even longer periods; weathering of rock to form a flambon may require more than a million years. The evidence for this last statement lies in the fact that flambons are found only on very old land surfaces which have been exposed to weathering since at least the Tertiary period.

Further evidence regarding the speed of horizon differentiation when complete weathering takes place, is gained by a study of the interglacial soils in the central U.S.A. where there are some strongly weathered horizons known as 'gumbotil'. These occur either at the surface or are sandwiched between two glacial deposits and many have formed during a single interglacial period extending over 50,000–70,000 years (see page 273). The rate of development of these horizons can be regarded as maximal since they have formed in a zone of permanent saturation where hydrolysis is most vigorous. A particularly important feature of soils is that they pass through a number of stages as they develop, culminating in deep pedounits with many well differentiated horizons. The principles underlying the development of soils in relation to time are considered in the section on soil evolution given below.

Soil evolution

Initially pedologists tended to interpret most soil features as the result of the interaction of the prevailing environmental conditions at the time when examinations were made. However, it soon became evident that most places have experienced a succession of different climates which induced changes in the vegetation and in the soil genesis. Therefore, most soils are not developed by a single set of processes but undergo successive waves of pedogenesis. Furthermore, each wave imparts certain features that are inherited by the succeeding phase or phases. In a number of cases these properties are developed so strongly that they are evident thousands and even millions of years after their formation. The outcome is that soils are regarded as having developmental sequences which manifest not only the present factors and processes of soil formation, but also a varying number of preceding phases.

Although changes in the external environmental factors are usually responsible for progressive changes in the soil, vegetation and landscape, there are a number of cases when the environmental factors remain relatively constant but continuing soil development leads to the formation of new features within the soil. Some of these can become very prominent and may themselves influence the further course of soil formation. The progressive development of an impermeable horizon can cause waterlogging at the surface and changes in the soil characteristics with corresponding changes in the vegetation. These various progressive changes in soils are known as *soil evolution*. Changes produced by variation in external factors are *exogenous evolution* and those due to internal changes are *endogenous evolution*. Examples of these changes have been reported from every major land surface and, set out below in Tables 10 and 11 are some of the major events that have produced evolutionary changes in soils. This is followed by an outline of their effects, while in Chapter 8 details are given of a number of the more interesting and widespread types of soil evolution.

King (1967) suggests that the oldest land surfaces, such as those in parts of Africa and Australia, developed in the mid-Tertiary period, but one or two may date from the early part of the Tertiary or even the Cretaceous. If this is the case, then the soils which occur on these surfaces may be very old. Consequently, any consideration of soil development and change through time must commence by discussing the events that took place during the early part of the Tertiary period. The evidence afforded by plant remains shows clearly that the surface of the earth at that time did not have the extremes of climate that it experiences at present. The flora from the late Cretaceous to the Miocene period in the northern latitudes grew under warmer conditions as indicated by the occurrence of palms in Canada and Britain, where conditions could be regarded as subtropical. However, full tropical conditions existed in many of the same areas where they are found today, as well as in many of the present subtropical and arid areas, including such places as central Australia and the southern part of the Sahara desert. These warmer conditions caused the rocks to undergo profound weathering by hydrolysis in the cases of the silicates and by solution in the case of limestone and other soluble rocks. Considerable natural erosion also took place during this prolonged period to form the characteristic flat or gently sloping surfaces of planation associated with old landscapes. But weathering and soil formation kept pace with and even proceeded faster than erosion so that great thicknesses of soil and weathered rock were maintained. The processes of weathering were so complete in many places that rocks of all types were transformed into clay minerals, hydroxides and oxides of iron and aluminium together with their resistant residues such as quartz and zircon, but sometimes even these primary minerals have decomposed. Most of the bauxite deposits of the world are Tertiary phenomena resulting from profound chemical weathering.

The warm, humid conditions were maintained in tropical and subtropical areas throughout most of the Tertiary period, but in the poleward areas the climate became progressively cooler from the end of the Miocene so that by the end of the Pliocene the floras of Eurasia and North America bore many similarities to those of the present, but they contained more species particularly in western Europe, where the present flora is a poor reflection of the past. The gradual cooling of the climate culminated in the Pleistocene period with its repeated glaciations both in the northern and southern hemispheres but it is in the former with its greater circumpolar land surface that the effects on pedology are most significant and widespread. There

Table 10

GENERALISED GEOCHRONOLOGICAL DATA FOR NORTH CENTRAL EUROPE AND CENTRAL NORTH AMERICA, TERTIARY TO HOLOCENE

PERIOD	TIME (Years BP)	NORTHERN EUROPE	ALPINE REGION	CENTRAL NORTH AMERICA
HOLOCENE	Present	HOLOCENE	HOLOCENE	HOLOCENE
	10,000			
		WEICHSEL	WÜRM	WISCONSINAN
	125,000			
		Eemian	Riss/Würm	Sangamon
	235,000			
		SAALE	RISS	ILLINOIAN
	360,000			
		Holstein	Mindel/Riss	Yarmouth
PLEISTOCENE	670,000			
		ELSTER	MINDEL	KANSAN
(Glacial and inter-glacial stages; Glacial periods in block capitals)	780,000			
			Gunz/Mindel	Afton
	900,000			
			GUNZ	NEBRASKAN
	1,150,000	Various deposits of 'cold' and 'warm' phases		
			Donau/Gunz	
	1,370,000			
			DONAU	
	1,600,000			
			Pre Donau	
	2,500,000			
TERTIARY		PLIOCENE (Development of man)		
	12,000,000			
		MIOCENE		
	25,000,000			
		OLIGOCENE		
	32,000,000			
		EOCENE		
	60,000,000			
		PALEOCENE		
	70,000,000			

were five distinct glacial episodes, the fourth or Saale (Illinoian) glaciation, being the most extensive and covered large parts of Eurasia and North America. These glaciations together with their associated periglacial conditions effectively removed most of the deep soils and weathered rock that formed during the Tertiary period. However there is still sufficient present to leave no doubt about its previous ubiquity or about its contribution to the various superficial deposits of the Pleistocene. Much of the so-called boulder clay (glacial drift) contains fine material derived from a previously weathered soil, mixed with fresh rock fragments.

Perhaps the most impressive soil evolution has taken place in Australia and various parts of central Africa particularly those bordering the Sahara where soils that developed under hot humid conditions are now partially fossilised in an arid environment.

Most other tropical and subtropical areas experienced periods of higher rainfall or *pluvial conditions*

Table 11

GENERALISED GEOCHRONOLOGICAL DIAGRAM OF NORTH CENTRAL EUROPE, LATE WEICHSEL TO THE PRESENT

PERIOD	TIME (Year BP)	CLIMATIC PHASES	VEGETATION	SOIL	CULTURE
	Present				
		SUB-ATLANTIC Cool – Dry	Beech, Spruce and Pine in the north	Altosols, Argillosols, Peat, Placosols, Podzols, Subgleysols	HISTORICAL IRON AGE
	2,500				
HOLOCENE (Post glacial)		SUB-BOREAL Warm – Dry	Oak, Ash, Spruce and Pine in the north	Progressive leaching	BRONZE AGE (Start of Agriculture) NEOLITHIC
	5,000	ATLANTIC Warm – Moist	Oak, Elm	Spread of peat	MESOLITHIC (Hunters and Collectors)
	7,500	BOREAL Warm – Moist / Moister	Oak, Hazel / Pine, Hazel	Altosols, Podzols, Subgleysols	
	9,000	PRE-BOREAL Cool	Birch, Pine	Altosols, Podzols, Subgleysols	
	10,000	UPPER DRYAS Sub-Arctic	Grasses, Sedges	Crysols, Solifluction	PALEOLITHIC
LATE WEICHSEL (Late Glacial)	11,000	ALLERØD Warm Sub-Arctic LOWER DRYAS Sub-Arctic	Birch Grasses, Sedges	Rankers Cryosols, Solifluction	
	15,000				

during the Pleistocene. Many workers regard these periods as coincident with the glacial episodes but the evidence for this is fragmental and the chronology is not as firmly established as that for the glaciated areas. It is certain that erosion by water did reduce the thickness of soil and weathered mantle in many tropical and subtropical areas but the effect was small as compared with widespread and deep stripping caused by glacial and periglacial processes.

The rapid climatic amelioration at the end of the Pleistocene possibly induced the most dramatic set of changes to be experienced by the earth's surface. This caused wave after wave of vegetation to pass rapidly over many areas particularly in western Europe and North America. These were quickly followed by the influence of man, so that soil development in many places in the middle latitudes has been subject to climatic conditions varying from full glacial to marine in the relatively short period of 10,000 years. As a result, many of the soils in this area have diverse properties inherited from the Tertiary and the Pleistocene periods.

In historical times many evolutionary paths have been directed by man's activity: the polders of Holland, the rice soils of eastern countries and the catastrophic erosion in many of the Mediterranean countries and North America quickly spring to mind. An interesting feature to emerge from recent investigations is that many plagioclimaxes in the vegetation may be due to the activity of man. Excellent examples are afforded by the *Ericaceous* heaths of Europe and many of the savannas of the tropics.

3

PROCESSES IN THE SOIL SYSTEM

Soils are complex, dynamic systems, in which an almost countless number of processes are taking place. Generally these processes can be classified as chemical, physical or biological, but there are no sharp divisions between these three groups; for example, oxidation and reduction are usually regarded as chemical processes but they can be accomplished by microrganisms.

Similarly the translocation of mineral particles can take place either in suspension or in the bodies of organisms such as earthworms. In this chapter the processes are discussed in general terms whereas in Chapter 6 they are related to the formation of specific soils. Some processes have been discussed in the previous chapter but they are mentioned again.

CHEMICAL PROCESSES

The main chemical processes taking place in soils are:

Hydration	Clay mineral formation
Hydrolysis	Oxidation
Solution	Reduction

Hydration

This is the process whereby a substance absorbs water to form a new compound which differs only slightly from the original state. Few of the primary minerals undergo direct hydration, therefore very little takes place during the early stages of weathering and soil formation. The principal exception is biotite which absorbs water between its layers, expands and finally splits apart (Fig. 34). Hydration is more often a secondary process following the hydrolysis of primary minerals and the formation of decomposition products which are then hydrated to varying degrees. The principal compounds affected in this way are iron and aluminium oxides.

Hydrolysis

This is probably the most important process participating in the destruction of minerals and determining the course of soil formation. Because of its complexity it is not well understood, however when defined in chemical terms it appears quite simple and straightforward. Hydrolysis of minerals is the replacement of cations in the structure of the primary silicates, by hydrogen ions from the soil solution, and eventually leading to the complete decomposition of the minerals.

Due to variations in the structure of the primary minerals (see page 13 *et seq.*), the course of hydrolysis does not follow the same path for each type, however it is always initiated by the replacement of basic cations such as Na, K, Ca and Mg, by hydrogen ions. The next stage for the ortho- silicates like olivine and the chain silicates (pyroxenes and amphiboles) is the removal of iron which links together the individual tetrahedra or the tetrahedral chains. This is sufficient to cause a high degree of decomposition of these minerals since these ions are responsible for holding their structure together.

During the hydrolysis of felspars with their framework structure the removal of the basic cations does not have the same marked influence upon their structure but the subsequent removal of aluminium from the tetrahedra causes severe weakening which leads to complete decomposition of the mineral. This takes place because aluminium is in a position of 4 coordination in the felspar structure so that its removal leaves a large number of broken bonds each of which is then individually satisfied by a hydrogen ion. However there is only weak bonding between the hydrogen ions, and the structure eventually collapses. The rate and efficiency of this reaction is influenced by the following factors:

1. Surface area
2. pH
3. Volume and speed of water flowing through the soil
4. Temperature
5. Chelating agents
6. Removal of substances by precipitation.

Surface area

Unconsolidated deposits have a greater surface area than consolidated rocks and therefore decompose much faster. The initial stages of hydrolysis of con-solidated rocks destroy the 'cementing' agent between the minerals, cause disaggregation so that the constituent minerals are separated one from the other, and produce a rapid increase in the surface area. A further increase is caused by weathering along the cleavages of the minerals, thus hydrolysis effectively increases the surface area which in turn increases the rate of weathering.

pH

The lower the pH values the more vigorous is the ensuing hydrolysis but since soil acidity seldom falls below pH 3·5 the intensity of the reactions is small when compared with that in mineral acids. The lowest pH values usually occur in the surface horizons in association with acid decomposition products of organic matter.

500μ

FIG. 34.   Splitting apart of biolite following hydration.

Volume and speed of water flowing through the soil

When water flows through the soil the soluble products of hydrolysis are removed; this prevents the establishment of an equilibrium between the minerals and the soil solution, and thus perpetuates conditions for hydrolysis. At the same time the pH of the soil solution is maintained at fairly low values by the steady inflow of water charged with CO_2 and acid decomposition products from the organic matter.

Temperature

It has already been stated on page 27 that as temperatures increases so does the rate of hydrolysis.

Chelating agents

The rate of decomposition is increased greatly in the presence of chelating agents. This explains the rapid loosening of minerals beneath lichens and may be a very important factor in releasing plant nutrients by plant exudates in the rhizosphere.

Removal of substances by precipitation

If substances released by hydrolysis remain in solution an equilibrium may be established thus preventing further hydrolysis; on the other hand if they are removed from solution by precipitation then the reaction can proceed. This applies particularly to iron and aluminium which precipitate as oxides and hydroxides.

Solution

There are only a few substances found in soils that are soluble in water or carbonic acid. Those that are very soluble in water include chlorides, nitrates and sulphates, but these are of restricted distribution in parent materials. However they may accumulate in the soils of arid areas. Those substances soluble in carbonic acid include calcite and dolomite which are widespread and form the major component of limestone, chalk, and some other parent materials. These substances are somewhat unique since they are almost completely soluble and therefore supply only a very small residue after solution. Consequently soils developed on these materials are normally quite shallow. On the other hand most other materials, particularly the silicate rocks furnish a considerable residue of

primary minerals and secondary products. Less soluble is apatite which can persist for thousands of years in some soils of humid areas developed in drift deposits.

Some minerals such as quartz and ilmenite that are usually considered to be inert and insoluble do dissolve eventually, this accounts for the small amount of primary material $< 50\ \mu$ found in many very old tropical soils. The rate of solution is influenced in the same way by those factors given above that effect hydrolysis namely, surface area, pH, volume and speed of water flow through the soils, temperature, chelating agents and removal of substances by precipitation.

Whereas some processes are confined largely to the upper horizons, solution and hydrolysis occur throughout the soil and extend to great depths, and far below what is usually regarded as soil. This is common in many tropical areas where the considerable thicknesses of decomposed rock might be regarded more correctly as a geological phenomenon rather than as soil but it is necessary to consider them as part of the system, since the soil often grades into this material or it may form parent material.

Resistance of minerals to transformations

The minerals in the soil have varying degrees of resistance to transformation; the simple salts like gypsum are soluble, and are lost comparatively quickly from the soil system in areas of high rainfall; carbonates are less soluble and are removed more slowly. The stability of the silicates is considerably greater but within this group some are more resistant than others.

As indicated above, minerals like olivine and the pyroxenes are among the easiest to be hydrolysed whereas clay minerals are very resistant with the felspars in an intermediate position. The most resistant minerals include quartz, zircon, magnetite and titanite, which constitute part of the residue but they usually succumb in time. Resistance to weathering also varies with the size of the mineral particles and environment. Since the small soil particles with their high surface area are more important in hydrolysis, the modified stability sequence of Jackson and Sherman (1953) for fine particles is given below in order of increasing resistance:

1. Gypsum
2. Calcite, dolomite, apatite
3. Olivine, hornblende, augite, diopside
4. Biotite
5. Albite, anorthite, microcline, orthoclase
6. Quartz

7. Muscovite, clay-mica
8. Vermiculite and interstratified 2:1 layer silicates
9. Montmorillonite
10. Kaolinite, halloysite
11. Gibbsite, boehmite, allophane
12. Hematite, goethite,
13. Anatase, zircon, rutile, titanite

Products of hydrolysis and solution

The end products of hydrolysis and solution are:

> Weathering solution
> Resistant residue
> Alteration compounds

Weathering solution

This contains the basic cations together with some iron, aluminium and silicate ions, all of these solutes are lost or redistributed in the soil system.

Resistant residue

This includes minerals such as quartz, zircon, rutile and magnetite which alter only very slowly but can decompose when present in a very finely divided state.

Alteration compounds

These are principally hydroxides and oxides of iron and aluminium, silica and clay minerals; also included are manganese dioxide and titanium compounds.

Ions of aluminium and iron only remain in solution at pH 2·5 or lower, but this degree of acidity is seldom found in soils therefore these ions precipitate out of solution very quickly as amorphous hydroxides. This reaction can take place immediately after hydrolysis so that some hydroxides may form pseudomorphs after many minerals, then follows slow crystallisation to form goethite, hematite, gibbsite or boehmite, or there may be a recombination of aluminium hydroxide with silica to form clay minerals. Indeed it would appear that all secondary crystalline substances are derived from a pool of amorphous materials but in some cases the amorphous phase is only of short duration and gives the impression that crystalline substances are the initial products. Precipitation also takes place after translocation away from the site of release: for example on the surfaces of other mineral grains or peds, or within pores.

Iron oxides

These occur mainly as ferric hydroxide gel, goethite $FeOOH$ or hematite Fe_2O_3. Ferric hydroxide gel is amorphous with yellowish-brown colour and there is some evidence to show that it closely resembles ferric hydroxide precipitated in the cold in the laboratory. Goethite is crystalline with reddish-brown colour but changes to yellowish-brown as it becomes hydrated. It has a wide distribution ranging from the tropics to the arctic and is one of the main colouring substances in soils. Hematite is bright red and occurs chiefly in soils of tropical and subtropical areas or in old geological formations. The other iron oxide is lepidocrocite which is bright orange in colour but up to the present has been found only in soils subject to waterlogging where it is responsible for the bright colouration of many of the mottles. Apparently iron is present as amorphous ferric hydroxide in young soils, becoming crystalline to form goethite or hematite. Thus, the little weathered soils of temperate areas mainly have amorphous ferric hydroxide and a small amount of goethite whereas those of older landscapes have goethite or hematite or a mixture of these latter two. Within soils ferric hydroxide and ferric oxides occur as discrete particles, as coatings on sand grains and on peds, or they form part of the soil matrix as microaggregates visable only with the electron microscope.

Aluminium oxides

In a crystalline form aluminium oxide occurs mainly in the soils of tropical areas as gibbsite — $\gamma Al(OH)_3$ but it can occur in limited quantities in soils of temperate areas. In very old and highly weathered soils boehmite — $\gamma AlO \cdot OH$, and diaspore $AlOOH$ are the principal forms. These minerals form hexagonal plates which are held together by hydrogen bonding and together or separately they can accumulate to form the main constituents of bauxite. A considerable proportion of gibbsite in some soils is associated with silica gel to form allophane, particularly in temperate areas or young soils.

Silica

Amorphous or hydrous silica — $SiO_2 \cdot nH_2O$ in addition to forming part of allophane, may be lost in the drainage water or redistributed within the soil system. For example in Australia there are certain layers known as silcrete that are formed principally by the accumulation of amorphous silica. Elsewhere there are a few instances of crystallization to form secondary quartz and it has been shown that the clay fraction in the zolon of the so-called Kauri podzol in New Zealand is composed largely of cristobalite – SiO_2. Many plants absorb and accumulate large quantities of silica within their structure to form masses of opaline silica. When the plants die the silica is returned to the soil as phytoliths which have been observed on the Island of Reamie (East Africa) to comprise the uppermost horizon with a thickness of 5–30 cm (Riquier 1960).

Manganese dioxide

At present this is considered to be of rather restricted distribution being found as a blue-black coating on the surface of peds or within peds or associated with iron in concretionary deposits and certain massive materials. Usually these features form in soils subject to periods of waterlogging as shown in Plate Vb.

Titanium compounds

These occur in the form of needles as secondary titanite or as crystals of anatase but both are very rare. The white coating of leucoxene – TiO_2 around ilmenite and other minerals is more common but it is also infrequent.

Clay Mineral Formation

Early workers visualised a simple transformation from primary minerals such as felspars to clay minerals like kaolinite but as Yaalon (1959) has pointed out this requires aluminium to move from a position of 4 co-ordination in the felspar to one of 6 co-ordination in the gibbsite sheet in kaolinite. This is not possible by simple rearrangement of the lattice. For some situations many workers now favour a hypothesis that requires a complete breakdown of the felspar structure to form amorphous materials or allophane followed by a synthesis of clay minerals. It was thought at one time that allophane was associated mainly with soils derived from volcanic ash but now it appears to be much more widespread and represents an early stage of hydrolysis. However there is experimental and field data to show that the decomposition of felspars in an acid environment can lead to either kaolinite or amorphous material. Evidence for the build up of clay minerals from amorphous materials is seen in the

weathering zone in many tropical soils that contain vermiform kaolinite (Fig. 35), which does not appear to form pseudomorphs after primary minerals but grows in pores. On the other hand in some rocks, felspar crystals and micas are seen to be replaced entirely by an intricate intergrowth of vermiform kaolinite. In attempting to resolve this problem Fields and Swindale (1954) have suggested that under conditions of vigorous decomposition there is breakdown to the hydroxides and allophane followed by synthesis of clay minerals whereas, where conditions for hydrolysis are slower, there is direct formation of clay minerals.

Even without a clear picture of the mechanism of clay formation there is sufficient field evidence to show that kaolinite is formed in an acid environment from which basic cations and iron are constantly being lost. The nature of the rock can be important; since alkali felspars lose silica by solution more readily than plagioclase, granitic rocks are kaolinised more readily than mafic or Ca–Mg rocks but these latter are readily transformed into kaolinite in an environment of vigorous hydrolysis. On the other hand montmorillonite is formed in the presence of large amounts of basic cations especially magnesium and iron, as in mafic rocks and volcanic ash of intermediate composition. The formation of montmorillonite can result from granitic rocks if basic cations accumulate in the system as happens in depressions and in arid and semi-arid areas. The micas either form in an intermediate type of environment by the transformation of felspars or some other mineral, or are merely fragments of muscovite and biotite of clay dimensions.

Although clay minerals are relatively stable they can undergo transformation. There is now a large body of evidence to show that montmorillonite can be transformed into mica and this into kaolinite. It also seems that under very acid conditions, clay minerals may be broken down to hydroxides which are either lost from the soil system or translocated and deposited at some point within the pedounit.

Hydrolysis of individual minerals

The general picture which seems to be emerging from studies in many parts of the world is that the type of breakdown and nature of the alteration compounds are determined largely by the soil climate and more particularly by the microenvironment in which the mineral is found. It is common to find similar minerals yielding different end products in contrasting environments. Similarly contrasting minerals can produce identical substances. Another important feature is that the number of decomposition products is relatively few as compared with the wide range of primary minerals. By far the greatest amount of work on mineral transformation has been conducted on the micas. Walker (1949) has shown that during hydrolysis, biotite goes from black through brown to dull brown and at the same time the structure changes from biotite to mixed layered vermiculite-biotite and then to vermiculite. Biotite can undergo other transformations and may be changed to chlorite and then to vermiculite (Stephen 1952) or to chlorite and then to kaolinite or straight to gibbsite. Similarly felspars

can be changed to mica, kaolinite, halloysite or gibbsite (Figs. 36 and 37). Hornblende goes to chlorite, vermiculite or gibbsite. Minerals having a high content of magnesium and iron such as olivine and many of the pyroxenes often form pseudomorphs of hydroxides and allophane as initial decomposition products or may have an intermediate stage of chlorite.

Weathering indices

A number of indices have been produced in an attempt to assess the rate and type of weathering under different conditions but none has been found to be of

FIG. 35.   Macrocrystalline kaolinite in partially crossed polarised light.

general applicability. One approach is to assume that a particular constituent of the parent material remains constant and unchanged while the others are lost, altered or redistributed in the soil system. Many workers have used either the total zirconium content or the mineral zircon in weathering studies. The former is the least reliable of the two since it may occur in very small zircons or within a mineral and therefore it can be lost from the soil by weathering. Zircons in the size range 20–200 μ are a fairly reliable guide to mineral transformation because of their relatively large size and consequent stability, but even these are partly decomposed in some situations. However this mineral is present only in relatively small amounts, therefore it has to be used as an indicator of weathering with caution because it may be unevenly distributed in the parent material.

Quartz is extremely resistant and has an important advantage as an index mineral primarily because it is usually present in large amounts in most parent materials and therefore is less liable to be affected by variations in distribution but small grains of quartz weather slowly hence the low silt content in many tropical soils.

During the initial stages of hydrolysis there appears to be little change in volume despite considerable changes in composition. Therefore it is possible to calculate the amount of material lost and redistributed. At a later stage the structure of the rock is destroyed and some of of the most resistant minerals become weathered therefore it is not possible to calculate with a high degree of accuracy the relative gains and losses.

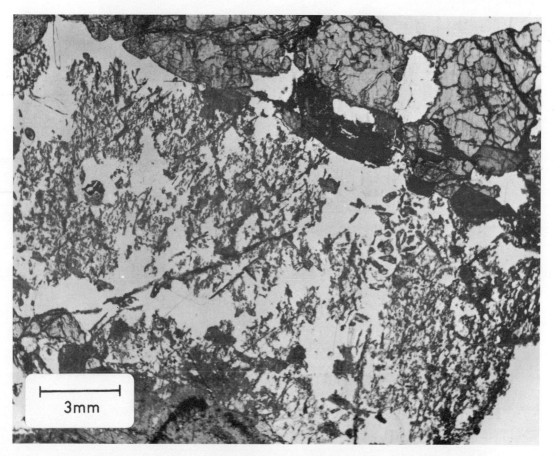

FIG. 36. Thin section of weathered granite in which the felspars have been replaced by gibbsite. In some cases the outlines and cleavages of the original felspar crystals have been preserved.

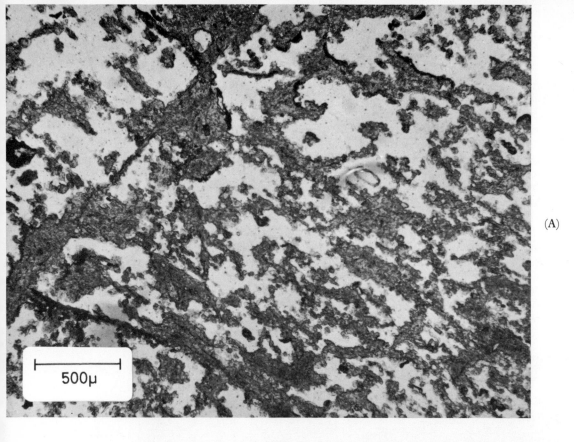

(A)

FIG. 37. Enlarged position of Fig 36. (A) Showing the finely crystalline gibbsite forming a network which is mainly a pseudomorph of the outline of the original felspar crystals and cleavages. (B) Same section in crossed polarised light showing that gibbsite is birefringent.

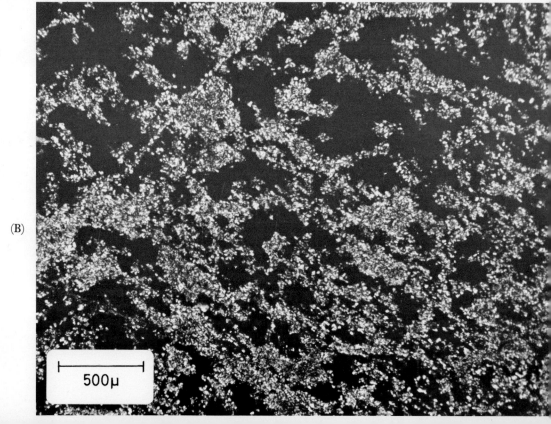

(B)

Oxidation and reduction

It is convenient to consider these two processes together since one is the reverse of the other. Iron is the principal substance affected by these processes for it is one of the few elements that is present in the reduced state in the primary mineral. Consequently when it is released by hydrolysis and enters an aerobic atmosphere it is quickly oxidised to the ferric state and precipitated as ferric hydroxide. Then it is transformed slowly to crystalline goethite or hematite, the latter forming in some soils of hot countries. If, on the other hand the iron is released into an anaerobic environment it stays in the ferrous state hence the greyish-blue colour of many constantly waterlogged soils. In addition reduction of iron from the ferric to the ferrous state is possible as in cases where soil formation is taking place under conditions of prolonged anaerobism and the parent material contains ferric iron. Dramatic examples of this are seen in areas of red sediments which have been changed to bluish-grey and even blue. These two processes can also be accomplished by microorganisms (see p. 37).

Rock weathering

The amount of rock weathering which has taken place during the Holocene period can be regarded as relatively insignificant as compared with the considerable amount of rock decomposition which took place during the Tertiary period. Therefore, in order to determine the full changes that take place when rocks are weathered it is necessary to examine old soils such as some of those found in tropical areas. Probably the most outstanding contribution to rock decomposition has been made by Harrison (1933) who has described weathering sequences for a number of different types of rocks. He found that the end product of weathering as well as the course of weathering varies with the type of rock and with the specific environmental conditions. With small modifications his findings have been confirmed for many other tropical countries therefore his results can be used to illustrate weathering in a humid tropical environment. Probably the weathering of a basic igneous rock is the best illustration of the profound changes that take place when rocks weather by hydrolysis. Therefore the decomposition of dolerite will be considered. This rock has the following composition:

Labradorite and small amounts of alkali felspar	50·0%
Augite	42·0%
Enstatite	2·0%
Biotite	1·0%
Ilmenite and titaniferous magnetite	5·0%
Olivine	<1 %
Quartz	<1 %

This sequence of weathering starts at the base of the profile with unweathered dolerite above which there is up to 2 m of weathered rock containing boulders or core stones some more than a metre in diameter. This has been formed by weathering along joint planes which were initially the only surfaces exposed to percolating water. Each core stone has five concentric rings of progressively more weathered rock. Above this layer there is mottled clay 45 cm thick, followed by a bright red layer 3 m thick, a grey sandy subsoil 25 cm thick and finally blocks of ironstone on the surface. In thin sections the greatest alteration of the rock is seen to be taking place within the innermost ring over a distance of not more than 4 mm. The first indication of alteration is shown by the occurrence of yellow staining of iron oxide and small scales of gibbsite along the cleavages of the felspars. This changes into highly corroded felspars which are replaced by crystalline gibbsite. Indeed the same felspar crystal may vary from relatively fresh at one end to completely decomposition and replaced by gibbsite at the other (cf Figs. 36 and 37). At the same time the augite is replaced by chlorite and iron oxide and there is a little secondary silica. Within the next 23 mm any remaining felspars are changed to gibbsite and the chlorite is decomposed to iron oxides. These changes are accompanied by dramatic changes in chemical composition, the most important being the considerable reduction in the amount of combined silica and basic cations and the large relative increase in the amount of aluminium and iron as shown in Table 12. Within the third and fourth layers when the structure of the rock has almost disappeared equally important changes have taken place. Apparently the gibbsite is resilicated to form kaolinite which represent 18% and 39% respectively of these two layers; also there is a concentration of iron. The fifth layer or crust does not have as much kaolin as the underlying two but there is a further increase of iron. Thus within 66·5 mm dolerite has been transformed into a weathered product containing mainly kaolinite, iron oxides and gibbsite. The mottled clay and red layers have a similar chemical composition with a higher content of silica indicative of resilication and the formation of more kaolinite. Also present are iron and aluminium oxides. The data for the sandy subsoil are interesting and do not appear to be due to any chemical alteration but to differential erosion at the surface, removing the fine material. This appears to be a common phenomenon in the humid tropics.

These data indicate the degree of alteration and reorganisation of the constituents that can take place when rocks undergo profound hydrolysis.

Although the overall picture is the same for all rocks i.e. the complete hydrolysis of the primary silicates and loss of some of the constituents in solution, particularly the basic cations, the details seem to vary somewhat from place to place and from rock type to rock type. Whereas basic rocks are transformed within a relatively short distance to highly altered compound, intermediate and acid igneous rocks sometimes show a more gradual decomposition. The primary decom-

position products also vary somewhat. Harrison has demonstrated that gibbsite does not form during the alteration of granite instead there is a fairly direct transformation to kaolinite but in other tropical countries one of the initial products of the weathering of granite can be gibbsite.

The moisture regime of the environment is a factor of some importance for it appears that while dolerite will form gibbsite in a humid subsurface horizon this does not happen when weathering is near to the surface. Here the weathered dolerite loses a considerable proportion of the aluminium and no gibbsite is formed. The reason for this is not clear but it has been shown that some amorphous hydroxides will not form in the presence of organic matter. This is a very important feature and should be borne in mind when examining a full set of data from a pedounit because differences in the chemical composition of the various layers might be due to the initial type of weathering rather than to pedogenetic reorganisation.

Table 12
CHEMICAL COMPOSITION OF DOLERITE AND WEATHERED PRODUCTS

A		Unaltered dolerite
B	0–4 mm	First layer of concentric weathering
C	4–27 mm	Second layer of concentric weathering
D	27–37 mm	Third layer of concentric weathering
E	37–64 mm	Fourth layer of concentric weathering
F	64–66·5 mm	Fifth layer (crust) of concentric weathering
G	330–375 cm	Mottled clay (pale colour)
H	330–375 cm	Mottled clay (dark colour)
I	375–675 cm	Red layer
J	675–765 cm	Grey sandy subsoil
K	765 cm	Ironstone on surface

	A	B	C	D	E	F	G	H	I	J	K
Quartz	<·1	9	8	13	10	9	8	8	5	81	1
SiO_2	51	8	5	10	19	6	29	29	26	6	3
Al_2O_3	16	34	40	19	23	24	24	24	21	3	21
Fe_2O_3	2	21	18	34	30	36	20	19	29	2	65
FeO	9	4	4	1	2	1	1	1	2	<·1	<·1
H_2O	<·1	18	20	15	11	20	13	13	14	5	9
TiO_2	2	4	4	7	5	4	5	5	4	2	2
MnO	<·1	Nil	Nil	Nil	Nil	Nil	<·1	Nil	Nil	Nil	<·1
MgO	8	1	<·1	<·1	<·1	<·1	<·1	<·1	<·1	<·1	<·1
CaO	9	1	<·1	<·1	Nil	<·1	Nil	Nil	Nil	<·1	Nil
K_2O	2	<·1	<·1	<·1	<·1	<·1	<·1	<·1	<·1	<·1	<·1
Na_2O	2	<·1	<·1	<·1	<·1	<·1	<·1	<·1	<·1	<·1	<·1
P_2O_5	<·1	<·1	<·1	<·1	<·1	<·1	<·1	<·1	<·1	<·1	<·1

PHYSICAL PROCESSES

The main physical processes are translocation, aggregation and fragmentation of material, but the agencies responsible are very varied, thus there are many specific ways in which these three processes can be achieved. For example, freezing and thawing can bring about specific aspects of all three processes and are responsible also for certain topographic forms. Therefore it seems more appropriate to treat all aspects of processes such as freezing and thawing together rather than to present them separately in three or more places. The physical processes can be considered under the following headings:

Aggregation

Translocation

Freezing and thawing

Wetting and drying

Expansion and contraction

Exfoliation

Unloading

Aggregation

This is the process whereby a number of particles are held or brought together to form units of varying but characteristic shapes. In many cases the details of the mechanisms are poorly understood but there appear to be certain correlations between the type and degree of aggregates and other soil properties. Therefore for most situations it is possible only to indicate the general relationship that exists between the aggregates and other features. For example, it is easy to understand how cementing substances will bind particles together, but it is difficult to explain how the characteristic and often regular shapes are formed. The type and degree of aggregation is usually referred to as structure which is discussed on page 84. The principal agencies responsible for the formation of aggregates are:

1. Clay and humus
2. Cementing substances
3. Mesofauna
4. Plant roots
5. Expansion and contraction
6. Freezing and thawing
7. Microorganisms
8. Exchangeable cations
9. Cultural operations

Clay and humus

Clay and humus are capable of binding particles together and are the main factors responsible for much of the aggregation in the upper horizons of soils.

Cementing substances

Many products of hydrolysis particularly compounds of iron and aluminium can cement small groups of particles together. In some cases progressive cementation can form very hard and massive horizons. Also the continuous deposition of calcium carbonate can lead to the formation of massive horizons.

Mesofauna

Earthworms, enchytraied worms, termites and other organisms that ingest organic and mineral material produce large amounts of faecal material. Where the activity of any one of these is particularly vigorous entire horizons may be composed of faecal pellets with their characteristic granular or ovoid shapes (Fig. 104).

Plant roots

The fine roots of many plants, especially those of grasses bind together small groups of particles to form a crumb or granular structure.

Expansion and contraction

In certain soils particularly those containing a high proportion of expanding lattice clay, expansion and contraction take place in response to the gain and loss of water, resulting in extensive cracking and the formation of large columns or wedge shaped peds with slickensided surfaces. When only a small amount of expanding lattice clay is present the volume change with moisture change is small but is usually sufficient to form a prismatic or blocky structure.

Freezing and thawing

These two processes can produce massive, platy and sub-cuboidal structures. See page 63.

Microorganisms

It appears that microorganisms may aid the formation of structure by secreting various mucilaginous compounds or gums. These substances bind the particles together to form crumbs or granules which are more frequent and better formed in the presence of organic matter and an active microflora.

Exchangeable cations

In some cases the nature of the exchangeable cations can have an influence, for example the presence of large amounts of calcium causes flocculation and the formation of a crumb or granular structure in the upper horizons whereas sodium causes a dispersion of the system and the formation of a massive or columnar structure.

Cultural operations

Structure is a somewhat ephemeral property which can be altered easily by cultivation; usually, and through mismanagement, the change is from a naturally good crumb or granular structure to poor blocky or massive. Probably a better example is the formation of a ploughpan through repeated cultivation to a constant depth with heavy implements. However ploughing and other tillage operations are conducted to improve the structure and aeration of the soil. In fact it is probably true to say that any great increase in productivity in the future will come through the improvement and management of structure rather than through improved fertiliser techniques the principles of which are now fairly well understood.

Translocation

Some or all of the material in the soil system migrates by a variety of methods from one place to the other. In fact many of the processes of soil formation are concerned primarily with the reorganisation and redistribution of material in the upper 2 m or so of the earth's crust.

The principal types of translocation are:

1. Translocation in solution
2. Translocation in suspension
3. Translocation due to organisms
4. Translocation *en masse*
5. Translocation through freezing and thawing
6. Translocation through expansion and contraction

Translocation in solution

In a humid environment much of the soil solution normally migrates in response to gravity by moving vertically down through the soil to underground waters and finally to the river system. As the solution moves some of the material such as compounds of iron and aluminium may be precipitated to form horizons such as sesquons, while the most soluble materials which include bicarbonates, sulphates and nitrates are lost completely from the soil system. In soils on slopes there is some lateral movement of moisture through the soil and down the slope; this attains very high proportions when the soils contain impermeable horizons over which the moisture can move. As a result ions are transferred laterally and enrich the soils in the lower part of the slope. If the volume of water is large or if the impermeable layer comes close to the surface the entire upper part of the soil can become saturated and there may be free water on the surface. Such situations are known as *flushes* and usually carry a plant community indicative of a moist habitat.

In a semi-arid environment there is only a partial loss of material from the soil system. The easily soluble simple salts including nitrates, chlorides and sulphates are completely removed whereas carbonates are usually precipitated in the middle or lower positions in the soil. In an arid climate there is only a short period following rainfall during which some of the most soluble materials become dissolved and are capable of movement. Since the rainfall under these conditions is small, any downward flow is quickly reversed by intense evaporation which induces upward movement of the soil solution and a deposition of salts at or near to the surface. The soil solution in humid areas will also have a period of temporary upward movement if there is a marked dry season.

Translocation in suspension

Fine particles and colloidal materials are often transported from one place to the other within the soil system. Perhaps the most important manifestation of this process is the removal of particles $< \cdot 5 \, \mu$ from the upper horizons of some soils followed by their deposition in the middle position to form cutans (see page 99). Sometimes the fine material may be redistributed within an horizon and in some cases it would appear that particles up to $50 \, \mu$ in diameter can be redistributed as in the formation of isons (page 155). These phenomena are considered further under thin section morphology on page 96 *et seq.*

Translocation by organisms

Many members of the mesofauna and some mammals are responsible for the movement of much material within the soil system; these processes are discussed on pages 38 *et seq.*

Translocation en masse

Mass movement in response to gravity has already been mentioned as a mechanism for the formation of landscapes involving the movement of the weathered mantle down the slope. Since soil usually constitutes the upper part of the weathered mantle mass movement can remove the soil, exposing fresh material to pedogenetic processes.

In addition, mass movement can also take place within the body of the soil on flat sites and therefore not caused by gravity. There are two processes by which this can take place:

Expansion and contraction (see page 67)
Freezing and thawing (see page 42)

Freezing and thawing

Freezing and thawing take place to varying degrees over a wide area of soils but the fluctuations in temperature about 0°C are of little importance if water is not present. When a wet soil freezes a number of ice patterns develop and segregations occur as determined by the speed of freezing, texture of the material and moisture content. These segregations lead to the following secondary processes:

Frost heaving
Frost shattering
Solifluction
Ice-wedge formation.

Frost heaving

Often when the soil cools through a loss of heat to the atmosphere the surface freezes fairly rapidly forming a massive structure, but when freezing is slow, ice crystals with a characteristic needle form grow out from the surface carrying a small capping of fine material. Alternatively they may grow beneath stones. This process is responsible for heaving stones on to the surface and breaking down large clods to form a better structure, hence the reason for ploughing before the onset of winter. When freezing takes place slowly beneath the surface, a number of ice patterns develop

within the body of the soil, the exact type being governed principally by the texture of the material. In very sandy materials the ice forms fillings in the large pores, giving a massive structure. In loams, a marked platy structure develops with layers of pure clear ice alternating with layers of frozen soil, (Fig. 87), while in clays, ice segregates into both vertical and horizontal layers to isolate subcuboidal blocks of frozen soil.

The thickness of the ice segregations is a function of the speed of freezing as well as moisture content. When freezing is very slow and prolonged and there is a plentiful supply of moisture, the ice segregations are very large, since water normally migrates to the freezing front. This is best seen in loams in polar areas where lenses over a metre in thickness are encountered.

Ice segregations also cause a reorganisation of the soil by withdrawing water and by compacting the laminar or subcuboidal units through expansion upon freezing and crystal growth. These units are usually so well formed that they maintain their shape when the material thaws.

Also in many polar areas the surface soils freeze during the winter and thaw during the summer, creating an annual cycle which, coupled with the development of ice segregations, causes profound disturbance of the soil surface and the formation of a number of characteristic patterns including mud polygons (Fig. 38) and stone polygons (Fig. 39). When a cryon (permafrost) is present the process is enhanced for as the soil freezes from the surface downwards large volumes of moisture are trapped between the freezing front and the cryon and as the moisture cools from 4°C to 0°C it expands creating very high pressures which are often released by an upward buckling of the surface to form 'boils', 'thufurs' or 'pingos', (Fig. 40). Since stones have a lower specific heat than the surrounding soil they cool down and heat up more quickly, so that during the cooling process they attain 0°C before the surrounding soil, and form loci for the formation of ice which can become quite thick around the stone displacing the surrounding material. Repeated freezing and thawing resulting in the formation and disappearance of ice can cause reorientation of the stones which on flat situations steadily develop a vertical orientation of their long axes and are gradually forced towards the surface.

Frost shattering

The occurrence in polar and alpine areas of extensive surfaces strewn with angular rock fragments has

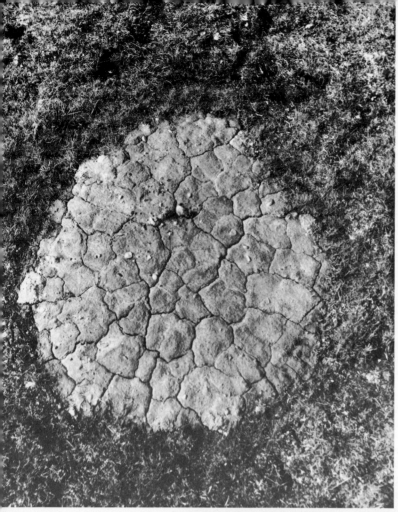

FIG. 38.

A mud polygon 120 cm wide from northern Alaska.

FIG. 39.

Stone polygons 2–3 m wide on Baffin
Island (photo by F.M. Synge).

been interpreted for a long time as due to alternate freezing and thawing. It is assumed that as water freezes in the pores in the rock, pressures of over 500 kg per sq. cm are created and the rock splits apart leaving a number of loose fragments when it thaws. Therefore the amount of fragmentation will increase with the number of times the rock freezes and thaws. This explanation however plausible does not stand up under careful analysis since many areas in the middle latitudes have more frost-thaw cycles than those of polar areas but do not have the same extensive occurrence of angular rock debris. Recently, Everett (1961) has suggested that the active agent is the steady growth of ice crystals within the rock during prolonged periods of low temperatures. This creates high pressure which severely weakens the rocks causing them to split apart, and it is claimed that, over a long period, rocks can be reduced to accumulations of silt size material. Frost-shattering of rock fragments also occurs in soils and unconsolidated sediments (Fig. 41).

Solifluction

On sloping situations repeated freezing and thawing causes the movement *en masse* of the material down the slope. The exact nature of the mechanism involved is not understood. However it appears that when the ground freezes, expansion takes place normal to the surface and when the soil thaws and contracts, it does not return to its original position but responds to gravity, thereby moving slightly down the slope. Alternatively the soil becomes very wet and even saturated upon thawing and being in a somewhat loosened state due to freezing, flows down the slope. This process can be accelerated in the presence of a cryon which is impermeable and causes water to accumulate within the surface layers. Where movement of material down a slope is gradual the stones become oriented with their long axis parallel to the slope. In addition the surface of such areas is usually characterised by a number of lobes or terraces produced by differential movement.

FIG. 40. A pingo about 20 m high on Svalbard.

FIG. 41.

Frost shattered material. In many cases adjacent stones can be fitted together, thus indicating that they were once part of a larger unit.

FIG. 42.

Fossil solifluction material overlain by loess in central Poland.

The characteristic arcs at depths of 1 to 3 m in materials of periglacial areas on slopes is attributed to movements resulting from freezing and thawing and therefore are regarded as a normal type of solifluction formation (Fig. 42). It is probable that these arcs are due to plastic deformation and slow flowage of the frozen mass down the slope. The depth of summer thawing in these areas seldom exceeds one metre and since the arcs occur below this depth it is difficult to explain their presence as due to freezing and thawing. However the temperature in this layer during the summer will only be a few degrees below freezing therefore it could flow by plastic deformation during this period of the year. The warmer and less viscous upper part will flow more rapidly than the cooler and more viscous lower portion, thus accounting for the formation of the arcs.

Another feature is that these arcs are destroyed when they are incorporated into the layer above, which has an annual freeze-thaw cycle. Therefore it seems necessary to differentiate between mass movement caused by freezing and thawing and that caused by plastic flowage.

Ice-wedge formation

During the winter of polar areas when temperatures fall to − 20°C or − 30°C the whole soil contracts sufficiently to cause the ground including the cryon to crack into a hexagonal or rectangular pattern about 10 to 15 m in diameter. Hoar frost then accumulates in these cracks so that during the following summer when the ground expands a thin vertical vein of ice is trapped in the cryon. Repetition of the process over many centuries leads to the formation of tundra polygons and large ice-wedges with their characteristic vertical lamination (Figs. 43 and 44).

In certain parts of Antarctica that are very dry the cracks become filled with soil material to form sand-wedges (Péwé 1959).

Wetting and drying

These processes have been mentioned already in connection with expansion and contraction but they can be considered in a much wider context. Few, if any, reactions in soils take place in the absence of water, therefore wetting initiates soil development. Similarly most reactions stop when the soil becomes dry. Therefore it is possible to regard soils that are subject to periods of wetting and drying as having

phases of activity followed by resting phases. Although this happens to a greater or lesser extent in most soils, it is only in extreme cases where the results are dramatic. A particularly good example is the formation of gilgai (see below). Another striking example is found in arid regions where nitrogen fixation takes place only during a wet period, thereby increasing the amount of nitrogen available to plants which begin to grow in response to the increased moisture.

Expansion and contraction

This is a very important phenomenon in soils containing a high proportion of clay with an expanding lattice and occurs principally in the soils of hot environments, with alternating wet and dry seasons. During the dry season the soil dries, contracts and eventually develops cracks which may be over 10 cm wide on the surface, extending down for over a metre. Invariably some of the upper horizon falls to the bottom of the crack where it stays until the wet season when the soil absorbs water and expands. Since there is now a greater amount of material in the middle or lower horizons they expand to a greater volume than before and in so doing create pressures which cause a displacement of material, which can only move upwards. In addition to the displacement of the material there is a considerable amount of minor sheering at an angle of about 45°. As one portion of soil glides over the other the sliding surfaces become polished to form *slickensides*.

With the annual repetition of this process a large number of areas of soil about one metre or more in diameter are raised above the general level of the surface and the process can proceed until the subsoil is brought to the surface and exposed. Such phenomena are known as *gilgai*. The precise pattern of the gilgai on the surface is determined by a number of factors, particularly slope. On flat sites gilgai are usually subcircular mounds whereas on gentle slopes they tend to be long and narrow running normal to the contour as shown in Figs. 45 and 46 (Hallsworth *et al.* 1955).

There are phenomena that develop on a micro scale and are visible only in thin sections. Some of these are mentioned on page 99. Freezing and thawing are also main agencies of expansion and contraction as discussed on page 65.

Exfoliation

Boulders or outcrops of rocks at the surface experience wide temperature fluctuations especially in polar

FIG. 43.

Oblique aerial view of tundra polygons in
northern Alaska. Each polygon is 15–20 m
in diameter.

FIG. 44.

An ice wedge 4 m deep on Gary island,
northern Canada.

FIG. 45.   Large normal gilgai on a flat surface in Australia.

FIG. 46.   Wavy gilgai on a gentle slope in Australia.

and desert areas where repeated expansion and contraction cause flaking or scaling of the rock. This is of some importance on sloping situations where the material that is released slips down the slope forming scree or talus cones, but generally it is of little importance as a process in the soil system.

Unloading

At present some importance is attached to unloading as a mechanism of physical weathering of rocks. It is suggested that as rocks are weathered and removed the resulting reduction in the thickness of the overburden allows the underlying material to expand and spall off in large sheets parallel to the surface. The main significance of this process is that it increases the surface area of the material and the depth of moisture penetration.

BIOLOGICAL PROCESSES

Probably the most important biological processes taking place in soils are humification of organic matter and the translocation of material from one place to the other. Other processes included are nitrification, nitrogen fixation and those given below that are discussed elsewhere:

1. Microbiological oxidation and reduction of inorganic substances, page 37.
2. Water and ion uptake, page 34.
3. Fragmentation, page 33.
4. Aggregation, pages 33, 37.

Translocation

Some emphasis must be given to those biological processes that bring about churning, and other soil disturbance. The most dramatic manifestation of this process is brought about by the soil inhabiting vertebrates (page 179) but probably the greatest amount is accomplished by earthworms and termites which mix the organic and mineral material and redistribute it in the soil system. These are extremely important processes because they are continually bringing the microbiological population into contact with a fresh food supply and therefore help to maintain a constant breakdown of the organic matter and release of ions for plant nutrition. Also they cause disaggregation and reaggregation of the soil with the result that the mineral grains are being exposed to regular changes of environment resulting in greater decomposition.

Humification

The breakdown of organic matter to form humus and to release various plant nutrients is an extremely complex and little understood process involving organisms as the principal agents. Relatively little work has been completed on the stages of breakdown of the litter falling on to the surface, therefore it is not possible to make many generalisations about these processes. The example given below is taken from the work of Kendrick and Burges (1962) who made a careful study of the decomposition of the needles of *Pinus sylvestris* and suggested the following sequence of events.

About 40% of the living needles on the tree are infected in the spring by the parasitic fungus *Lophodermium pinastri* which causes characteristic black spots. Other fungi are also present but they do not appear to attack the needles until they become old and turn brown, by which time over 90% of the needles are infected. The needles fall to the ground in autumn becoming part of the litter where they spend about six months and are then infected by another fungus (*Desmazierella acicola*) which replaces the previous group. The needles now become part of the fermenton and over a period of two years *Desmazierella acicola* spreads, particularly inside the needles where it deposits areas of black pigment. The outsides of the needles are attacked by two other fungi, *Sympodiella acicola* and *Helicoma monospora* which also produce the dark coloured materials that are a characteristic feature of fermentons. After two years the needles become compressed and fragmented and are attacked by the mesofauna. Mites, collembola and enchytraied worms eat the conidiophores of the fungi and simultaneously instars of mites attack the soft interior of the needles (Fig. 104). The remaining fragments are attacked by yet another group of fungi (Basidiomycetes) and for the next seven years the remains become the humifon where biological activity is at a very low level.

Further studies on the breakdown of organic material have been made by Saitô (1956, 1957) who has given an account of the decomposition of beech leaves. The litter consists of freshly fallen leaves and brown leaves only slightly subject to microbiological attack. In the fermenton, where the moisture content remains fairly constant, many leaves turn yellow following the attack of Basidiomycete mycelia associated with vigorous growth of bacteria. Here the infected leaves at first become much thinner without losing their structure, then mouldy from overgrowth of Basidiomycete mycelia. Later growth of other organisms takes place and gradually the leaves are transformed into amorphous debris. Not all leaves in the same layer become infected by Basidiomycete mycelia, furthermore decomposition does not always take place uniformly throughout a leaf. In the yellow infected leaves there is a marked disappearance of lignin followed by a rise in the quantity of water-soluble substances, finally the Basidiomycete mycelia in the decomposing leaves are broken down by bacteria.

Generally the early stages of decay are characterised by the loss of the water soluble materials, starches and proteins. This is followed by the decomposition of hemicelluloses and cellulose leaving a residue consisting largely of lignin and cuticularised cell walls. In the case of the mesofauna their soft proteinaceous material is rapidly decomposed leaving chitinous exoskeletons which accumulate at the junction of the organic and mineral material.

It is clear from the above work that the breakdown of the organic matter in soil is accomplished by successive waves of organisms some of which are symbiotic heterotrophs, like the fungi, which can be attacked by parasitic mites which in turn are decomposed by bacteria. This means that a given carbon atom may pass from the atmosphere to a plant leaf and then to a succession of soil organisms before it passes back to the atmosphere in CO_2; or it may become part of the soil humus.

There is still a considerable amount of doubt about the true character of the black material comprising the humifon. Many of the earlier workers regarded humus as a ligno-protein (Waksman and Iyer 1932) or a condensation product of tannins and proteins, (Handley 1954). It is now generally believed that part of the humus results from the reaction between polyphenolic compounds of plant or microbial origin, with amino acids, peptides and proteins also of plant or microbial origin. The result of the reaction is to produce insoluble heterogeneous polymers which are comparatively resistant to further microbiological decomposition,

(Flaig, 1960; Kononova, 1961). It is also suggested that humus is composed of a resistant core of complex polycyclic compounds to which are linked carbohydrates and polymers such as those mentioned above, (Cheshire et al. 1967). Resistance to decomposition is achieved through heterogeneity since a very wide range of enzymes are required for their decomposition. The decomposition of organic matter is also accomplished by other organisms including earthworms, termites and mites.

Nitrification

This is the process during which nitrate is formed by soil organisms. When heterotrophic organisms, including mesofauna, decompose organic residues, the principal and simplest nitrogen compound produced is ammonia; this can then be oxidised to nitrite and afterwards to nitrate, each stage being accomplished by specific autotrophic bacteria. Ammonia is oxidised by *Nitrobacter*, *Nitrosomonas* and *Nitrococcus* and the nitrite by *Nitrobacter*.

These organisms require fairly exacting conditions for their activity; the soils must be aerobic and must at the same time be moist with about 50% of the pore space saturated. The optimum temperature is 27–32°C but they can operate down to 4°C. Although they prefer conditions around neutrality they can also operate down to about pH 5. Under anaerobic conditions these organisms are inhibited, others take their place and reduce the nitrogenous compounds to nitrogen which is lost to the atmosphere. This process may be of relatively little significance in the direct formation of soils but it is of paramount importance in plant nutrition. Most plants in natural or seminatural plant communities obtain the greater part of their nitrogen in the form of nitrate produced by bacterial activity from the remains of organisms including parts of plants and the soil microorganisms. Thus nitrogen is constantly being cycled from the soil to plants and then back to the soil via the litter and its decomposition products (Figs. 50 and 51).

Nitrogen fixation

There are a number of free living heterotrophic bacteria including *Azotobacter*, *Clostridium pastorianum* and *Beijerinckia* sp, that are capable of utilising atmospheric nitrogen to form their body protein which upon the death of the organism, is decomposed to ammonia to form part of the nitrogen available to plants or to take part in nitrification. Other microorganisms capable of fixing atmospheric nitrogen include some algae.

THE WEATHERING UNIT – PRIMARY SOIL SYSTEMS

Weathering can be used as a comprehensive term to include all of the subaerial processes responsible for the transformation of mineral material both consolidated and unconsolidated. These diverse processes combine to produce several types of weathering units or primary soil systems which are formed either from superficial drifts or from consolidated rocks. Chemical and physical processes predominate, the biological processes are significant only at the surface but are of paramount importance where they occur, being responsible largely for the formation of the pedounit. The following are the principal mechanisms responsible for the formation of primary soil systems.

1. Hydrolysis of consolidated rock
2. Hydrolysis of unconsolidated material
3. Solution of consolidated rock
4. Solution of unconsolidated material
5. Fragmentation of consolidated rock
6. Fragmentation of unconsolidated material

Of these six mechanisms the three illustrated in Figs. 47 to 49 are possibly the most common.

Hydrolysis of consolidated rock – System 1

This mechanism produces a system which starts at

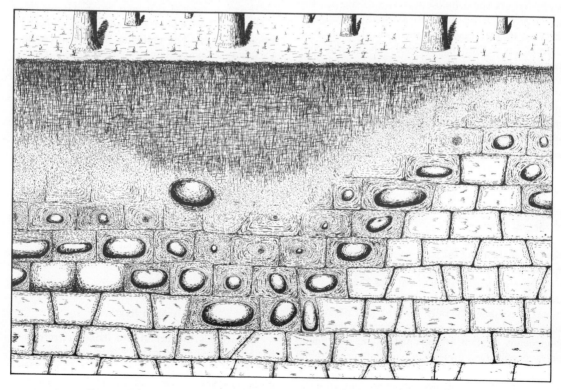

FIG. 47. Primary soil system produced by chemical weathering of a consolidated igneous rock.

the top with strongly weathered material composed of clay minerals, hydrous oxides and the resistant residue, and grades with depth through progressively less weathered material into fresh rock. In the lower parts of systems developed on crystalline rocks there are usually a number of core stones which increase in number and frequency with depth but on sedimentary rock the areas of more resistant materials often occur as strata rather than as core stones. A characteristic feature of this system is the uneven surface of the weathering front which may be caused by hydrological differences or variations in the rock structure (Fig. 47).

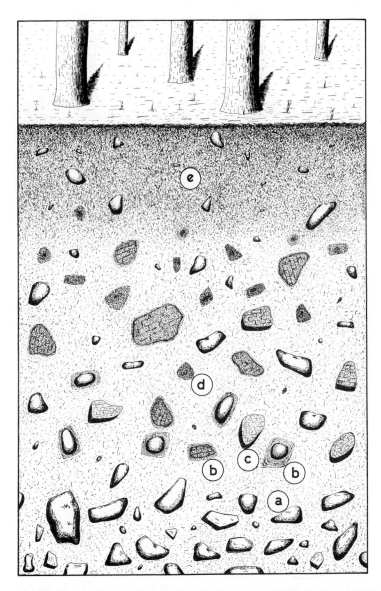

FIG. 48. Primary soil system produced by the hydrolysis of unconsolidated material. (a) Relatively unaltered material, (b) partially weathered boulders such as granite or basalt with an unweathered core, (c) very resistant boulder such as quartzite, (d) very decomposed boulder, a 'ghost' of granite or basalt, (e) soil containing very resistant boulders of quartzite.

Hydrolysis of unconsolidated material – System 2

This mechanism produces a system which has many similarities to that developed by the hydrolysis of hard rocks by having the zone of maximum decomposition at the surface. However in most cases the amount of weathering throughout the material is minimal since most unconsolidated deposits such as loess or alluvium are of recent age, but when an old system is examined the material becomes progressively less weathered with depth passing through a zone where the stones are almost completely decomposed forming 'Ghosts' (Fig. 48).

Solution of consolidated rock – System 3

The characteristic feature of solution is the manner

FIG. 49. Primary soil system produced by physical disintegration of rock. This is an example of disintegration by frost on a sloping site hence the movement of material down the slope by solifluction.

in which subterranean channels develop down which the fine soil particles can fall leaving bare rock surfaces or a shallow layer of residue. These channels gradually develop into depressions or caves (see Fig. 30).

Solution of unconsolidated material – System 4

Systems produced by this mechanism are characterised by extreme differential weathering to form pipes of altered material within the body of the relatively unweathered material.

Fragmentation of consolidated rock – System 5

The only agency of significance producing fragmentation is freezing and thawing. Systems produced by this mechanism have at the surface a matrix of fine material containing many angular rock fragments which become more frequent with depth until there is virtually no matrix. On sloping situations an additional effect of frost action is to cause the stones to become oriented parallel to the slope resulting in the development of pattern shown in Figs. 41 and 49.

Fragmentation of unconsolidated material – System 6

This mechanism produces systems that are similar to the one above with the exception that the amount of fine materials is uniformly distributed throughout the system but again the degree of frost shattering of stones and boulders diminishes with depth and is usually absent below 2 m.

These systems are somewhat schematic and do not specify the precise type of soil present. For example a chernozem or a podzol may develop when basic System 2 is operating. Similarly a vertisol or a krasnozem may form when basic System 1 is taking place. Further, as explained in Chapter 2, it is common for one set of processes to stop and be replaced by another set. This took place on a large scale during the Pleistocene period. Thus we find today that System 1 is best developed in tropical and subtropical areas whereas System 2 occurs mainly in the middle latitudes. Systems 5 and 6 now occur mainly in polar areas but during the Pleistocene they extended into the middle latitudes and were superimposed upon Systems 1 and 2 but now Systems 1 and 2 are superimposed upon Systems 5 and 6. For these reasons we find that in the humid areas outside the limit of the influence of Pleistocene glacial and periglacial processes the soils are deep and have developed by chemical processes whereas within that limit the soils are shallow and are mainly in System 2. This finds its maximum expression in the middle and early Pleistocene deposits of the central U.S.A. where the material, mainly loess and till has been transformed to a clay-rich soil up to 3 m thick (see page 275).

SUMMARY

Under natural conditions it is the rule rather than the exception to find several of the above processes operating simultaneously or superimposed one upon the other. For example, hydrolysis, oxidation and clay mineral formation usually proceed together in most freely drained soils; whereas reduction, hydrolysis and clay formation will take place together in wet soils. Therefore it is really quite impossible to give a comprehensive picture in the form of a single diagram without making it too complex. Figs. 50 and 51 are attempts to summarise much of what has already been said about the processes in the soil system in a humid climate where chemical and biological processes predominate. Similar illustrations can be constructed for arid and tundra environments.

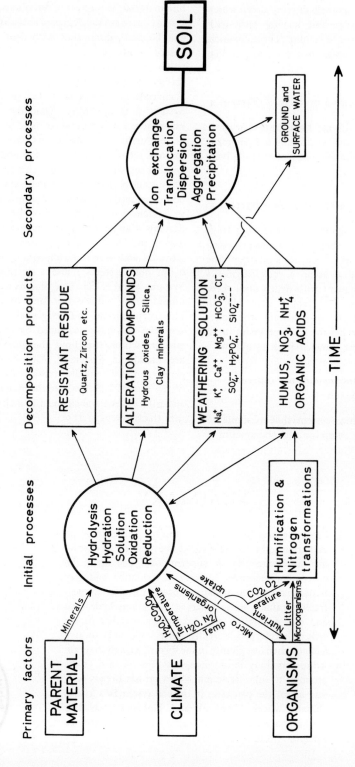

FIG. 50. Soil formation (adapted from Yaalon 1960).

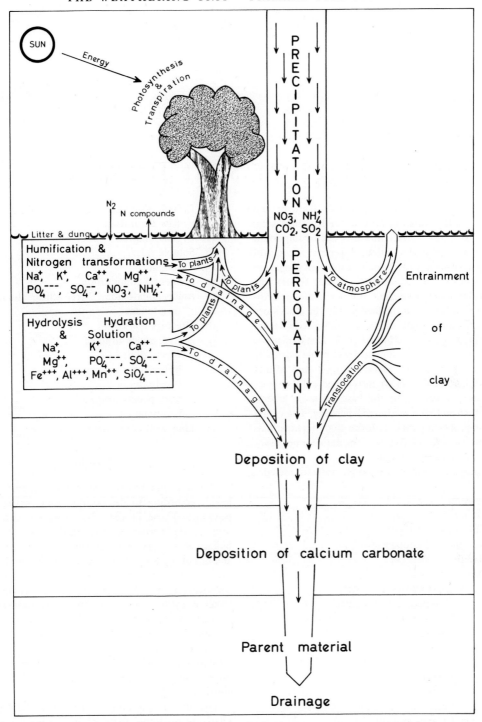

FIG. 51. The soil system with particular reference to those soils in which there is translocation and deposition of clay and calcium carbonate.

4

PROPERTIES OF SOIL HORIZONS

Some soil properties are distinctive features and important differentiating criteria while others seem to have little pedological significance but are important in relation to crop production. A general account of the character of the pedologically important properties is given below but no attempt is made to describe methods for their assessment. Analytical and descriptive methods for use in the field and in the laboratory are covered adequately in many publications particularly the *Soil Survey Manual* (U.S. Department of Agriculture 1951) and in standard text-books especially those by Piper (1947), Jackson (1958) and Milner (1962). Where new ideas are introduced a full characterisation is given as in the case of some aspects of structure and thin section morphology. The values for some properties vary with the analytical methods employed, therefore for some of the data it is necessary to state the technique adopted.

The properties used to categorise horizons include:

1. Position
2. Colour
3. Thickness and regularity
4. Sharpness of boundaries
5. Handling consistence
6. Mineral particles
   Larger separates > 2 mm
      Mineralogical composition
      Shape of particles
      Size and frequency
      Orientation
   The fine earth < 2 mm
      Texture
      Shape of particles
      Mineralogical composition
7. Structure and porosity
8. Bulk density
9. pH
10. Cation exchange capacity and percentage base saturation
11. Organic matter
12. Soluble salts
13. Carbonates
14. Elemental composition
15. Amorphous oxides – free silica, alumina and iron oxides
16. Moisture content
    Excessively drained
    Freely drained
    Imperfectly drained
    Poorly drained
    Very poorly drained
17. Concretions
18. Thin section morphology

1. Position

Horizons can be grouped into upper, middle, and lower, according to the position they occupy in the pedounit. Those in the upper position occur at or near to the surface and are strongly influenced by biological activity. Usually they contain the largest amount of organic matter and in a humid environment the greatest amount of water passes through them, consequently they lose significant amounts of material, either in solution or suspension. The middle position includes those horizons that are influenced less strongly by biological processes. On the other hand they receive and sometimes retain some of the material washed in from above. Where an upper horizon is very thick it may be regarded as extending into the middle position.

The lower position is occupied regularly by relatively unaltered material as in the case of many soils developed in Pleistocene or Holocene sediments, but in a dry environment there may be accumulations of calcium carbonate or calcium sulphate forming calcons or gypsons respectively.

Areas of older soils frequently have weathered materials in various stages of decomposition occupying this position. In the so-called bi-sequal soils a few horizons that normally occur in the upper or middle positions are found forming a second lower sequence or horizons. These situations are relatively rare but may be very important in certain local situations.

2. Colour

Generally, the colour of a soil is determined by the amount and state of the iron and/or the organic matter. Hematite is responsible for the red colour in many soils developed under free drainage in tropical and subtropical areas. This substance occurs in a very finely divided state and need be present only in fairly small amounts to impart intense coloration. The mineral responsible for most of the inorganic colouration of freely drained soils is goethite which has colours that range from reddish brown to yellow with an increasing degree of hydration. The highly hydrated yellow and yellowish-brown forms are sometimes referred to as limonite. Many grey, olive, and blue colours occur in the soils of wet situations and originate through the presence of iron in the reduced or ferrous state. Sometimes, substances with blue colours such as vivianite may form under these conditions and therefore contribute to the colouration. The colour of the upper horizons usually changes from brown to dark brown to black as the organic matter content increases and there is the tendency for organic matter to become darker in colour with increasing humification. Also the pH and cation status can be influences, for acid soils low in calcium and organic matter can often be pale in colour whereas in the presence of large amounts of calcium or sodium even small amounts of organic matter will give rise to dark colours. Dark colours are produced also through the presence of manganese dioxide or may be caused by elemental carbon following burning.

Pale grey and white originate through the lack of alteration of grey or white parent material, depositions of calcium carbonate, efflorescences of salts or as a result of the removal of iron leaving significant amounts of uncoated light coloured minerals such as quartz, felspars and kaolinite.

Each horizon usually has its own specific and fairly narrow range of colours. Furthermore most horizons exhibit some variability in colour within the same pedounit but this is usually ignored when small. On the other hand the marked colour patterns of some horizons are distinguishing characteristics. Colour patterns may be mottled, streaked, spotted, variegated or tongued.

Possibly the most common and important colour pattern is yellow or brown mottles on a grey background which is interpreted as resulting from seasonal wetting and drying of the horizon. The duration of the dry period is indicated by the yellow and brown areas which become progressively larger and more frequent as aeration increases.

Although the colours of most horizons are produced by pedogenic processes there are a number of instances where they are inherited from the parent material, for example, many sediments of Devonian or Triassic age are bright red owing to the presence of hematite. Unless pedogenic processes are very vigorous, soils developed from such parent materials will retain these colours for thousands of years through many contrasting phases of pedogenesis.

A very high proportion of the names of soils are based upon colour as this is the most conspicuous property and sometimes the only one that is easily remembered. For example, class names such as chernozems (page 178) and podzols (page 216) are derived from the colour of a particular horizon. This system of nomenclature is not entirely satisfactory and now other criteria are used to establish soil names, but some terms are so well entrenched in the literature that it would be impractical to introduce replacements. It should be pointed out that many inferences based upon colour are made about soils since there appears to be a correlation between it and other properties.

Many soils are darker when moist or wet, and in order to attain a measure of uniformity there is a tendency to report in the literature the colours of moist soils. Until recently many bizarre terms were used to describe soil colours, but with the introduction of the Munsell Soil Colour Charts a high degree of standardization has been achieved.

3. Thickness and regularity

Perhaps it is questionable whether the thickness of horizons should be considered as an important differentiating property but, in most circumstances, fully developed horizons have fairly well defined limits of thickness. Within these limits the exact thickness is used as a crude indication of the degree of development and age of the horizons. However, fully developed horizons vary widely in their thickness, some are as thin as 1 cm whereas others may be several metres thick. In any given pedounit the constituent horizons are seldom uniform in thickness, sometimes they vary by not more than one or two cm from the mean thickness but others are extremely variable particularly those that form tongues into the underlying horizons. Thin horizons, of which placons (Plate IIIc) are a good example are often very irregular in outline and present many difficulties with regard to delimiting pedounits.

4. Sharpness of boundaries

Horizons usually grade vertically from one to the other and the distance over which this takes place varies considerably with the nature of the horizon. Usually, when an horizon is thin the change is sharp or very sharp (Plate IIId) on the other hand as horizons increase in thickness they tend to have merging boundaries (Plates IIIa) but there are a number of situations where there are departures from these generalisations, particularly in the case of thick horizons some of which may have at least one sharp boundary.

5. Handling consistence

When soils are manipulated between the fingers and thumb they exert varying degrees of resistance to disruption and deformation as determined by their particle size distribution, degree of aggregation, and moisture content. Generally the pressure needed to disrupt a dry soil increases with the content of fine material, sands are usually quite loose while clays form very hard aggregates. Moist sands have a small measure of coherence whereas moist clays are plastic and become very sticky when wet, particularly if the content of montmorillonite is high. The presence of large amounts of humified organic matter in the soil is particularly important for it increases the plasticity of sandy soils but has the reverse affect on a clay by reducing the stickiness. The consistence of soils of medium texture does not change very much with variations in moisture content. In either the dry or moist state they are usually *friable*, i.e. firm with well formed aggregates that crumble easily when pressure is exerted on them. When wet, these soils

tend to become sticky but never to the same extent as clays.

Some horizons are massive and hard and offer a considerable degree of resistance to disruption as a result of cementation by substances such as iron oxides, aluminium oxides and calcium carbonate. Resistance to disruption can result also from physical compaction.

6. Mineral particles

Soil particles are divided initially into two size classes. The limit is normally set at 2 mm which delimits the 'fine earth' from the larger separates including gravel.

Larger separates > 2 mm

A knowledge of the nature and properties of the coarser particles in soils can lead often to important conclusions about the origin and formation of the parent material and about the soil itself.

The properties of the larger separates that are important in pedological studies are:

1. Mineralogical composition
2. Shape
3. Size and frequency
4. Orientation

Mineralogical composition

There is a substantial body of evidence to show that, in some superficial deposits the sand and silt fractions of soils have a similar mineralogical composition to the material > 2 mm. Therefore the composition of the larger separates can be used to assess the chemical composition of soils and parent materials. This is a common field practice in many glaciated areas.

Shape of particles

The shape of the stones often gives a clear indication of the processes which have influenced the formation of the parent material and/or the soil itself. Rounded stones are found in alluvium and beach deposits whereas in glacial drift the stones tend to be subangular to subrounded while angular stones are associated with frost action or exfoliation. Where stones or joint blocks are subjected to weathering for long periods they usually become rounded forming core stones (Fig. 47) but limestone tends to develop weird shapes due to differential solution.

Size and frequency

The size of the larger separates varies from material just larger than 2 mm to boulders one or more metres in diameter. In transported materials the size increases with the vigour of the transporting medium. In the case of water the carrying capacity of a river on a constant slope varies with the 3·2 power of the velocity. Therefore alluvial deposits associated with mountain streams normally have a high proportion of large rounded stones and boulders. Whereas those associated with meandering streams are free of large separates. Glacial deposits are usually characterised by a large number of subangular stones and boulders of various sizes. The size and frequency of frost shattered rock fragments are determined by the original nature of the material as well as by the intensity of frost action. Generally the size decreases as frost action increases, but this is more the case on laminated rocks such as slates and schists than on the crystalline rocks such as granite.

Orientation

Stones in superficial deposits are orientated in a number of specific patterns. The stones in alluvium usually have an imbricate pattern with their long axes aligned in the direction of river flow but they dip upstream. Glacial deposits often have stones aligned in the directions of ice flow, a feature that is used in glaciological studies to determine the direction of ice movement. In areas of vigorous frost action the stones become oriented by frost heaving to form a number of patterns at the surface as well as in the soil itself. On a flat site the stones within the soil are vertically oriented whereas on a sloping situation they become oriented parallel to the slope and normal to the contour. This feature is particularly important in interpreting the origin of slope deposits. A similar type of orientation on slopes is associated with certain forms of soil creep found in many tropical areas.

The fine earth < 2 mm

The principal properties of particles of this size range are:

1. Texture
2. Shape
3. Mineralogy

Texture

The material < 2 mm is divided further into various size fractions, the precise number depends upon the nature of the investigation, but normally either the International scheme or that proposed by the United States Department of Agriculture is adopted. The former has four fractions whereas the latter is more detailed and recognises seven fractions. A comparison of these two schemes together with that used in the U.S.S.R. is shown in Fig. 52. The relative proportions of each of these size fractions in any one soil is defined as the *soil texture* of which the following 21 textural classes are recognised by the USDA.

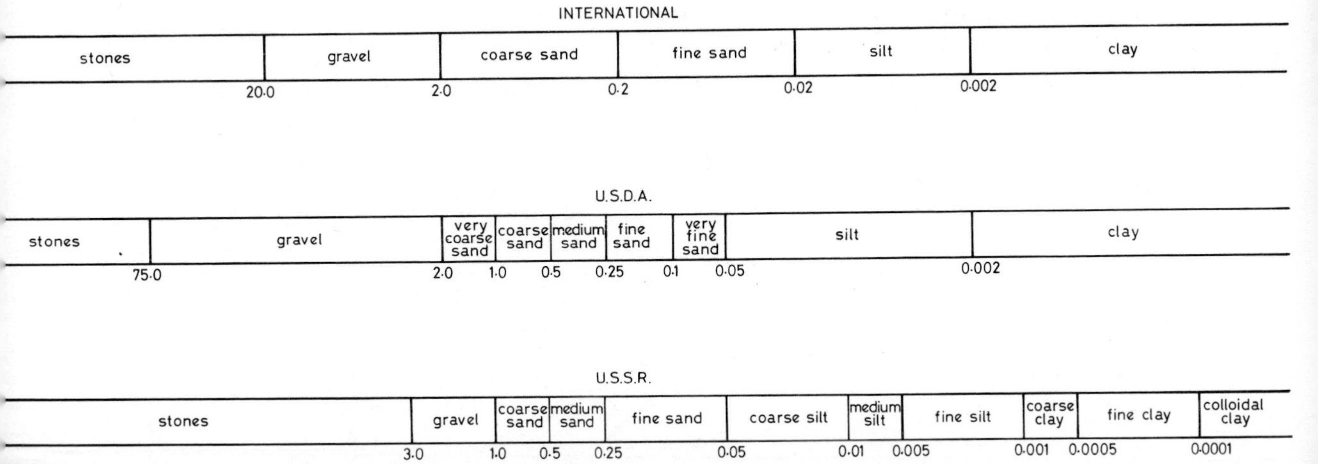

INTERNATIONAL

stones	gravel	coarse sand	fine sand	silt	clay

20·0   2·0   0·2   0·02   0·002

U.S.D.A.

stones	gravel	very coarse sand	coarse sand	medium sand	fine sand	very fine sand	silt	clay

75·0   2·0 1·0 0·5 0·25 0·1 0·05   0·002

U.S.S.R.

stones	gravel	coarse sand	medium sand	fine sand	coarse silt	medium silt	fine silt	coarse clay	fine clay	colloidal clay

3·0 1·0 0·5 0·25 0·05 0·01 0·005 0·001 0·0005 0·0001

FIG. 52. Comparison of three major schemes of particle size distribution in millimetres.

SANDS
  Coarse sand
  Sand
  Fine sand
  Very fine sand

LOAMY SANDS
  Loamy coarse sand
  Loamy sand
  Loamy fine sand
  Loamy very fine sand

SANDY LOAMS
  Coarse sandy loam
  Sandy loam

Fine sandy loam
Very fine sandy loam

LOAM
SILT LOAM
SILT
SANDY CLAY LOAM
CLAY LOAM
SILTY CLAY LOAM
SANDY CLAY
SILTY CLAY
CLAY

This is a rather large number of classes having fine shades of difference between adjacent classes with the result that many workers prefer to use fewer broader classes. These 21 classes can be grouped into twelve basic classes and presented in the form of a triangular diagram by using only three size fractions as shown in Fig. 53. This is a very convenient method of presenting the textural classes as well as being useful in comparative studies. This scheme can be simplified even further as shown in Fig. 54.

The estimation of texture can be carried out in the field or in the laboratory. A fairly precise determination of the various size fractions can be made in the laboratory but in the field only an approximate estimate is possible by manipulating the moist soil between the fingers. This is a very subjective technique and necessitates considerable experience but it is essential that it should be mastered by all field workers since the results can be more meaningful in agriculture

than a laboratory assessment, for it takes into account the content and state of organic matter and the type of clay minerals, not assessed in the laboratory method.

In very detailed studies, the fine earth is divided into 12 or more fractions of which the $< 0.5 \mu$ (or $< 0.2 \mu$) fraction is particularly important in categorising the soil for it is composed of the clay minerals, amorphous hydroxides and oxides.

The textures of soil horizons result from fairly specific processes. Soils developed in fairly recent deposits have inherited a considerable proportion of their textural characteristics from the parent materials but, as soils increase in age, more and more clay is formed and the texture gradually becomes increasingly fine. However the translocation of clay particles from one horizon to the other may cause the horizon losing the clay to become coarser and the receiving horizon to become finer.

As pointed out earlier the main influence of texture is upon permeability which gradually decreases with decreasing particle size but there are notable exceptions when there are well formed aggregates.

Shape of particles

The shape of the particles in the fine earth has an influence upon the consistence of the soil. In

FIG. 53.

Triangular diagram of texture classes according to the USDA. (Sand 2 to 0·05 mm, silt 0·05 to 0·002 mm, clay < 0·002 mm).

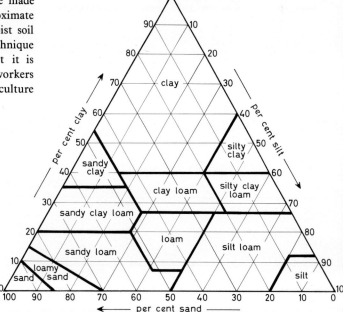

addition, the shape of the sand fraction is often a useful guide to the origin of the material. Sand varies in shape from smooth and round to very rough and angular. The smooth round shape is found in wind blown materials and beach sand; subrounded grains occur in alluvium and beach deposits; rough, angular sand occurs in glacial deposits.

Minerals in a strongly weathering environment develop varied forms, determined by the nature of the weathering and the crystal structure of the mineral. For example, as apatite of sand size weathers by solution a number of very irregular forms are produced, this contrasts with other minerals such as hypersthene which upon hydrolysis develops serrated edges. Minerals of silt size are either angular, having been formed by physical comminution or irregular as a result of differential decomposition and do not seem to become rounded in the same way as sand.

The shape of some particles is controlled very largely by their crystal structure. The primary micas and chlorite keep their marked platy shape in all size fractions and the amphiboles always tend to be fibrous. Some of the very hard and resistant minerals, such as zircon which has well formed crystal outlines loses its initial form only after prolonged decomposition and even then the larger grains maintain a considerable part of their original crystalline form.

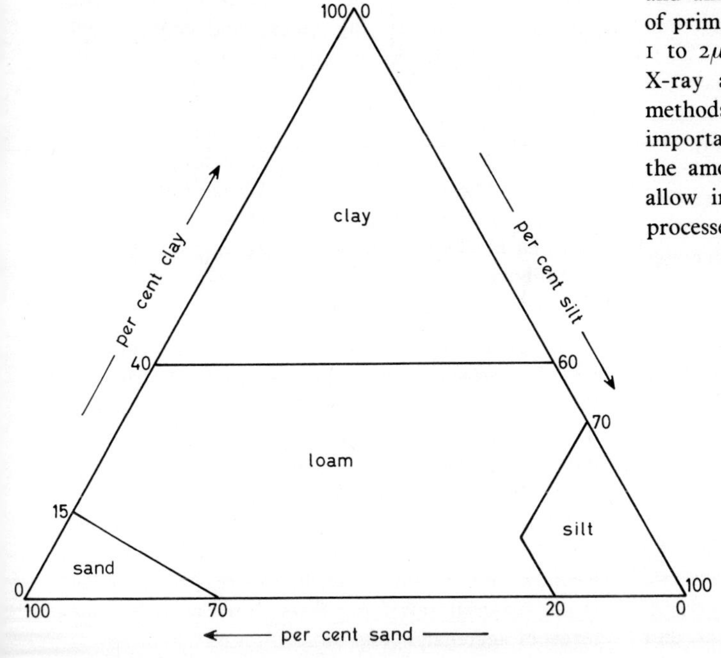

Finally it has been pointed out already, that most of the clay minerals have a platy shape except halloysite which is tubular.

Mineralogical compostion

The fine earth contains both primary and secondary minerals. The former dominate the sand and silt fractions but occasionally within these fractions there may be a large amount of various types of concretions including secondary minerals.

The mineralogical examination of the sand fraction is a very long and tedious process therefore it is customary to investigate only one size fraction, normally 50 to 100 μ, as this has been found to give a fairly accurate picture of the overall sand mineralogy. This fraction usually has large but varying amounts of quartz, felspars and muscovite and in order to obtain an accurate estimate of the minerals of low frequency the fraction is divided into two subfractions, the 'light minerals' and the 'heavy minerals' by specific gravity separation using a liquid (usually bromoform) of sp.gr. 2·9. The light fraction which floats is composed mainly of quartz, felspars and muscovite while the heavy fraction contains a wide range of ferromagnesian and accessory minerals and is often the important fraction in categorising soils and parent materials. The clay fraction is composed predominantly of crystalline clay minerals and amorphous materials except for a small amount of primary minerals mainly quartz that occurs in the 1 to 2μ size range. The clay fraction is studied by X-ray and D.T.A. (Differential Thermal Analysis) methods which give data that are usually more important than those of the sand mineralogy since the amount and type of the different clay minerals allow important conclusions to be made about the processes taking place in soils.

Fig. 54.

Simplified triangular diagram of texture classes, mainly for use in classifying parent materials.

7. Structure and porosity

These refer to the spatial distribution and total organisation of the soil system as expressed by the degree and type of aggregation and the nature and distribution of pores and pore space.

In many soils the individual particles do not exist as discrete entities but are grouped into aggregates with fairly distinctive shapes and sizes. This is found most often in horizons of medium texture. Soils that are composed predominantly of coarse sand are loose and without well formed aggregates, while those composed mainly of clay tend to be massive or to form fewer large aggregates. It is found that in soils with well-formed aggregates or when there is mainly sand the individual units may have only a few points of contact but generally are surrounded by a continuous pore phase. On the other hand in massive soils it is the mineral material that forms the continuum but it may and often does contain discrete pores. Thus in some instances there is a continuous phase of pore space whereas in others the mineral material forms the continuum. Frequently an intermediate situation is encountered when both the pores and the mineral material form continuous phases by having interlocking systems.

The degree and type of aggregation determines aeration and permeability and, therefore, the infiltration capacity and moisture movement. It often determines the volume of the pore and pore-space and therefore the volume of the soil atmosphere. This is usually saturated with moisture in a humid climate and has the following composition: 79–80% nitrogen, 15–20% oxygen and 0.25–5% carbon dioxide. The above ground atmosphere differs by having only 0.03% carbon dioxide. Structure also influences the erosive potential since the presence of surface horizons with a massive structure reduces infiltration, which increases run-off thereby increasing the erosion hazard.

There are at present several good systems for classifying aggregates but there is no general agreement about the classification of pores. However the system recently presented by Brewer (1964) has merit. Nevertheless, in some respects it seems superfluous to have two systems of classification – one for aggregates and the other for pores since they are largely interdependent or the complement of each other, except when discrete pores occur within the aggregates. It seems more appropriate to have one system based on the type of aggregates since aggregation is the more common type of arrangement, and

to describe pores in terms of their size and shape.

Most systems of structure classification used in pedology are based mainly on field characteristics and consequently are somewhat crude.

A fuller understanding of the undisturbed structure can be obtained by impregnating the soil with a resin containing a fluorescent dye. The impregnated block is cut open to expose the undisturbed central part which is studied or photographed using ultra-violet light. This latter technique gives photographs in which the black areas are soil and the light areas are pores and pore space. Furthermore it is possible to study these specimens at many levels of magnification thus obtaining a detailed insight into the architecture of the soil.

By this method it has been found that much of the aggregation observed in a hand specimen in the field is caused by disturbance. Many soils that appear to have aggregates are seen to be poorly aggregated when impregnated and examined under UV light. Therefore it is necessary to differentiate between those soils that have discrete aggregates and those in which the aggregates are poorly formed or partially joined to each other. When the aggregates are separate the structure is termed *complete* and when the aggregates are partially joined the structure is *incomplete*.

Alternatively structure can be studied in thin sections as discussed on page 96 *et seq*. By using thin sections, fluorescing blocks and conventional field methods the following types of structure are recognised.

Single grained Fig. 55	Columnar Fig. 61
Crumb Fig. 56	Lenticular Fig. 62
Granular Fig. 57	Wedge Fig. 63
Subangular blocky Fig. 68	Massive Fig. 64
Angular blocky Fig. 58	Spongy Fig. 65
Subcuboidal Fig. 59	Alveolar Fig. 66
Prismatic Fig. 60	Vermicular Fig. 67

In a number of cases the structure of a horizon may be *compound* or *complex*. Compound structure usually occurs in soils with prismatic or columnar structure in which the large peds are made up of incomplete smaller peds. Complex structure occurs when two or more contrasting processes are operating simultaneously in the same horizon. This is common when the soil has a fairly vigorous fauna such as termites and worms resulting in areas of granular faecal material ramifying through areas with other forms of aggregates (Fig. 68).

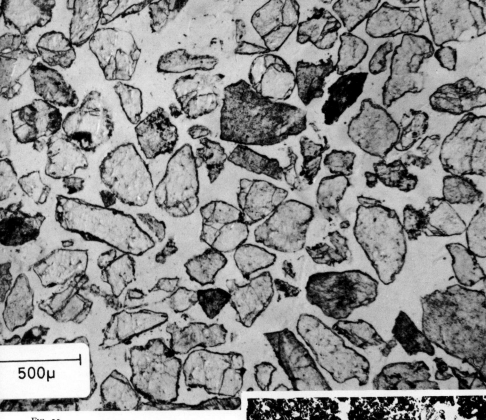

500μ

FIG. 55.

Thin section of single grain structure, there are no aggregates and the particles are separate one from the other.

FIG. 56.

Thin section of crumb structure – composed of numerous loose porous aggregates.

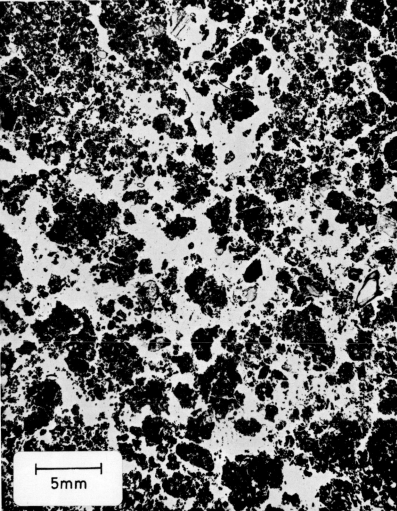

5mm

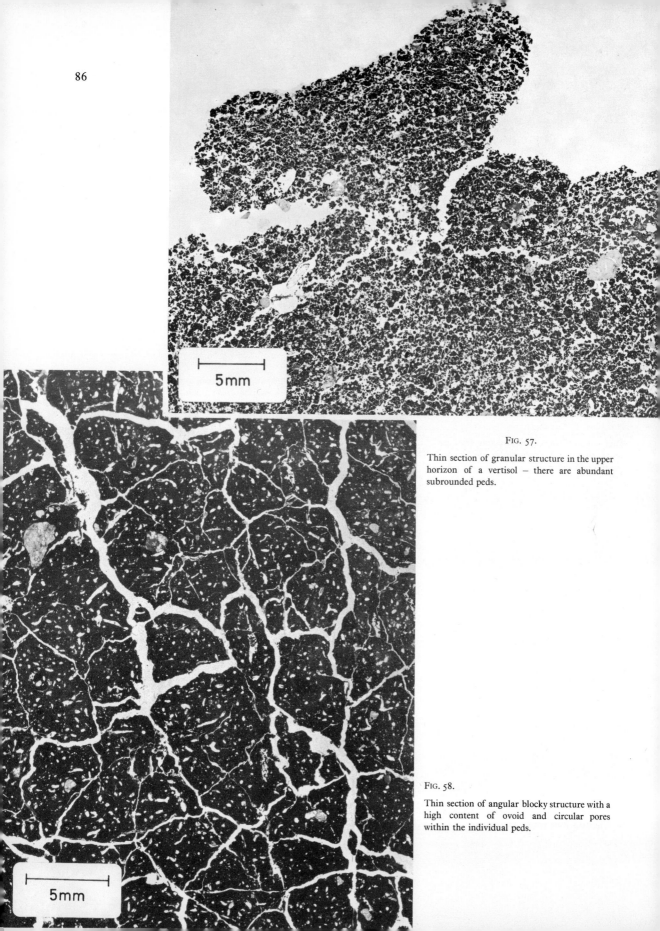

FIG. 57.

Thin section of granular structure in the upper horizon of a vertisol — there are abundant subrounded peds.

FIG. 58.

Thin section of angular blocky structure with a high content of ovoid and circular pores within the individual peds.

4cm

FIG. 59.
Subcuboidal structure formed by freezing.

FIG. 60.
Prismatic structure in the middle horizon. Note the sharp edges and flat faces of the prisms.

10 cm

FIG. 61.

Columnar structure in a soloretz.

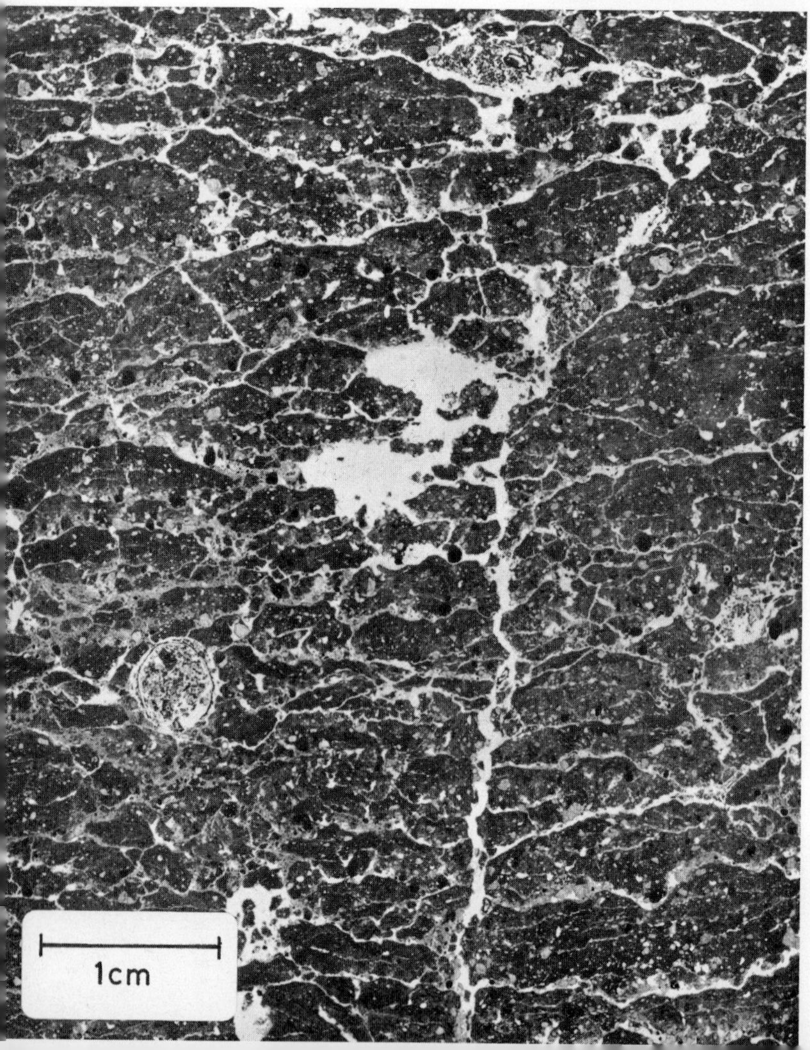

1cm

FIG. 62.

Thin section of lenticular structure.

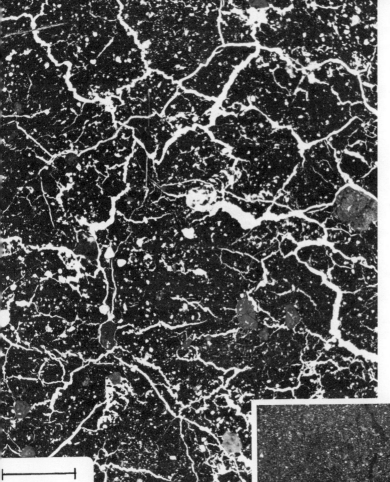

5 mm

FIG. 63.

Thin section of wedge structure characterised by pore space running and intersecting at an angle of about forty-five degrees.

FIG. 64.

Thin section of massive structure with contraction cracks caused by drying.

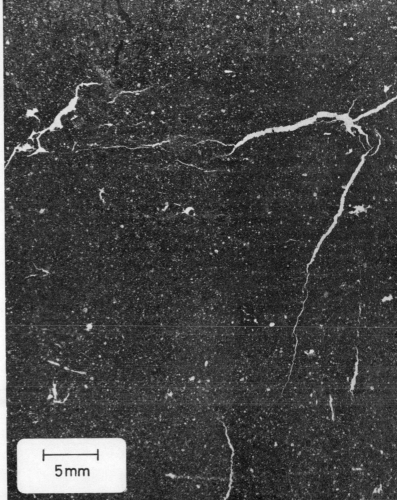

5 mm

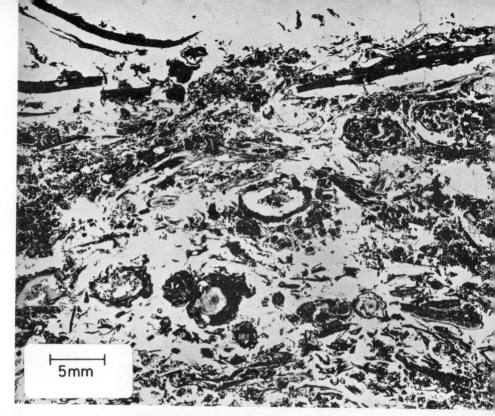

FIG. 65. Thin section of spongy structure, characteristic of many surface organic horizons.

FIG. 66. Thin section of alveolar structure.

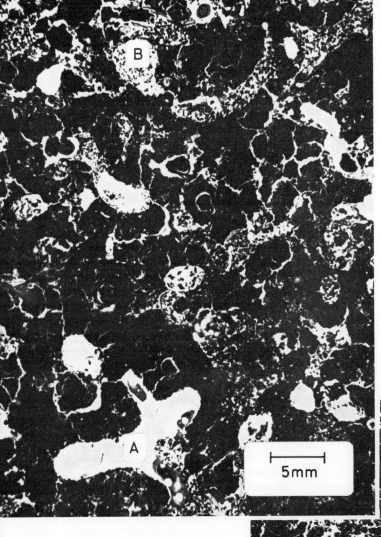

Fig. 67.

Fluorescent block of vermicular structure produced by termites. A is a termite passage and B is faecal material of termites.

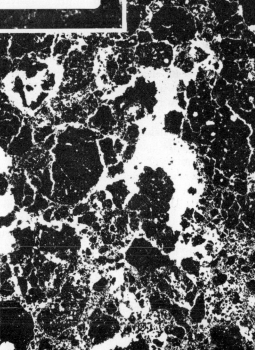

Fig. 68.

Thin section of complex structure of complete subangular blocky and granular.

Considering now the distribution of structure within soils, the surface horizons usually have a crumb or granular structure with peds up to about 3 mm diameter. This normally changes to blocky or prismatic in the middle horizon where the peds may vary from 1–10 cm in diameter and up to 30 cm high; finally to massive in the lower position, but there are very wide departures from this simple outline.

8. Bulk density

Increasing attention is being given to the bulk density of soils since it is a significant and distinctively characteristic feature of some horizons. Whereas the density of most cultivated soils is about 1·3 the extremes can vary from 0·55 for soils developed on volcanic ash to 2·0 for some strongly compacted lower horizons. This property is also attaining some importance in fertility studies because continuous cultivation by heavy implements is inducing compaction which reduces percolation and root penetration. Oertel (1961) has used bulk density in certain fundamental studies to establish whether the clay content of certain horizons has been increased by translocation or weathering *in situ*.

9. pH

The reaction of the soil normally ranges from pH 3 to 9 but occasionally values outside these limits are encountered. Very low values are found in soil of marshes and swamps that contain pyrite or elemental sulphur. At the other extreme very high values result from the presence of sodium carbonate. Within the normal range the two principal controlling factors are organic matter and the type and amount of cations. Large amounts of organic matter induce acidity except when counterbalanced by high concentrations of base cations. Although it is normal for peats to have low pH values of 3 to 4 they can be near to neutrality if there is a high content of bases, mainly calcium bicarbonate in the water influencing their formation.

Aluminium often contributes to soil acidity for when it is released by hydrolysis or upon moving into solution from exchange sites, each Al^{+++} ion combines with three OH^- ions and is precipitated leaving three free H^+ ions which reduce the pH value. Generally pH values about neutrality are associated with high contents of exchangeable calcium and magnesium sometimes supplemented by free carbonates.

10. Cation exchange capacity and percentage base saturation

The only exchange property which seems to be important for use as a differentiating criterion in categorising soils is the cation exchange capacity (C.E.C.) of the whole soil. This property is a measure of the exchange capacity of the constituent clay minerals, allophane and humus, expressed as milligram equivalents per 100 gm of soil i.e. me%. The range is from about 5 me% for some lower horizons up to 100 me% for upper horizons containing high percentages of organic matter, vermiculite or montmorillonite. In soils containing a large amount of allophane the data should be interpreted with caution since they can vary between wide limits depending upon the technique used to measure cation exchange capacity.

The silt fraction sometimes contributes substantially to the cation exchange capacity especially when it is present in high proportions.

A further property which is of some importance is the percentage base saturation (B.S.) which is a measure of the extent to which the exchange complex is saturated with basic cations.

The general trend is for the amount of exchangeable bases to increase with decreasing rainfall and for calcium to be dominant but the nature of parent material may have a local influence. As the cation content increases there is also the tendency for sodium to become increasingly important and may even be dominant in many soils of arid and semi-arid areas. Conversely, low figures for the percentage base saturation are used as a criterion of intense leaching. However since the amount and type of cations can be changed rather easily by cultivation they are used only occasionally as differentiating criteria.

11. Organic matter

Despite decades of intensive and painstaking research, a detailed categorisation of the humified organic matter occurring separately at the surface or mixed with mineral material has not been achieved. Therefore in routine analyses for categorising soils only the total amounts of carbon and nitrogen are determined. The results are normally presented as percentages by weight but it might be more realistic to present the data on a volume basis so as to take account of the very wide difference in density between mineral material and organic matter.

The carbon percentages usually are multiplied by the conversion factor of 1·72 to give an indication of the total amounts of organic matter present. The ratio of carbon to nitrogen (C/N ratio) is used as a measure of the degree of humification and ranges from > 100 to 8. The former high figure is for the fresh and little decomposed material found in litter and peat while the latter occurs in many upper horizons containing a mixture of mineral material and well humified organic matter and shows that there is a reduction of carbon relative to nitrogen upon humification. The C/N ratio usually decreases with depth and a few middle horizons have ratios as low as 4 which is difficult to explain and may be caused by a high content of ammonium ions fixed by the clay. Another important characteristic of the C/N ratio is that it tends to increase with increasing soil acidity.

The total amount of organic matter in soils as measured by the total amount of carbon × 1·72 varies from <1% up to >90%. This latter figure is for the relatively unaltered material that occurs at the surface. Normally however, except in peats, upper horizons contain <15% organic matter and a very large number contain <2% even when the supply of litter to the surface is high as in certain humid tropical areas.

The properties of humus are unique and very largely determine the character of many upper horizons. Firstly, humus is capable of absorbing large quantities of water thus increasing the water holding capacity of soils and is therefore of importance in crop production. Also it has a cation exchange capacity of about 300 me/100 g and it can be dispersed or flocculated like clays depending upon the nature of the cations present, and as already stated on page 80 it influences the handling consistence.

Under natural conditions the humus content of a virgin soil is usually higher than in adjacent cultivated areas. This is caused by a higher rate of addition of organic matter by the natural vegetation accompanied by a lower rate of biological activity and lower temperatures.

12. Soluble salts

Soluble salts occur in significant proportions only in the soils of arid and semi-arid areas where they accumulate because the annual precipitation is insufficient to leach the soils or because the water-table is at a shallow depth and moisture is drawn to the surface by capillarity bringing with it dissolved salts which are left behind as the moisture evaporates. Flooding by sea water also causes salinity in soil but this is of minor importance except in countries such as Holland which depends upon large areas reclaimed from the sea. The predominant anions are bicarbonate, carbonate, sulphate and chloride while the cations include sodium, calcium, magnesium and small amounts of potassium.

These ions occur in widely varying proportions and depending upon the particular ratio they impart a number of properties to the soil, some of which are detrimental to plant growth. Soils can be divided broadly into saline and sodic depending upon the proportions of these cations as follows:

saline: E.C. (electrical conductivity) of saturated extract > 4 mmhos/cm
exchangeable sodium < 15 me%, pH < 8·5
sodic: E.C. (electrical conductivity) of saturated extract < 4 mmhos/cm
exchangeable sodium > 15 me%, pH > 8·5

Although the above classification covers most situations there are a number of soils that are not adequately accommodated, therefore some workers (Miljkovic, 1965) prefer to have separate scales for salinity and alkalinity as follows:

Degree of salinity

Slightly saline with E.C.	2–4 mmhos/cm
Moderately saline with E.C.	4–8 mmhos/cm
Strongly saline with E.C.	8–15 mmhos/cm
Very strongly saline with E.C.	> 15 mmhos/cm

Degree of alkalinity

Slightly alkaline	< 20%	exchangeable sodium
Moderately alkaline	20–50%	exchangeable sodium
Strongly alkaline	> 50%	exchangeable sodium

In neither of these two schemes is an account taken of the proportions of the different ions present as done by some workers, particularly the variations in the amounts of anions. A further method of sub-division of the soluble salts taking into account the proportions of the anions can be achieved using a ternary system in which the coordinates are percent chloride, percent sulphate and percent carbonate and bicarbonate as given in Fig. 69.

A common feature of many soils containing soluble salts is that micro elements accumulate to toxic proportions. In the case of boron, concentrations over 1·0 ppm are toxic.

13. Carbonates

Carbonates of calcium and magnesium particularly the former, are widely distributed in soils, occurring separately or they may be associated with soluble salts. The most important properties of carbonates are (1) They are relatively easily soluble in water containing dissolved carbon dioxide, and therefore can be quickly lost or redistributed within the soil. (2) When present in as small amounts as 1% of the soil they can dominate the course of soil development, because this amount is sufficient to raise the pH value over neutrality and sustain a high level of biological activity. (3) Carbonates, particularly calcium carbonate, are the first substances to start accumulating as the climate becomes arid. (4) Both calcium and magnesium are essential plant nutrients.

14. Elemental composition

Only in very detailed studies is it necessary to determine the total amount of each element present in the soil. Where soil formation has proceeded for a long period and where there has been a considerable amount of hydrolysis and solution it is usual to perform a partial ultimate analysis to estimate the ten or so dominant elements in each horizon as well as in the underlying parent material in order to determine the predominant chemical changes that have taken place during soil formation. These elements include silicon, aluminium, iron, sodium, potassium, calcium, magnesium, manganese, titanium, zirconium, nitrogen and phosphorus. However the determination of the total amount of certain single elements is performed very frequently, particularly nitrogen, because of its immense importance as a plant nutrient.

It is claimed that the molecular ratio of silica to alumina and iron oxide (silica: sesquioxide ratio or SiO_2/R_2O_3 where $R = Fe + Al$) is an important criterion in the study of tropical soils. Therefore the total amount of these three substances is often determined, however the validity of this ratio has still to be demonstrated convincingly.

The silica:sesquioxide ratio is also determined for the clay fraction isolated from soils such as altosols and podzols. In these soils the amount of hydrolysis is relatively small and a partial ultimate analysis does not show strong contrasts between their horizons. However the data from the clay analysis are used as an aid in establishing distinguishing criteria for by this means it is possible to detect differential translocation of these substances. As well as the silica:sesquioxide ratio, silica:alumina, silica:iron oxide and alumina:iron oxide ratios are used.

At present most laboratories conducting these investigations employ long and tedious methods, but it is hoped that more rapid and sophisticated techniques such as X-ray fluorescence will be used in the future so that greater amounts of data will become available, further to test the validity of some of these criteria.

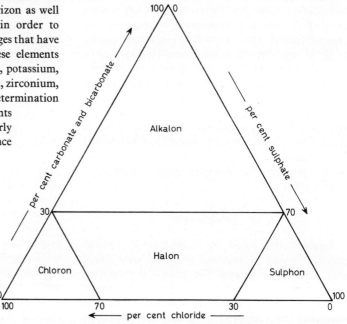

FIG. 69.

The subdivision of saline soils based on the proportions of the four principal anions. Each class is a horizon and shows the range of ions in the saturation extract.

15. Amorphous oxides – free silica, alumina and iron oxides

It has been stated previously that the initial stages of hydrolysis may form amorphous materials which may then crystallise gradually to produce a number of new substances or they may be translocated within the soil. It is assumed therefore that the type, amount and distribution of amorphous materials can be used as criteria for measuring the degree and type of soil formation. Also they are regarded as being formed during the current phase of pedogenesis and, therefore, are valuable criteria for differentiating between relic and contemporary phenomena. When performing the determinations it is customary to treat the soil with an extracting reagent such as sodium dithionite, citric acid or sodium pyrophosphate and the amount of each constituent in the extract is determined. However, published data pertaining to these constituents should be treated with caution as the suitability of many of the techniques can be challenged but certain trends have been discovered which seem to be reasonably valid.

Although chemical techniques have been used almost exclusively for the study of these oxides, other methods including the study of soils in thin sections, by the electron microscope and electron probe are yielding much new and interesting data. For example it has been found that amorphous materials are isotropic in cross polarised light, but probably the information supplied by the electron microscope is more revealing for it has been shown that silica often forms small aggregates while alumina and iron oxides form coatings on clay minerals and other surfaces.

16. Moisture content

There is a great need for more quantitative data about the variations in moisture regimes of soils in different environments and to relate these data to other soil characteristics. At present the annual variations in the content of moisture of soils is inferred from morphological characteristics and on limited observations made at various times of the year. On this basis the following five classes of drainage for the whole soil are usually recognised.

Excessively drained

Water moves rapidly through the soils which have bright colours due to oxidising conditions.

Freely drained

Water moves steadily and completely through the soils with little tendency to be waterlogged. These soils also have bright colours due to oxidising conditions.

Imperfectly drained

These soils are moist for part of the year with one or two horizons showing mottling owing to extended periods of wetness and reduction of iron to the ferrous state. The wetness in these soils as in the other two classes below may be caused by an impermeable horizon, high water table or high precipitation.

Poorly drained

These soils are wet for long periods of the year with the result that many of the horizons are mottled, with at least one that is blue or grey resulting from intensely reducing conditions.

Very poorly drained

These soils are saturated with water for the greater part of the year so that most of the horizons are blue or grey caused by reducing conditions. Peat may also be present as a consequence of the high degree of wetness.

Recently Stewart and Adams (1968) have shown that there is a correlation between soil morphology and the percentage waterlogging measured three days after rain and defined as a 'field capacity day'. The percentage waterlogging is the amount of moisture present in the soil on a field capacity day expressed as a percentage of the total pore space. They showed that horizons, mottled grey and yellowish brown (gleysons), were more than 70% waterlogged on such a day. Brown, middle horizons (altons) were usually less than 40% waterlogged while yellowish brown horizons with faint greyish areas (marblons) were about 60% waterlogged. Massive surface horizons (lutons) were over 70% waterlogged. This is very interesting since there are few morphological features in lutons that could be ascribed to wetness apart from a little rusty mottling along root channels. These horizons usually have a large population of anaerobic bacteria correlating well with the high percentage waterlogging which in these horizons is due not only to rainfall but to higher retentivity resulting from poor structure.

Perhaps it should be pointed out that a given figure of ca. 50% waterlogged does not mean that all the pores are half-full of water but that the larger pores are almost empty while the small ones are waterlogged. Thus, in a moist environment, soils that are more than 50% waterlogged three days after rain are likely to show some evidence of waterlogging either through their morphology or in their bacterial population. The concept of percentage waterlogging cannot be over stated since soils with many small pores and pore spaces are likely to have higher waterlogging percentages for longer periods following rain. This will influence the microflora and determine very largely the space available for root penetration. Although these results indicate a valuable potential method for the study of soil moisture in a humid environment its usefulness has to be demonstrated for other environments.

17. Concretions

These are discussed under thin section morphology given below.

18. Thin section morphology

Resulting from the pioneering work of Kubiëna (1938) thin sections of soils are being used in an increasing number of investigations, and it seems probable that in the future this technique will form one of the major methods for categorising soils.

Thin sections of soils are prepared by a technique similar to that employed by geologists for making thin sections of rocks, but soils must be impregnated with a resin and hardened before they can be cut, ground and polished. In the earlier studies, sections about 3 × 2 cm were produced but the tendency at present is to prepare larger sections, thus a considerable amount of thin section data given in this book is derived from thin section 10 × 5 cm ground to the standard thickness of 25 to 30μ, except where stated otherwise.

With this technique, investigations can be conducted on small details of the type and degree of organisation or *fabric* of the soil, for thin section studies reveal that particles of various sizes react differently to pedological processes and become organised to form many specific patterns. Therefore thin section morphology can be regarded as an extension of structure being concerned with the identification and study of the organisation of the soil constituents including the pores and pore space. Although many workers have attempted to give a specific definition for fabric it is used here as a general term to refer to the organisation and nature of material in the soil. This technique is used also in studies of weathering and other mineralogical investigations and has a high potential for microbiological investigations, allowing observations to be made on organisms in their natural habitats.

The aspects of soils usually studied in thin sections are given below.

1. Structure – the overall arrangement of the constituents including the pores and pore space.
2. Passages – faunal and root
3. Faecal material
4. Soil matrix
5. Organisation of soil constituents
   a) Organic matter d) Silt
   b) Stones e) Clay
   c) Sand
6. Domains
7. Cutans
8. Luvans
9. Papules
10. Concretions
11. Salt efflorescences and crystals
12. Microorganisms

There are a number of properties that are applicable to many of the phenomena given above. These properties are listed briefly below before discussing each phenomenon separately.

Distribution pattern: banded, clustered, concentric, radial, uneven, uniform, zoned.

Frequency: percentage of volume:

Term	Symbol	Range in %
Dominant	D	> 50
Very abundant	VA	25–50
Abundant	A	15–25
Very frequent	VF	10–15
Frequent	F	5–10
Occasional	O	2–5
Rare	R	0·5–2
Very rare	VR	< 0·5

Boundaries: abrupt, clear, diffuse, sharp.

Orientation: horizontal, oblique, random, vertical.

Shape: acicular, blocky, circular, compound forms, curved, dendritic, granular, irregular, lenticular, linear, ovoid, rectilinear, sinuous, tubular.

Roundness: angular, subangular, subrounded, rounded.

Size:

measurements of the phenomena are according to the length of the longest dimension.

Very large	> 10,000 μ (1 cm)
Large	2,000–10,000 μ (2 mm–1 cm)
Medium	200–2,000 μ
Small	50–200 μ
Very small	2–50 μ
Micro	< 2 μ

Surface characteristics: accordant (as between peds), mammillated, rough, scaly, serrated, slicken-sided, smooth.

1. Structure

By using large thin sections it is possible to examine the complete range of structures for most soils, starting with the unaided eye followed by examination under the microscope at various magnifications. The structural forms seen under the microscope are similar in shape to those seen with the unaided eye and in some cases the macro-peds are the only structural units present. Therefore the terms given on page 84 can be used for the descriptions of thin sections. However a conspicuous additional feature of peds and massive soils is the presence of discrete pores which are described in terms of their frequency, size, shape, etc. There are a number of recurring shapes of discrete pores such as small circular or ovoid pores which are attributed to the formation of gas bubbles. Figs. 58 and 63 illustrate a number of different types of intra-pedal pores.

2. Passages – faunal and root

Organisms such as earthworms and termites that burrow through the soil create passages which in cross section are usually circular or ovoid but they differ from discrete pores by containing faecal material and also by having interconnecting pores or pore space Figs. 67, 68, 83. When these passages occur in the subsoil the contrast between the material they contain and the surrounding soil is usually striking. Tree roots also form passages which may become filled with material when the tree dies and the roots decompose. Good examples of this are seen in semi-arid and arid areas where pipes of calcium carbonate may form in the spaces left by the roots. Also in wet soils local oxidation around the sides of the old root passages forms tubes of material cemented by iron oxide.

3. Faecal material

The mesofauna produce large amounts of faecal material particularly in the upper organic horizons where there is a plentiful supply of food. Faecal material also occurs at depths of over a metre and is then a useful indication of the depth of faunal penetration. Figs. 67, 68, 82, and 104 illustrate some types of faecal material and its location.

4. Soil matrix

The soil matrix is composed of material less than about 20 μ which appears in plane polarised light as a continuous phase enclosing larger grains, pores and other material but when examined in cross polarised light it may vary from isotropic to strongly birefringent with a high degree of organisation. Therefore it is only in soils with an appreciable amount of fine material that there is much matrix. As the content of sand increases there is a tendency for the matrix to be arranged around the sand grains or to form bridges between the grains. However, there are some very sandy soils that are very compact with the matrix occurring in the spaces between the sand grains. The principal form of organisation within the matrix are domains (Emerson 1959) which are discussed below.

5. Organisation of soil constituents

a. Organic matter

Fresh or partially decomposed organic matter is easily recognised by its cellular structure but there is as yet no positive optical test for identifying humified material apart from its isotropic characteristics and dark colours but some confusion may arise when dark coloured soils contain allophane and ferric oxide which are also isotropic. When a horizon is composed of an intimate blend of organic and mineral material it is even more difficult to recognise the organic matter since it forms a clay-humus complex which is also isotropic. An indirect indication of the presence of organic matter is through the occurrence of faecal material. Figs. 67, 68, 82, and 104 illustrate the nature of organic material in some horizons.

b. Stones

One of the advantages of large thin sections is that the larger separates including stones can be studied, thus very useful information can be obtained about their mineralogy, orientation, shape, frequency, degree of weathering, relationship to other stones and to the surrounding soil. Except in certain specific cases stones are thought to have random orientation and distribution but very few investigations along these lines have been conducted.

c. Sand

The nature of the organisation of the sand is very variable. In most soils the sand grains occur separate one from the other and embedded in the matrix, then they are described in terms of size and frequency. In some cases the matrix is small in amount and the sand grains are separate one from the other forming a single grain structure. A third type of fabric is composed of closely packed sand grains with a small amount of matrix. This is a characteristic feature of many soils having a high content of silt and fine sand but is most marked in soils containing a high proportion of small calcite grains which may be inherited from the parent material or formed *in situ*. In addition the sand fraction may have a specific distribution pattern such as occurring in bands, clusters, or related to some other feature. A random orientation pattern is striking when there is a high proportion of mica present. This is seen as a false diamond arrangement in cross polarised light since it is at the 45° position that mica flakes show their maximum birefringence (Fig. 70).

The arrangement of the sand grains may be related also to a weathering pattern, for it is found that in some horizons formed by strong weathering the quartz grains may retain a very similar relationship to each other in the weathered material as was present in the original rock, thus allowing certain deductions to be made about the original rock structure (Fig. 89).

d. Silt

At present the main optical studies on the silt fraction are conducted on detrital grains and it appears that they have the same general characteristics as the sand fraction but as most thin sections vary between 20–50 μ it is impossible to see clearly the individual silt grains. Perhaps in future, the use of high magnifications, very thin sections and phase contrast tech-

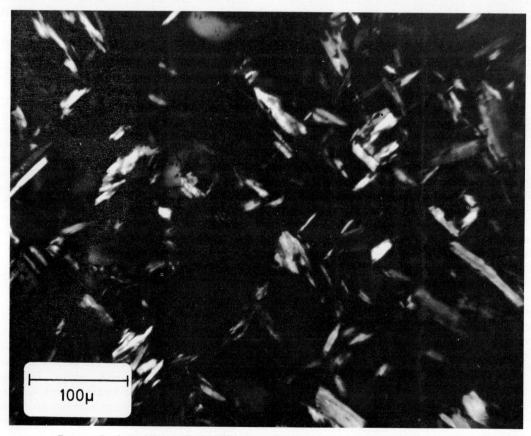

100μ

FIG. 70. Random orientation of mica flakes which show maximum birefringence in the 45° position: hence the apparent lattice like arrangement.

niques may demonstrate some specific properties, since this is the size range that is weathering most rapidly. In the meantime the silt is regarded as forming part of the matrix.

e. Clay

The clay fraction is considered to comprise all material $< 2\ \mu$, including the clay minerals and oxides which together constitute a considerable part of the matrix. The various patterns of organisation displayed by the crystalline clay are among the most important distinguishing and differentiating criteria to be seen in thin sections. Of these the two most common are domains and cutans both of which are discussed below.

A conspicuous feature of many lower horizons in tropical and subtropical soils is the presence of macrocrystalline kaolin, forming pseudomorphs of primary minerals or growing elsewhere within the soil (Fig. 35).

6. Domains

These are segregations of clay sized material within the matrix. They are not visible in plain light but are seen in crossed polarised light as small birefringent areas with diffuse boundaries – properties which serve to differentiate them from mica flakes. Domains are usually randomly orientated within the central part of peds but they tend to be aligned parallel to the surfaces of peds and around pores. They are formed by the coalescence and similar alignment of clay particles hence their birefringence. However, the exact mechanism is obscure but most probably expansion and contraction are responsible (Fig. 71). Domains vary considerably in length and width and can be classified on their size characteristics as set out in Table 13 below.

Table 13
CLASSIFICATION OF DOMAINS

	LENGTH		WIDTH	
Type	Length(μ)	Type	Width(μ)	
Very short	$< 10\ \mu$	Very narrow	$< 1\mu$	
Short	$10–50\ \mu$	Narrow	$1–5\ \mu$	
Long	$50–100\ \mu$	Broad	$5–10\ \mu$	
Very long	$> 100\ \mu$	Very broad	$> 10\ \mu$	

7. Cutans

These are deposits or concentrations of small particles and colloidal material on the surfaces of peds, pores, sand grains or stones and can be classified according to their composition and method of formation.

a. Depositional clay cutans

This type of cutan is the most common and probably the most important. In plane polarised light they are recognised by their layered structure with an absence of coarse material while in cross polarised light they are birefringent being formed by the progressive deposition of clay sized particles on ped or pore surfaces in such a way that the particles gradually build up like mica flakes; hence their birefringence. They are easily differentiated from primary micas by their lower birefringence and undulose extinction which often takes the form of a black extinction band (Fig. 72; pp. 102–3). Depositional cutans are relatively common phenomena and therefore they are regarded as significant only when present in excess of about 3%.

b. Diffusion cutans

Cutans with similar optical properties to depositional clay cutans are commonly found on the surfaces of large sand grains within the matrix where their formation is attributed to expansion and contraction and they are regarded as being formed by diffusion.

Just beneath the surface of many peds or next to the surface of some pores there are features that closely resemble depositional clay cutans but they contain larger particles. These features are probably caused by the diffusion of material to the surface and are also regarded as diffusion cutans.

c. Stress cutans

On the surfaces of slickensides and some other ped surfaces clay becomes orientated in response to pressure to form stress cutans.

d. Silt cutans

Certain horizons have local concentrations of material mainly of silt size. These may be randomly distributed through the soil but usually they occur on the upper surfaces of stone (Fig. 103) or peds or may form linings in fissures and cracks.

e. Other cutans

Many other substances form cutans, one of the most common being manganese dioxide but this also occurs

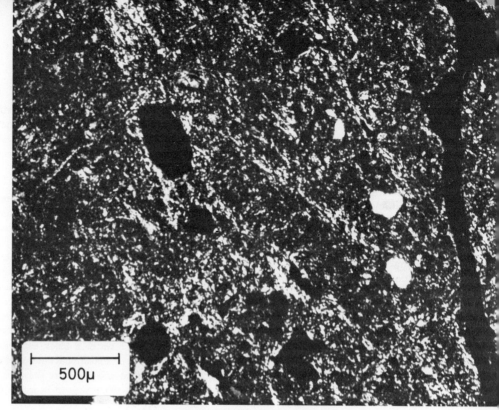

500μ

A: Abundant short random domains.

FIG. 71.   DOMAINS

B: Abundant short random domains and stress cutans on either side of the oblique pore space.

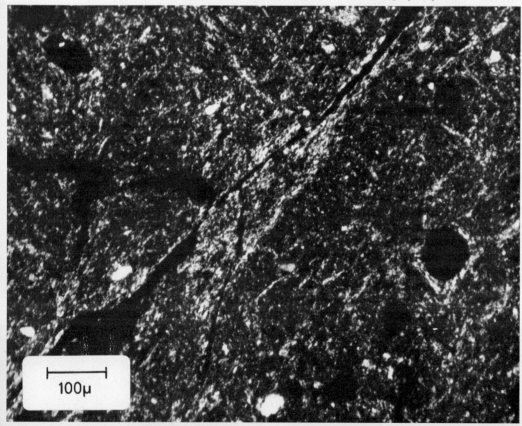

100μ

within the matrix as irregular dendritic areas.

8. Luvans

These are thin coatings of sand grains that occur on the surfaces of some peds and have formed by differential removal of the matrix through leaching.

9. Papules

These are small areas of clay size material which represent weathered mineral grains or fragments of cutans. They are usually weakly birefringent with uniform extinction or they may be composed of domains (Brewer 1964).

10. Concretions

Under certain specialised conditions some constituents form local concentrations which may become very hard. For example, iron may accumulate to form concretions in certain soils of the tropics and some periodically waterlogged soils. Similarly calcium carbonate forms concretions or massive horizons in arid and semi-arid areas. Figs. 90 and 116 illustrate the micro-structures of two types of concretions.

11. Salt efflorescences and crystals

Efflorescences of salts on the surface of peds and the growth of crystals within the matrix are a common feature. This is restricted largely to arid and semi-arid situations where crystalline gypsum and calcite are very common (Fig. 113). On the other hand pyrite forms in waterlogged soils of tidal marshes.

12. Microorganisms

By using suitable stains it is possible to investigate the type and occurrence of microorganisms in their natural habitats in the soil but as yet only a very few studies of this type have been conducted. Fig. 77 shows the occurrence of fungal mycelia on decomposing organic matter.

See overleaf for Figs. 72 A and B

A: in plain polarised light

FIG. 72. DEPOSITIONAL CLAY CUTANS

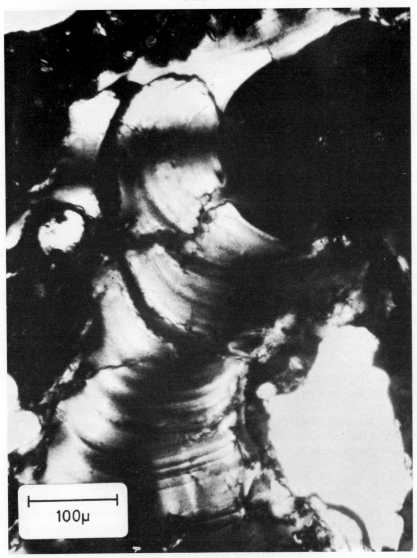

B: in cross polarised light showing the characteristic black extinction bands.

BRIDGING PORE SPACE

5

NOMENCLATURE AND CLASSIFICATION
HORIZON NOMENCLATURE

Soil horizons are normally designated by symbols composed of letters, following in some measure the classical system introduced by Dokuchaev (1883) who used the letters A, B, and C to indicate the upper, middle and lower positions respectively, in the chernozem pedounit. Since then, many attempts have been made to amplify and refine this system to fit other soils, mainly by adding qualifying letters, numbers and subscripts to form symbols such as A_1, A_2 and Bfe which are then used to designate specific horizons. However these attempts have not been very successful because workers in different parts of the world have used the same symbol such as A_2 for quite different horizons as well as using different symbols for similar horizons. This is clearly shown in Table 14, the full explanation of which is to follow.

Apparently Müller (1879, 1884) initiated soil horizon nomenclature by introducing a small number of names for specific surface horizons; these included the terms mor, torf, insect mull, mull-like mor, and muld. This approach was followed by many workers particularly Kubiëna (1953) and it is found in the 7th Approximation, which is a system of classification recently devised by the Soil Survey Staff of the United States Department of Agriculture, but no attempt has been made to name all soil horizons. Some confusion has arisen with regard to naming horizons; Kubiëna and many other workers have adopted some of the terminology of Müller, plus the addition of a few new names of their own, whereas in the 7th Approximation completely new terms have been introduced. Most of these duplicate many of those already in existence but there is also an addition of a small number of new names which have been added for horizons which have not been named previously.

However, all workers use the A-B-C system of horizon designation and consequently it is found that for some horizons, names and symbols are used simultaneously and synonymously. In recent years the trend is to use horizon names particularly in general discussions and to ascribe symbols only when writing profile descriptions. The result of these various attempts to designate horizons is that nearly every worker has his own particular set of symbols and names. Furthermore there is seldom any relationship between the horizon name and its symbol. Notable exceptions to this include the fermentation and humification layers designated by the letters F and H respectively.

Jenny (1941) states there exists a great need for rigorous criteria for horizon identification, because all scientific systems of soil classification as well as theories regarding soil development rest on horizon interpretation. An attempt to meet this need and to produce an orderly system of names and symbols for horizons has recently been published (FitzPatrick 1967) and it is this approach that is used in this book. In this new system, each and every standard horizon that has been recognised is given a name and a symbol. The name is based on some conspicuous or unique property of the horizon; this may be its colour or some substance present in significant proportions. The symbols are formed by using the first letter of the horizon name plus one other which is usually a part of the name.

All of the names of horizons are coined words ending with -on the same as for the word horizon. When creating these new words every attempt was made to modify words that already form part of pedological terminology so that the new names appear to be somewhat familiar. For example the name mullon is the well known term mull with an -on ending; fermenton and humifon are derived from fermentation and humification respectively. It has been necessary to introduce a number of new names such as zolon and cerulon. Wherever possible these are based on a conspicuous property and the derivation of each term is given at the end of each definition in Table 17. In addition two groupings of horizons are given in Tables 15 and 16 which are intended to aid the identification of unknown horizons.

It has been shown in previous chapters that horizons are not discrete entities but grade from one into the other. This phenomenon of continuous variation imposes almost insurmountable difficulties with regard to establishing and naming horizons. The diagram below (Fig. 73) demonstrates the principles used to create separate horizons and ascribe names to them. It illustrates the situation in which three horizons, a sesquon, Sq, an alton, At, and an argillon, Ar, are intergrading horizontally. Since it is virtually impossible to communicate information in terms of continuous variation unless lengthy descriptions are used, some attempt must be made to impose boundaries in the continuum, both with regard to the areas of apparent uniformity as well as with regard to the intergrading situations, even although the divisions are of necessity somewhat arbitrary. There would seem to be two methods of creating divisions in a continuum such as soil. The first is to set boundaries half-way between Sq and At, and At and Ar at points 2 and 5, in Fig. 73. Each horizon could then be defined in terms

of its central part as well as in terms of the intergrading situation on either side. This might be a practical solution if each horizon intergraded only into two other horizons, but in reality horizons intergrade vertically as well as horizontally. Furthermore, in another situation an alton might intergrade laterally into two other quite different horizons. The result is that horizon definitions would have to be written in terms of all the possible intergrading situations which would be cumbersome and could even be meaningless. The second and more realistic approach is to create what would appear to be three standard horizons and two intergrades, by placing boundaries at points 1, 3, 4, and 6. However, it must be stated that these boundaries have no fundamental meaning because any two adjacent points situated on either side of a boundary are more like each other than each is to distant points within its own horizon irrespective of whether it is an intergrade or standard horizon. This method produces a relatively small number of standard horizons but there are a very large number of intergrades since any one horizon may intergrade into two or more other horizons. Creating horizon boundaries and attempting to introduce the concept of standard horizons is not as simple as it might appear at first, since each horizon has a wide variety of properties and usually one or more of these properties are intergrading towards those of other horizons. Therefore the first step in horizon nomenclature is to recognise all the various standard horizons which are

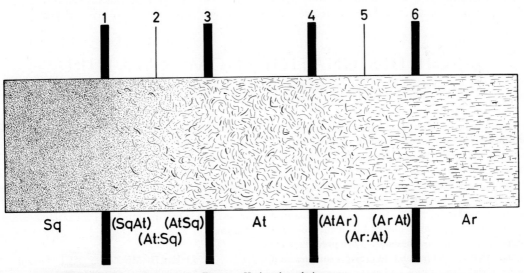

FIG. 73. Horizon boundaries.

then defined in terms of all their known intrinsic properties although in a few cases a comparison with another horizon is necessary. The lack of intrinsic differentiating criteria for some horizons is probably due to a lack of knowledge rather than to lack of criteria. It is hoped that further research will supply these criteria. The names of the horizons and their symbols are given alphabetically in Table 14 together with conventional and 7th Approximation equivalents. The page number after each horizon name refers to the location of the description of the properties of that horizon in Table 17. Most of these properties have a wide range of variability; therefore it is necessary to set arbitrary limits for the range of each property within each horizon. At present the limits are set on a purely subjective basis but wherever possible horizons have been created using coordinate principles which would seem to be the only satisfactory way in which to divide up a continuum. Figs. 69, 141, 142, 143 and the accompanying discussions illustrate the manner in which a number of horizons were created. But with present computing methods involving the use of ordination techniques it may be possible to set objective standards and create more exact limits in the future.

The intergrades are not given names but are designated by the use of the horizon symbols. Referring again to Fig. 73 the intergrades are designated by placing the appropriate symbols in round brackets according to the proportion of each set of properties, the dominant set being placed first. The intergrade (SqAt) displays properties of a sesquon and an alton but is more like a sesquon. In cases where there are equal proportions of the properties from two horizons their symbols are separated by a colon and written in alphabetical order thus (At:Sq) and (Ar:At). Where intergrades are formed through physical mixing by ploughing or faunal activity the symbols are separated by a stroke, placing the symbol of the receiving horizon first. One of the best examples is found in chernozems which have krotovinas. Horizons of this type are designated (Ch/CZ), Plate IIb. Horizon designations of this type are particularly well suited to soils with polygenetic evolution, such as those found in certain parts of East Africa where red horizons (krasnons) developed in a humid climate are now in an arid environment and are accumulating calcium carbonate. These horizons should really be regarded as compound horizons but they are treated as krasnon-calcon intergrades and designated (KsCk).

This system of notation is an attempt to recognise and preserve the reality of the soil continuum. It must be emphasised that horizon symbols are merely a means of communicating information in a concise form and are *not* a substitute for a full and comprehensive characterisation which involves a detailed description of the soil and supporting analytical data. In many cases a specific horizon may be very thick and it may be necessary for it to be subdivided for analytical purposes in which case the subhorizons are designated with the aid of Arabic numbers — a krasnon which has three subdivisions would be symbolised as follows: 1Ks, 2Ks, 3Ks.

It has been found necessary to use a small number of suffices to cope with the wide variability found in some horizons. These suffices are given in Table 17.

SOIL DESIGNATION, NOMENCLATURE
AND CLASSIFICATION

Soils are composed of horizons and therefore also form continua in space and time, which implies that any divisions or boundaries created within soils are similarly as arbitrary as the divisions that separate horizons, therefore they do not delimit discrete entities. This situation defies classification. The best that is attainable for soil, is to give each pedounit a designation, then to have one or two higher levels of grouping using *ad hoc* methods. This procedure recognises that it is impossible to create a fundamental system of soil classification and at the same time places greater emphasis on the pedounit. The designation of the pedounit itself is achieved by using a formula which is produced by writing in order, the horizon symbols as they occur in vertical sequence starting with the uppermost horizon. The thickness of each horizon in centimetres is also indicated by means of a subscript number. Two examples are given below:

1. $Lt_2Fm_3Hf_2Mo_7Zo_5Sq_{10}AST$ Podzol
2. $Lt_2(Ch/CZ)_{100}Ck_{20}CZ$ Chernozem

Each set of symbols in each formula characterises a *group* of soils, all of which have the same horizon sequence but there may be variations in the thickness of the horizons, particle size distribution and many other features. However, all the members of the group are sufficiently alike to be placed in the same category. Each distinctive member of each group is given a name, usually a local place name which is also given to the series (mapping unit) in which this group member is dominant. Perhaps it should be stated that mapping units are not necessarily taxonomic units. Most mapping units are defined as having a certain range of variability within each property. Sometimes the range may fall within the defined limits of the constituent horizons but as often happens the range overlaps the arbitrary divisions separating the horizons. Therefore the mapping unit may span two taxanomic units in the same way that mapping units regularly span two arbitrarily created textural classes.

The groups are arranged in *subclasses* and *classes* both of which, particularly the latter, are established on an *ad hoc* basis.

The subclasses are established on the basis of two, three or four prominent horizons in the same pedounit. They are not given names but are designated by writing the symbols of each horizon and placing them in square brackets. Four characteristic subclasses of the podzol class are:

[ZoSq] [ZoHd] [ZoHs] [ZoSqZoAr]

The classes are given names and are characterised by the presence of one prominent horizon in the pedounit or by a unique combination or relationship between the constituent horizons. When a class is established on the basis of a single horizon it often derives its name from that horizon. The argillosols are created because of a prominent and well developed argillon. On the other hand the supragleysols are established because of the unique occurrence of a markedly gleyed horizon or horizons in the middle position of the soil; the upper and lower horizons normally being freely drained and lacking any characteristic associated with wetness. Sometimes difficulties may be encountered in ascribing a group to a given class particularly if two prominent horizons are present. In such cases the group is placed in the class with which it seems to have the greatest affinities.

This system possesses a high level of versatility, allowing classes to be created on the basis of any prominent horizon or upon any unique combination of horizons. Therefore groupings can be produced to meet the demands and requirements of local situations. It also has advantages when trying to accommodate soils that have complex polygenesis or those that have been truncated by erosion. Furthermore the horizon symbols can be used in the construction of maps and map legends. This makes it possible to put more information on to maps and to have a single set of symbols to cover most pedological work.

As names are given only to the classes and to the groups, it is possible to derive a system of binomial nomenclature for soils by combining the class name and the series names of each soil. The former gives a general indication of the fundamental properties of the soil and the latter is a specific characterisation. In order to facilitate this system of nomenclature single words are used for the names of classes as well as for the names of the members of the groups. Furthermore classical pedological names or slight modifications of these are used wherever possible. These include the names podzol, chernozem and rendzina. The binomial name of two soils discussed in Chapter 6 are Placosol Charr and Podzol Corby.

SUMMARY

The principles of soil designation, nomenclature and classification are summarised as follows:

1. Each horizon is defined, named and given a symbol which is made up of the first letter of the horizon name, plus one other.

2. Each soil is given a formula made up by placing in order, the symbols for each horizon followed by a subscript number which gives the thickness of each horizon in cm and so the depth of the whole soil. Each set of symbols represents a different group which will contain a number of members.

   Each distinctive member of each group is given a name, usually a local place name. This name is also the name of the series (mapping unit) of which the group is the dominant member.

3. The groups are arranged in subclasses which are established on the basis of two or three prominent horizons in the same soil – an upper, a middle and a lower horizon.

4. The subclasses are arranged in classes using *ad hoc* methods, and are sometimes based on prominent horizon or upon a unique combination or relationships between certain horizons. The classes are given names.

5. A binomial system of nomenclature is possible by combining the names of the classes and series.

HORIZON DATA

This section (pp. 109–156), composed mainly of Tables 14–17 contains information about a large number of soil horizons. Table 14 is a list of the horizons and their symbols and includes a number of equivalent horizon symbols and names used in other systems of classification and is intended to aid in the conversion from one system to the other. Tables 15 and 16 are groupings of the horizons on the basis of colour and position. It is always difficult to produce groupings of this type without creating some ambiguities. This is partly true for Table 15–'Horizon Grouping According to Colour', for in this Table it has been necessary to reduce the wide range of soil colours to a relative few therefore each colour term should be used in a fairly broad sense.

Tables 15 and 16 can be used together to aid in the identification of horizons. A positive identification cannot be achieved by such a simple system but an unknown horizon can be narrowed down to one of about three or four. Then, by consulting Table 17 it should be possible to identify it as a standard or intergrade horizon.

In Table 17, most of the common properties of the horizons are given. Those properties which differentiate and distinguish one horizon from the other are underlined so that it is possible at a glance to see the criteria used to establish each one.

Table 17 should be regarded as a summary of the most important information of the given standard horizons. It is hoped that fuller information for each horizon will be published eventually. Fuller data for a number of horizons have been prepared and on pages 154, 155 and 156 ISONS are given as an example of the forthcoming treatment but it will be some time before the complete set is ready.

In addition the intention is to produce a key for the identification of the standard horizons and also to give some indication about the principal intergrades. Since most horizons intergrade in many different directions it is difficult if not impossible at this stage to include all intergrades into a simple key.

Table 14
SOIL HORIZONS AND SYMBOLS

Names	Page No.	Symbols	Traditional approximate equivalent symbols and names	7th Approximation approximate equivalent names
Alkalon	116	Ak	A, B, C,	Salic
Alton	116	At	B_1, B_2; B_3, (B)	Cambic
Amorphon	116	Ap	A_0. Amorphous peat	Histic
Andon	116	An	B_1, B_2, B_3, (B)	Cambic
Anmooron	118	Am	A_1. Anmoor	Umbric
Arenon	118	Ae	B, C.	Cambic
Argillon	118	Ar	B_1, B_2, B_3, Bt	Argillic
Buron	118	Bu	A_1, A_2.	Umbric
Calcon	120	Ck	Ca, Cca.	Calcic
Candon	120	Co	A_2g.	Albic
Celon	120	Cs	Bg, Cg, g	Cambic
Cerulon	120	Cu	Bg, Cg, g	Cambic
Chernon	122	Ch	A_1, A_2, Ap.	Mollic
Chloron	122	Ci	A, B, C.	Salic
Clamon	122	Cn	B_1, B_2, Bt	Argillic
Crumon	122	Cw	A_1	Mollic
Cryon	124	Cy	C, Cf. Permafrost	
Cumulon	124	Cm	B	
Dermon	124	De	A_1	Umbric
Duron	124	Du	Bm. Duripan	Duripan
Fermenton	126	Fm	A_0, F, O_2 Fermentation	
Ferron	126	Fr	B_1, Bfe, Bir	Spodic
Fibron	126	Fi	A_{00}, A_0. Fibrous peat	Histic
Flambon	126	Fb	C. Laterite	Plinthite
Flavon	128	Fv	B	Oxic
Fragon	128	Fg	B_3, Cx. Fragipan Indurated layer	Fragipan
Gelon	128	Gn	A, B.	Umbric, Ochric
Gleyson	128	Gl	Bg, Cg, g	Cambic
Glosson	130	Gs	B, Bg	Cambic
Gluton	130	Gt		Plinthite
Gypson	130	Gy	Ccs.	Gypsic
Gyttjon	130	Gj	A_0, Gyttja	Histic
Halon	132	Hl	Bsa, Csa, Sa.	Salic
Hamadon	132	Ha	Hamada	
Hudepon	132	Hd	B_1, B_2, B_3, Bh	Spodic
Humifon	132	Hf	A_0, H, O_2 Humification.	
Husesquon	134	Hs	Bhfe. Ortstein	Spodic
Hydromoron	134	Hy	A_0, H	
Ison	134	In	B_3, Cx. Fragipan Indurated layer	Fragipan

Table 14 continued

Names	Page No.	Symbols	Traditional approximate equivalent symbols and names	7th Approximation approximate equivalent names
Jaron	134	Ja	Bg, Cg,	Cambic
Kastanon	136	Kt	A_1, A_2	Umbric
Krasnon	136	Ks	B_1, B_2, B	Oxic
Kurón	136	Ku	A_1.	Umbric, Mollic
Limon	136	Lm	Ca	
Lithon	138	Lh	C, D. Stoneline	
Litter	138	Lt	A_{00}, L, O_4, Litter	
Luton	138	Ln	A_1	
Luvon	138	Lv	$A\epsilon$, E, A_2.	
Marblon	138	Mb	Bg, Cg.	Cambic
Minon	140	Mi	A_2g	Albic
Modon	140	Mo	A_1, Moder.	Umbric
Mullon	140	Mu	A_1, A_2, Ap. Mull	Mollic, Umbric
Oron	140	Or	B. Bog iron ore	
Pallon	142	Pl	Pallid layer	
Pelon	142	Pe	B, B_1, B_2,	Cambic, Argillic
Pesson	142	Ps	B	
Placon	142	Pk	B_1, B_2. Thin iron pan	Placic
Plaggon	144	Pg	A_1	Plaggen
Planon	144	Pn	Bg, Cg.	Argillic
Primon	144	Pr	A/C	
Proxon	144	Px	B_1, B_2,	
Pseudofibron	146	Pd	A_{00}, A_0, Pseudofibrous pea	Histic
Rosson	146	Ro	B_1,	Cambic
Rubon	146	Ru	B_1.	Cambic
Rufon	146	Rf	B_1, B_2	Cambic, Argillic
Sapron	148	Sa	C,	Saprolite
Seron	148	Sn	A_1, A_2	Ochric
Sesquon	148	Sq	B_1, B_2, B_3, Bfe, Bir, Ortstein	Spodic
Sideron	148	Sd	Amorphous peat	Histic
Solon	150	Sl	B, B_1, Bt	Natric
Sulphon	150	Su	A, B, C,	Salic
Tannon	150	Tn	A_1, Ap.	Ochric
Thion	150	To	Bg, Cg.	Cambic
Verton	152	Ve	A_1, A_2	Mollic
Veson	152	Vs	Laterite	Plinthite
Zhelton	152	Zh	B_1, B_2	Cambic, Argillic
Zolon	152	Zo	A_2, Ae. Bleicherde	Albic

Table 15

HORIZON GROUPING ACCORDING TO COLOUR

Black, Very dark brown, Very dark grey	Dark brown, Reddish brown	Brown, Yellowish brown	Dark grey, Greyish brown	Brownish grey, Grey, Pale brown	Mottled	Bluish grey, Olive brown, Olive	Red, Brownish red, Yellowish red
—	—	—	ALKALON	—	ALKALON	ALKALON	—
—	—	ALTON	—	—	—	—	Alton
AMORPHON*	—	—	—	—	—	—	—
—	ANDON	ANDON	—	—	—	—	—
ANMOORON	—	—	ANMOORON	—	—	—	—
—	—	ARENON	—	—	—	—	ARENON
—	ARGILLON	ARGILLON	—	—	—	—	Argillon
—	BURON	—	—	—	—	—	—
—	—	—	CALCON	CALCON	—	—	—
—	—	—	—	—	CANDON	—	—
—	—	—	CELON	—	—	CELON	—
—	—	—	—	—	—	CERULON	—
CHERNON	—	—	—	—	—	—	—
—	—	—	CHLORON	—	CHLORON	CHLORON	—
—	CLAMON	CLAMON	—	—	—	—	CLAMON
CRUMON	—	—	—	—	—	—	—
—	—	—	CRYON	—	—	CRYON	—
—	CUMULON	—	—	—	—	—	CUMULON
DERMON	—	—	—	—	—	—	—
—	DURON	DURON	—	—	—	—	—
FERMENTON*	—	—	—	—	—	—	—
—	—	FERRON	—	—	—	—	—
FIBRON*	—	—	—	—	—	—	—
—	—	—	—	—	FLAMBON	—	—
—	—	FLAVON	—	—	—	—	FLAVON
—	—	FRAGON	—	FRAGON	Fragon	Fragon	—
—	—	GELON	GELON	—	—	—	—
—	—	—	—	—	GLEYSON	—	—
—	—	—	—	—	GLOSSON	—	—
—	GLUTON	—	—	—	—	—	GLUTON
—	—	—	GYPSON	GYPSON	—	—	—
GYTTJON*	—	—	—	—	—	—	—
—	—	HALON	HALON	HALON	—	—	—
HAMADON	—	HAMADON	HAMADON	—	—	—	—
HUDEPON	—	—	—	—	—	—	—
HUMIFON*	—	—	—	—	—	—	—
HUSESQUON	—	—	—	—	—	—	—
HYDRO-MORON*	—	—	—	—	—	—	—
—	—	—	—	ISON	Ison	Ison	Ison
—	—	—	—	—	JARON	—	—
KASTANON	KASTANON	—	—	—	—	—	—
—	—	—	—	—	—	—	KRASNON
KURON	—	—	—	—	—	—	—

Table 15 continued

Black, Very dark brown, Very dark grey	Dark brown, Reddish brown	Brown, Yellowish brown	Dark grey, Greyish brown	Brownish grey, Grey, Pale brown	Mottled	Bluish grey, Olive brown, Olive	Red, Brownish red, Yellowish red
—	—	—	—	LIMON	—	—	—
—	—	LITHON	—	—	—	—	—
—	—	LITTER	—	—	—	—	—
LUTON	LUTON	—	LUTON	—	LUTON	—	Luton
—	—	—	—	LUVON	—	—	—
—	—	—	—	—	MARBLON	—	—
—	—	—	—	MINON	MINON	—	—
MODON	—	—	Modon	—	—	—	—
MULLON	MULLON	Mullon	—	—	—	—	—
ORON	—	—	—	—	—	—	—
—	—	—	—	—	PALLON	—	—
—	PELON	PELON	—	—	—	—	PELON
—	PESSON	—	—	—	—	—	PESSON
PLACON	PLACON	—	—	—	—	—	—
PLAGGON	PLAGGON	PLAGGON	—	—	—	—	—
—	—	—	—	—	PLANON	—	—
—	—	—	PRIMON	—	—	—	—
—	PROXON	PROXON	—	—	—	—	PROXON
PSEUDO-FIBRON*	—	—	—	—	—	—	—
—	Rosson	—	—	—	—	—	ROSSON
—	Rubon	—	—	—	—	—	RUBON
—	Ruffon	—	—	—	—	—	RUFFON
—	SAPRON	SAPRON	—	—	SAPRON	—	SAPRON
—	SERON	SERON	—	—	—	—	Seron
—	SESQUON	SESQUON	—	—	—	—	—
SIDERON	—	—	—	—	—	—	—
SOLON	SOLON	—	—	—	—	—	—
—	—	—	SULPHON	—	SULPHON	SULPHON	—
—	TANNON	TANNON	—	—	—	—	—
—	—	—	THION	—	—	THION	—
VERTON	—	—	—	—	—	—	—
—	VESON	—	—	—	VESON	—	VESON
—	—	ZHELTON	—	—	—	—	—
—	—	—	—	ZOLON	—	—	—

*Predominantly organic horizons
Lower case indicates lower frequency.

Table 16
HORIZON GROUPING ACCORDING TO POSITION

UPPER	MIDDLE	LOWER
ALKALON	ALKALON	ALKALON
—	ALTON	—
AMORPHON	AMORPHON	AMORPHON
ANMOORON	ANDON	—
—	ARENON	Arenon
—	ARGILLON†	Argillon†
BURON	—	—
—	Calcon	CALCON
CANDON	Candon	—
—	Celon	CELON
—	Cerulon	CERULON
CHERNON	Chernon	—
CHLORON	CHLORON	CHLORON
—	CLAMON†	—
CRUMON	—	—
—	Cryon	CRYON
Cumulon	CUMULON	—
DERMON	—	—
—	Duron	DURON
FERMENTON	—	—
—	FERRON†	Ferron†
FIBRON	FIBRON	FIBRON
—	Flambon	FLAMBON
—	FLAVON	Flavon
—	—	FRAGON
GELON	Gelon	—
—	GLEYSON	GLEYSON
—	GLOSSON	GLOSSON
—	GLUTON	—
—	Gypson	GYPSON
GYTTJON	GYTTJON	GYTTJON
Halon	HALON	HALON
HAMADON	—	—
—	HUDEPON	—
HUMIFON	—	—
—	HUSESQUON†	—
HYDROMORON	—	—
—	—	ISON

UPPER	MIDDLE	LOWER
—	JARON	—
KASTANON	—	—
—	KRASNON	Krasnon
KURON	—	—
—	LIMON	—
—	LITHON	—
LITTER	—	—
LUTON	—	—
LUVON*	Luvon*	—
—	MARBLON	MARBLON
MINON*	Minon*	—
MODON*	—	—
MULLON	—	—
—	ORON†	—
—	Pallon	PALLON
—	PELON	PELON
—	Pesson	PESSON
—	PLACON†	PLACON†
PLAGGON	Plaggon	—
—	PLANON†	—
PRIMON	—	—
—	PROXON	—
PSEUDOFIBRON	PSEUDOFIBRON	PSEUDOFIBRON
—	ROSSON†	ROSSON†
—	RUBON†	RUBON†
—	RUFON	—
—	Sapron	SAPRON
SERON	—	—
—	SESQUON†	—
—	SIDERON	SIDERON
—	SOLON†	—
SULPHON	SULPHON	SULPHON
TANNON	—	—
—	Thion	THION
VERTON	VERTON	—
Veson	Veson	VESON
—	ZHELTON	Zhelton
ZOLON*	Zolon*	Zolon*

*Contain less sesquioxides or clay than the horizon below.
†Contain more sesquioxides or clay than the horizon above or below.
Lower case indicates lower frequency.

PROPERTIES OF SOIL HORIZONS

Table 17

PROPERTIES OF SOIL HORIZONS

The terminology used in the descriptions is as follows:

1. Colour is normally given for the soil in the moist state and is described according to the Munsell Soil Colour Charts (1954).

2. Texture, structure, stones, boulders and consistence are described according to the U.S.D.A. Soil Survey Manual (1951). Stones and boulders are given on a weight (wt), and volume (v) basis.

3. The frequency of the clay minerals is according to the scale on page 96.

4. Thin section morphology: the terms given on page 96 *et seq.* are used.

5. The abbreviations used are as follows:
   OM = Organic matter, expressed as a percentage by weight of the < 2mm fraction
   C/N = Carbon:nitrogen ratio
   CEC = Cation exchange capacity, expressed in me per 100 g of the < 2 mm fraction
   BS = Base saturation, expressed as a percentage of the CEC

6. Carbonates and salts expressed as percentages of the < 2mm fraction

7. The abbreviations for the origin of the words are:

C	= Celtic	Gk	= Greek	OE	= Old English
E	= English	I	= Italian	R	= Russian
F	= French	J	= Japanese	S	= Swedish
G	= German	L	= Latin		

8. The asterisks and other symbols in the upper part of the Table are explained under "Other distinguishing features".

QUALIFYING SUFFIXES

c – containing calcium and/or magnesium carbonate from the parent material.

h – high base status.

d – hardened.

Table 17
PROPERTIES OF SOIL HORIZONS

HORIZON Name and symbol	ALKALON Ak	ALTON At	AMORPHON Ap	ANDON An
Position	Upper, middle or lower	Middle	Upper, middle or lower	Middle
Colour	Dark greyish brown, olive brown	Brown, yellowish brown, browish red	Black, very dark grey, very dark brown	Brown, yellowish brown, reddish brown
Thickness (cm)	> 20	15–50	5–30	15–50
Texture	Variable	Loam, silt	—	Loam, silt
Stones and boulders (%)	Variable	< 60wt < 50v	—	< 60wt < 50v
Structure	Massive, prismatic	Crumb, blocky	Massive	Granular, blocky
Consistence	Firm, hard	Friable, firm, loose when clay is low	Plastic, amorphous, crumbly	Greasy, fluffy, friable
OM (%)	< 2	1–3	> 60	3–8
C/N	12–20	8–12	20–35	8–12
pH	8·5–10·5	5·0–7·0*	3–5·0*	5·0–7·0
CEC (me%)	variable	10–30	80–100	15–80*
BS (%)	100	30–100	10–30*	10–50
Carbonates (%)	> 1*	0–50 from parent material	Absent	Absent
Salts (%)	< 0·2*	Absent	Absent	Absent
Clay minerals	Montmorillonite F – A Mica O – A	Mica F – D Vermiculite O – A Kaolinite O – A Goethite O – A Montmorillonite < F	—	Allophane D Iron and aluminium hydroxides O – F Halloysite < O Manganese dioxide < R
Weatherable minerals (%)	Variable	10–90	0–10	50–90
Thin section morphology		Isotropic matrix or few small random domains.	Isotropic, occasional cellular plant fragment	Isotropic matrix

HORIZON Name and symbol	ALKALON Ak	ALTON At	AMORPHON Ap	ANDON An
Associated horizons Above: Below:		Mullon, tannon Fragon, ison parent material	Litter, pseudofibron Candon, cerulon celon	Kuron Fragon, duron, parent material†
Intergrades and variants		(AtSq) (AtAr) (AtGl) Atc	(ApHy) (ApPd) (ApCk) Aph	(AnSq) (AnKs)
Illustrations: Plates Figs.	69	Ia 75, 143		Ib
Other distinguishing features	* $< 40\%$ CO_3^{--} and HCO_3^- in the saturation extract EC < 2 mmhos	* Higher when parent material contains carbonates. SiO_2/R_2O_3 similar to horizon above.	* Higher when developed in base rich water. When wet $< 10\%$ hard fibrous material. Permanently saturated with water.	* Varies widely with technique employed. †Volcanic ash May be thixotropic
Genesis	Accumulation of CO_3^{--} and HCO_3^-, mainly of sodium	Weathering and *in situ* deposition of oxides of iron and aluminium with little addition from outside.	Accumulation and decomposition of organic materials under anaerobic conditions.	Weathering and *in situ* deposition of hydroxides of iron and aluminium with little addition from outside.
Derivation	Alkaline	L. *Alter* = other	Gk. *Amorphos* = shapeless.	J. *An* = soil *do* = dark

Table 17 continued

HORIZON Name and symbol	ANMOORON Am	ARENON Ae	ARGILLON Ar	BURON Bu
Position	Upper	Middle or lower	Middle or lower	Upper
Colour	Black, very dark grey, very dark brown	Yellowish brown, yellowish red, brown	Brown, reddish brown, red	Dark brown
Thickness (cm)	10–20	25–200	1–70	25–35
Texture	Loam, silt	Loamy sand, sand *	Loam to clay *	Loam to silt
Stones and boulders (%)	Rare to absent	Rare to absent	< 60wt < 50v	< 60wt < 50v
Structure	Massive, blocky	Single grain, massive	Blocky, massive, cuboidal.†	Granular, blocky, prismatic with depth
Consistence	Firm, friable plastic, hard	Loose to firm	Friable, firm, hard.	Firm, friable
OM (%)	10–30	< 2	0·5–3	2–3*
C/N	12–20	8–15	5–10	8–12
pH	4·5–7·5	4·0–7·0	4·5–7·5‡	7–8·5
CEC (me%)	20–40	< 5	15–30	12–40
BS (%)	30–100	< 20	30–100	80–100
Carbonates (%)	0–50 from parent material *	Absent	0–5 from parent material	1–3†
Salts (%)	Absent†	Absent	Absent	< 0·5
Clay minerals	Mica F – D Kaolinite F – D Vermiculite < F	Kaolinite F – A Mica < O Goethite O – F	Mica F – D Vermiculite O – VF Montmorillonite < F Kaolinite F – D	Mica A – D Montmorillonite A – D
Weatherable minerals (%)	0–50	< 5	0–60	10–50
Thin section morphology	Numerous faecal pellets, occasional plant fragment.	Thin isotropic or weakly birefringent coatings on sand grains	> 10% clay cutans on ped surfaces, lining pores and within the matrix. Variable domains.	

Table 17 continued

HORIZON Name and symbol	ANMOORON Am	ARENON Ae	ARGILLON Ar	BURON Bu
Associated Above: horizons Below:	Litter Gleyson, cerulon, parent material.	Tannon Parent material	Modon, mullon, luvon, alton. Fragon, ison, calcon parent material.	Litter Calcon, argillon, parent material
Intergrades and variants	(AmMu) (AmMo) Amc	(AeZh) (AeFv) (AeKs) (Aeh)(AeRo)	(ArAt) (ArSq) (ArKs) (ArRo)	(BuKt) (BuSn) (BuHl)
Illustrations: Plates Figs.		55, 143	Ic, Id 72	
Other distinguishing features	*When present Anc †When present (AnHl) Often wet and saturated with water < 10% brown mottles	* < 10% silt plus clay	*Usually contains more clay than the horizons above and below. †May have cutans on ped surfaces. ‡Often has lower pH than horizon above.	*7–15 kg OM per m² †May be higher when derived from parent material Pseudomycelia and concretions in lower part
Genesis	Rapid breakdown and incorporation of organic matter in a moist environment.	Preferential removal of fine material. Some may form directly from coarse textured material.	Accumulation of clay translocated from the horizon above.	Rapid decomposition and incorporation of organic matter in a dry environment.
Derivation	Kubiëna 1953	L. *Arena* = sand	L. *Argilla* = clay	R. *Buryi* = brown

Table 17 continued

HORIZON Name and symbol	CALCON Ck	CANDON Co	CELON Cs	CERULON Cu
Position	Upper, middle or lower	Upper or middle	Middle or lower	Middle or lower
Colour	Pale brown, greyish brown	Grey, olive, with faint mottling	Greyish brown,* olive brown	Greyish brown,* olive brown
Thickness (cm)	5–100	5–50	20–200	20–200
Texture	Variable	Loam to silt*	Clay†	Sand, loam, silt
Stones and boulders (%)	Variable	Absent to frequent	Absent to rare	< 60wt < 50v
Structure	Blocky, prismatic, massive	Blocky, massive	Massive, prismatic, blocky	Massive, prismatic
Consistence	Firm, friable, hard*	Friable, firm, hard.	Hard, plastic	Friable, firm, plastic†
OM (%)	1–2	0·5–5	< 1	< 1
C/N	8–10	10–20	8–15	8–15
pH	7·5–8·5	4·5–7·0	4·5–7·0 §	4·5–7·0 §
CEC (me%)	Variable	10–20	40–80	10–40
BS (%)	100	10–30	40–100	40–100
Carbonates (%)	2–25†	Absent	0–20 from parent material‡	0–20 from parent material
Salts (%)	< 0·5	Absent	Absent	Absent
Clay minerals	Variable	Mica F – D Kaolinite F – A Vermiculite < F Montmorillonite < F	Mica F – D Montmorillonite F – D Kaolinite < F	Mica O – D Vermiculite < F Kaolinite O – D Montmorillonite < F
Weatherable minerals (%)	Variable	10–60	5–20	5–50
Thin section morphology	Needles of calcite in pores and comprising *pseudomycelia.* Crystals of calcite occurring singly, in clusters, lining pores and surrounding the larger separates.	Abundant random domains. Clay cutans are absent to frequent	Birefringent matrix, abundant small random domains.	Birefringent matrix, abundant small random domains.

Table 17 continued

HORIZON Name and symbol	CALCON Ck	CANDON Co	CELON Cs	CERULON Cu
Associated horizons Above: Below:	Chernon, buron, seron, verton, argillon Gypson, halon, parent material	Modon, mullon, luton, hydromoron. Marblon, glosson, placon, gleyson	Gleyson, verton Parent material	Gleyson, anmooron. Parent material
Intergrades and variants	(CkAr) Ckd	(CoZo) (CoLv)	(CsCu) (CsVe) (CsHl)	(CuCs) (CuGl) (CuHl)
Illustrations: Plates Figs.	Id, IIb, IIc, IVd. 78, 84	IIIc, Vb		Va
Other distinguishing features	* When hard Ckd As powdery fillings concretions, pseudomycelia, pendants below stones and sheets	May be thixotropic. Usually wet or saturated. Sesquioxidic concretions may be present. Similar clay content to horizon below	* < 5% prominent yellow or brown mottles † > 60% clay § Higher when carbonates present ‡ When present Cec. May be thixotropic	* < 5% prominent yellow or brown mottles †Sometimes loose when sand content is high. § Higher when carbonates present.
Genesis	Accumulation of calcium carbonate and growth of calcite crystals.	Reducing conditions in upper part of soil due to an impermeable horizon or excessive surface wetness	Reduction of iron to the ferrous state under conditions of complete saturation.	Reduction of iron to the ferrous state under conditions of complete saturation.
Derivation	L. *Calx* = lime	L. *Candela* = to shine	L. *Caelum* = sky	L. *Caeruleus* = blue

Table 17 continued

HORIZON Name and symbol	CHERNON Ch	CHLORON Ci	CLAMON Cn	CRUMON Cw
Position	Upper	Upper middle or lower	Middle	Upper
Colour	Black, very dark grey, dark greyish brown	Greyish brown, mottled, olive brown	Brown, reddish brown, brownish red	Black, very dark grey, dark brown
Thickness (cm)	50–150	> 20	10–50	2–15
Texture	Silt to loam	Variable	Loam, clay*	Clay loam to clay
Stones and boulders (%)	Usually absent	Variable	Absent to frequent	Usually absent
Structure	Crumb, granular, subangular blocky	Massive, prismatic	Blocky, prismatic	Granular
Consistence	Friable, firm	Firm, hard	Firm, hard	Friable, firm, plastic
OM (%)	3–15*	< 3	< 2	< 2
C/N	8–15	12–20	8–12	8–12
pH	6·5–8·0	8·0–9·0	5·5–7·5	5·5–7·0
CEC (me%)	15–40	Variable	15–30	40–70
BS (%)	50–100	100	40–100	60–100
Carbonates (%)	Pseudomycelia and concretions in lower part, 0–60 from parent material†	Variable	—	0–50 from parent material
Salts (%)	< 0·1	> 0·5*	—	Absent
Clay minerals	Mica F – D Vermiculite < VF Montmorillonite F – D	Montmorillonite F – A Mica O – A	Mica F – D Kaolinite < A Montmorillonite < F	Montmorillonite A – D Mica F – D Kaolinite < F
Weatherable minerals (%)	30–70	Variable	10–50	20–40
Thin section morphology	Numerous faecal pellets and worm casts. Isotropic. In lower part surfaces covered with calcite crystals also needles and concretions.		Numerous small and medium domains Rare to occasional clay cutans	Dense isotropic matrix. Occasional large ovoid pore.

Table 17 continued

HORIZON Name and symbol	CHERNON Ch	CHLORON Ci	CLAMON Cn	CRUMON Cw
Associated horizons Above: Below:	Litter Calcon, gleyson, halon		Luvon, candon, tannon, hamadon, mullon. Calcon, halon, gypson, parent material	— Verton
Intergrades and variants	(ChMu) (ChKa) (Ch/CZ) Chc		(CnHl) (CnCk)	(CwVe) (CwDe) (Cwc) (CwCk) (CwHl)
Illustrations: Plates Figs.	IIa, IIb 83	69	IIc	57
Other distinguishing features	* > 20 kg OM per m² † When derived from parent material Chc Krotovinas	* > 80% Cl⁻ in the saturation extract EC > 4 mmhos	*Contains more clay than the horizon above or below	Wide cracks develop upon drying
Genesis	Rapid breakdown and incorporation of organic matter in a dry environment. Constant churning by soil fauna.	Accumulation of chloride ions.	Weathering in situ to increase the content of clay	Repeated wetting and drying in a dry environment.
Derivation	R. Cherni = black	Chlorine	OE. Clam = clay	OE. Cruma = crumb

Table 17 continued

HORIZON Name and symbol	CRYON Cy	CUMULON Cm	DERMON De	DURON Du
Position	Middle or lower	Upper or middle	Upper	Middle or lower
Colour	Dark olive, greyish brown, olive brown	Red, reddish brown, yellowish brown dark brown	Black, very dark grey, dark brown	Brown, reddish brown*
Thickness (cm)	> 100	10–50	2–15	10–50
Texture	Variable	Loam	Clay loam to clay	
Stones and boulders (%)	Variable	> 30 gravel and stones	Usually absent	
Structure	Massive, platy or cuboidal, pore space filled with ice*	Crumb, granular, massive.	Massive to coarse, platy	Massive
Consistence	Hard to very hard	Loose to firm	Hard, plastic	Hard and cemented
OM(%)	< 30	< 3	< 2	< 2
C/N	10–40	8–12	8–12	8–12
pH	5·5–8·5	4·5–6·0	5·5–7·0	
CEC (me%)	10–50	5–20	40–70	
BS (%)	20–100	10–80	60–100	
Carbonates (%)	Variable	Absent	0–50 from parent material	Absent
Salts (%)	Usually < 2	Absent	Absent	Absent
Clay minerals	Variable	Kaolinite F – D Mica < F Goethite O – F Gibbsite F – A	Montmorillonite A – D Mica F – D Kaolinite < F	
Weatherable minerals (%)	5–100	0–10	20–40	
Thin section morphology	Dense isotropic matrix. Ice crystals oriented normal to ped surfaces.	Abundant quartz gravel and concretions with a thin coating of matrix containing abundant small domains.		Frequent thick birefringent cutans.

Table 17 continued

HORIZON Name and symbol	CRYON Cy	CUMULON Cm	DERMON De	DURON Du
Associated horizons Above: Below:	Gelon, gleyson (GlFm) alton, lithon Parent material	Mullon, tannon Krasnon, zhelton, chemically weathered rock.	— Verton	
Intergrades and variants		(CmKs) (CmZh)	(DeVe) (DeCw) Dec (DeCk)	
Illustrations: Plates Figs.	IId 44, 87	96		
Other distinguishing features	*Massive in sands, platy in loams, cuboidal in clays. Lenses and wedges of ice.		Wide cracks develop upon drying.	*Vertical grey tongues with hexagonal pattern often present.
Genesis	Gradual freezing of a wet mass of soil.	Concentration of quartz gravel and concretions by differential erosion, mass movement and termite activity.	Repeated puddling of the surface in a dry environment.	
Derivation	Gk. *Kuros* = frost	L. *Cumulus* = heap	Gk. *Derma* = skin	L. *Durus* = hard

Table 17 continued

HORIZON Name and symbol	FERRON Fr	FERMENTON Fm	FIBRON Fi	FLAMBON Fb
Position	Middle or lower	Upper	Upper, middle	Middle or lower
Colour	Brown	Very dark brown	Black, very dark brown	Mottled red, reddish brown and cream.
Thickness (cm)	20–60	1–20	50–300	100–500
Texture	Sand, loam	—	—	Clay loam to clay
Stones and boulders (%)	(< 60 wt < 50 v)	—	—	Absent to rare (< 60 wt < 50 v)
Structure	Single grain, crumb	Spongy*	Spongy, massive	Massive
Consistence	Loose, firm	Fibrous	Fibrous, hard	Firm, plastic*
OM (%)	< 3	> 75	> 75	< 1
C/N	12–20	25–30	25–40	6–10
pH	4·5–5·5	3·5–5·0	3·5–5·0*	5–7
CEC (me%)	10–20	80–120	80–100	< 15
BS (%)	10–20	20–40	10–30	10–100
Carbonates (%)	Absent	Absent	Absent	Absent
Salts (%)	Absent	Absent	Absent	Absent
Clay minerals	Mica A – D Hydroxides VF – VA	—	—	Kaolinite F – D Mica < F Gibbsite A – D Goethite F – VA
Weatherable minerals (%)	5–60	< 5	0–10	< 10
Thin section morphology		Partly decomposed plant fragments, numerous faecal pellets, occasional fungal mycelia.	Isotropic, abundant cellular plant fragments.	Partly maintains the original rock structure. Occasional to abundant clay cutans, occasional to rare concretions in process of formation. Occasional to rare macrocrystalline kaolinite and gibbsite

Table 17 continued

HORIZON Name and symbol	FERRON Fr	FERMENTON Fm	FIBRON Fi	FLAMBON Fb
Associated horizons Above: Below:	Modon, zolon, sesquon. Parent material	Litter Humifon, modon	Litter Pseudofibron, amorphon, cerulon, celon.	Krasnon, rosson, pesson, veson Chemically weathered rock.
Intergrades and variants	(FrSq)	(FmHf)	(FiPd), Fih, (FiCk)	(FbKs) (FbRo) (FbVs) (FbAkw)
Illustrations: Plates Figs.		IId, IIId, 65, 104		IIIa, 35, 89, 90, 91, 92,
Other distinguishing features	SiO_2/R_2O_3 less than horizon above by > 0.5	$* < 20\%$ amorphous plant material. Occasional fungal mycelia.	*Higher when developed in base rich water. When wet $> 50\%$ hard fibrous material. Permanently saturated with water.	*Sometimes hardens irreversibly on exposure to the atmosphere, then Fbd
Genesis	Accumulation of hydroxides of aluminium and iron translocated from the horizon above.	Aerobic decomposition of litter by the meso and microorganisms.	Accumulation and decomposition of organic materials under anaerobic conditions.	Hydrolysis and clay mineral formation in a moist soil in a humid tropical climate.
Derivation	L. *Ferrum* = iron	L. *Fevere* = to boil	L. *Fibra* = fibre	F. *Flambe* = flame

Table 17 continued

HORIZON Name and symbol	FLAVON Fv	FRAGON Fg	GELON Gn	GLEYSON Gl
Position	Middle or lower	Lower	Upper or middle	Middle or lower
Colour	Yellowish brown, brown, yellowish red, reddish brown	Yellowish brown, olive	Brown, greyish brown, yellowish brown.	Grey with yellow or brown mottles: reticulate mottling
Thickness (cm)	50–200	10–100	10–100	20–200
Texture	Clay loam to clay†	Sand, loam	Loam, silt	Loam, sand, clay
Stones and boulders (%)	Absent to rare	Absent to very abundant (< 60 wt < 50 v)	Variable (< 60 wt < 50 v)	Variable (< 60 wt 50 v)
Structure	Subangular blocky§	Massive, very coarse platy, prismatic	Platy, massive, cuboidal	Massive, prismatic
Consistence	Firm to friable	Firm to very firm*	Loose, firm.*	Friable, firm, plastic*.
OM (%)	< 5	< 2	< 5	< 2
C/N	8–12	8–12	12–25	8–15
pH	4·5–5·5	5·5–7·0	4·5–7·5	4·5–7·0
CEC (me%)	> 20 (clay)	3–15	Variable	10–40
BS (%)	< 20	30–100	Variable	20–100
Carbonates (%)	Absent	Absent	Variable	0–50 from parent material†
Salts (%)	Absent	Absent	Absent	Absent
Clay minerals	Kaolinite F – A Goethite O – F Gibbsite O – F Mica O – F	Mica A – D Kaolinite A – D	Variable	Mica F – D Kaolinite F – D Montmorillonite < F
Weatherable minerals (%)	< 5	5–100	20–100	10–100
Thin section morphology	Massive, incomplete subangular blocky, abundant random domains, cutans absent to rare.	Isotropic matrix, sometimes without pores. Rare to occasional cutans on ped surfaces and pores.	When fine or medium platy; finer texture at top of peds sometimes with thin silt cutans.	Birefringent matrix with abundant random domains. Sesquioxidic staining around pores and on ped surfaces. Clay cutans absent to rare.

Table 17 continued

HORIZON Name and symbol	FLAVON Fv	FRAGON Fg	GELON Gn	GLEYSON Gl
Associated horizons Above: Below:	Tannon Pesson, chemically weathered parent material§	Sesquon, alton, andon, zolon. Gleyson, argillon, parent material.	Litter, humifon Parent material	Modon, mullon, anmooron Cerulon, ison, pelon
Intergrades and variants		(FgGs)	(GnHa) (GnZo) (GnSq) (GnAm) (GnMo)	Numerous intergrades Glc
Illustrations: Plates Figs.	143			IId, Va
Other distinguishing features	*5YR, 7·5 YR, 10YR † > 30% silt plus clay Silt: clay ratio < ·2 § Smooth shiny ped surfaces	*Softens when wet. Bulk density greater than horizon above sometimes similar to horizon below. Erupts suddenly when pressed between finger and thumb.	*Hard when frozen. Pore space occupied by ice during winter	*Sometimes loose when sand content is high. †When present Glc > 5% prominent mottles. Similar clay content to horizons above and below.
Genesis	Progressive hydrolysis through a very long time.	Weak cementation by aluminium hydroxide silica and clay.	Frozen during winter thawed during summer	Reduction of iron to the ferrous state in the matrix. Oxidation of iron around pores and pore space.
Derivation	L. *Flavus* = yellow	L. *Fragilis* = fragile	L. *Gelare* = to freeze	R. *Glei* = grey sticky loam.

Table 17 continued

HORIZON Name and symbol	GLOSSON Gs	GLUTON Gt	GYPSON Gy	GYTTJON Gj
Position	Middle or lower	Upper or middle	Middle or lower	Upper, middle
Colour	Brown to red with light vertical streaks	Dark brown, reddish brown	Pale brown, brown, greyish brown*	Black, very dark brown
Thickness (cm)	30–150	10–50	5–50	20–200
Texture	Loam, silt.	Sand or loam	Variable	—
Stones and boulders (%)	Absent to frequent.	Occasional to frequent	Absent to frequent	—
Structure	Massive, prismatic.	Massive	Massive, blocky, prismatic.	Massive
Consistence	Firm, hard.	Hard	Firm, hard	Plastic
OM (%)	< 2	< 1	< 2	30–60
C/N	8–15		8–12	12–20
pH	5·5–7·5		7·0–9·0	3·5–7·5
CEC (me%)	15–30		Variable	50–100
BS (%)	40–100		100	10–30*
Carbonates (%)	0–20 from parent material	Absent	< 5	Absent
Salts (%)	Absent	Absent	≥ 5 $CaSO_4$	Absent
Clay minerals	Mica F – D Kaolinite F – O	Kaolinite F – A Goethite O – F Mica O – A Hematite < O	Variable	Variable
Weatherable minerals (%)	20–60	< 10	Variable	0–20
Thin section morphology	Numerous small domains in brown or red areas, isotropic matrix in lighter coloured streaks. Manganiferous concretions may be present.	Sand grains coated and cemented by goethite and manganese dioxide.	Clusters of well formed gypsum crystals	Isotropic, occasional cellular plant fragment.

Table 17 continued

HORIZON Name and symbol	GLOSSON Gs	GLUTON Gt	GYPSON Gy	GYTTJON Gj
Associated Above: horizons Below:	Candon, gleyson, cerulon. Parent material	Arenon Arenon, parent material	Calcon Halon, parent material	Litter Cerulon, celon
Intergrades and variants	(GsIn) (GsFg) Gsc	(GtAe)	(GyCk)	Gjh, Gjc.
Illustrations: Plates Figs.	Vb		113	
Other distinguishing features	Often mottled with small black concretions.		*Varies with colour of parent material.	*Higher when developed in base rich water.
Genesis	Reduction and removal of iron from cracks caused by contraction or previously occupied by ice.	Cementation of sandy material by material translocated laterally	Gradual growth of gypsum crystals in the soil.	Accumulation of organic and mineral material under anaerobic conditions.
Derivation	Gk. *Glossa* = tongue	L. *Gluten* = glue	Gk. *Gypsos* = gypsum	S. *Gyttja*

Table 17 continued

HORIZON Name and symbol	HALON Hl	HAMADON Ha	HUDEPON Hd	HUMIFON Hf
Position	Upper, middle or lower	Upper	Middle	Upper
Colour	Yellowish brown* brownish grey	Black, dark grey, brown.	Black, very dark brown	Black, very dark brown
Thickness (cm)	10–100	5–15	20–60	1–10
Texture	Variable	Loam	Loam or sand	—
Stones and boulders (%)	Absent to frequent	> 30 surface covered	< 60 wt < 50 v	—
Structure	Massive, prismatic	Single grain	Single grain, sub-angular blocky, massive.	Amorphous, small granular.
Consistence	Firm, friable, hard	Loose, friable	Loose, firm, hard cemented	Plastic
OM (%)	< 5	< 1	15–25*	> 75
C/N	< 2	—	18–22	15–20
pH	7·5–8·5	—	4·5–5·5	3·5–4·5
CEC (me%)	Variable	—	15–40	80–100
BS (%)	100	—	10–20	20–50
Carbonates (%)	0–50†	0–100	Absent	Absent
Salts (%)	> 0·5		Absent	Absent
Clay minerals	Variable	Variable	Mica F – D Kaolinite F – D Hydroxides < R	—
Weatherable minerals (%)	0–80	5–100	0–50	< 10
Thin section morphology		—	Small, isotropic granules, isotropic coatings on sand grains.	Small granular, opaque, numerous faecal pellets.

Table 17 continued

HORIZON Name and symbol	HALON Hl	HAMADON Ha	HUDEPON Hd	HUMIFON Hf
Associated horizons Above: Below:		— Gelon, gluton, halon, seron	Zolon, modon, mullon†, tannon†. Husesquon, sesquon, fragon, ison, parent material.	Fermenton, litter. Modon
Intergrades and variants	(HlCu)		(HdHs) (HdAS)	(HfFm) (HfMo)
Illustrations: Plates Figs.	69			
Other distinguishing features	*Varies with the colour of the parent material †From parent material < 80% Cl⁻ and 60% SO₄⁻⁻ in the saturation extract <u>EC > 4 mmhos</u>	Vegetation absent or rare. > 30% of surface covered by larger separates randomly distributed.	*Contains more organic matter than the horizons above or below †Only when cultivated	< 30% fibrous plant material Numerous fine roots
Genesis	Accumulation of soluble salts in an arid or semi-arid environment	Removal of fine material by deflation or surface wash. Frost heaving of stones and boulders.	Accumulation of translocated organic matter.	Decomposition of organic matter by arthropods, bacteria and fungi.
Derivation	Gk. *Hals* = salt	Kübiena 1953	Hu-from humus and dep-from deposit.	E. Humification

In the "Other distinguishing features" column for HALON, the notation uses $< 80\% \ Cl^-$ and $60\% \ SO_4^{--}$, $EC > 4$ mmhos.

Table 17 continued

HORIZON Name and symbol	HUSESQUON Hs	HYDROMORON Hy	ISON In	JARON Ja
Position	Middle or lower	Upper	Lower	Middle
Colour	Black, very dark brown	Black, very dark brown	Pale brown, olive brown	Grey, olive with yellow mottling*
Thickness (cm)	20–60	5–20	10–200*	10–50
Texture	Loam or sand	—	Sand or loam†	Loam, clay
Stones and boulders (%)	< 60wt < 50v	—	Absent to dominant	Absent to rare
Structure	Single grain, blocky, massive.	Massive, blocky	Medium and fine lenticular or massive§†	Massive, prismatic, angular blocky
Consistence	Loose, firm, hard* cemented.	Plastic, firm.	Firm to very hard	Firm, hard, plastic
OM (%)	10–25	> 50	< 2	< 3
C/N	18–22	15–25	8–12	—
pH	4·5–5·5	3·0–5·0	5·5–7·0‡	3–4·5
CEC (me%)	15–40	50–100	3–15	20–40
BS (%)	10–20	20–40	30–100	30–50
Carbonates (%)	Absent	Absent	Absent to dominant	Absent
Salts (%)	Absent	Absent	Absent	< 0·5
Clay minerals	Mica F – D Kaolinite F – D Hydroxides > F†	—	Mica F – A Kaolinite O – F	Mica F – A Montmorillonite < O Kaolinite F – A
Weatherable minerals (%)	0–50	< 10	0–100	5–30
Thin section morphology	Small, isotropic granules; isotropic coatings on sand grains.	Massive, opaque, occasional faecal pellets.	Isotropic matrix, rare to occasional small circular or irregular pores. Rare to occasional cutans on ped surfaces and pores. Silt cutans on upper surfaces of stones and boulders.	

Table 17 continued

HORIZON Name and symbol	HUSESQUON Hs	HYDROMORON Hy	ISON In	JARON Ja
Associated Above: horizons Below:	Zolon, modon, mullon§ tannon. Sesquon, fragon, ison, parent material.	Litter, fermenton. Modon, candon.	Sesquon, alton, zolon, mullon Parent material	Anmooron, hydromoron Thion
Intergrades and variants	(HsSq) (HsHd)	(HyMo) (HyFm)	Parent material	(JaTo)
Illustrations: Plates Figs.	IIId 106	IIIc, Va	IIIc 103	
Other distinguishing features	*Fine fraction has greasy feel. †SiO$_2$/R$_2$O$_3$ less than horizon above by ·> 0·5. § Only when cultivated.	Numerous fine roots. Usually moist or wet.	*Often with very sharp upper boundary †Faceted sand grains § Silt cutans on upper surfaces of stones and boulders. †When hard Ind ‡Higher when carbonates are present	*Mottling due in part to straw coloured jarosite
Genesis	Accumulation of translocated organic matter and amorphous oxides of iron and aluminium.	Accumulation of organic matter under wet but not waterlogged conditions.	Derived from a cryon by disappearance of the ice. Followed by slight cementation by aluminium hydroxide, silica, and carbonates	Drainage of a thion causing the oxidation of pyrite to form straw coloured basic ferric sulphate – jarosite.
Derivation	Hu-from humus sesq- from sesquioxide.	Hydromor Duchaufour 1960	OE. *Is* = Ice	Jarosite

Table 17 continued

HORIZON Name and symbol	KASTANON Kt	KRASNON Ks	KURON Ku	LIMON Lm
Position	Upper	Upper or middle	Upper	Middle or lower
Colour	Dark brown Very dark grey	Red, brownish red*	Black, very dark brown,	Grey to white
Thickness (cm)	35–50	25– > 3 m	10–25	3–100
Texture	Loam, silt	> 30% clay†	Loam, sand	Silt
Stones and boulders (%)	Usually absent	Absent to rare	Absent to rare	Absent
Structure	Crumb or granular to prismatic with depth	Massive, blocky	Granular, crumb, subangular blocky	Massive
Consistence	Firm, friable	Friable, firm, hard, plastic.	Friable, greasy	Firm, plastic.
OM (%)	2–5*	< 5	5–20	< 5
C/N	8–12	8–12	12–20	15–40
pH	7·0–8·5	4·5–5·5	4·5–6·6	7·5–9·0
CEC (me%)	10–40	< 15 (clay)	20–50	
BS (%)	100	< 15§	10–30	—
Carbonates (%)	Pseudomycelia and concretions in lower part. 0–60 from parent material†	Absent	Absent	> 30
Salts (%)	< 0·5	Absent	Absent	Absent
Clay minerals	Mica F – A Montmorillonite F – A	Kaolinite F – D Gibbsite F – D Goethite F – D Hematite < F Mica < F	Allophane A – D Halloysite < F	
Weatherable minerals (%)	20–60	< 5	5–80	Variable
Thin section morphology	Numerous faecal pellets and worm casts. Isotropic. In lower part, surfaces covered with calcite crystals also needles and concretions.	Abundant random very small domains	Isotropic matrix	

Table 17 continued

HORIZON Name and symbol	KASTANON Kt	KRASNON Ks	KURON Ku	LIMON Lm
Associated Above: horizons Below:	Litter Calcon, gypson, parent material	Tannon, mullon, flambon Pesson, veson, chemically weathered rock	Litter Andon, parent material	Amorphon, pseudofibron Parent material
Intergrades and variants	(KtCh) (KtBu)	(KsRf) (KsRo) Ksh	(KuMu)	
Illustrations: Plates Figs.		IIIa 141, 143	Ib	
Other distinguishing features	*15–20 kg OM per m² †When derived from parent material Ktc.	*10 R and 2·5 YR †Maximum content of clay in middle of horizon – not due to translocation. §When higher Krh. Silt: clay ratio < 0·15 on crystalline rocks <0·2 on sediments		Permanently saturated with water
Genesis	Rapid breakdown and incorporation of organic material in a dry environment.	Hydrolysis and clay mineral formation in a humid tropical climate.	Decomposition and blending of organic and mineral material.	Accumulation of precipitated calcium carbonate or skeletons of aquatic animals on the bottom of a lake.
Derivation	R. *Kastano* = chestnut	R. *Krasni* = red	J. *Kuroboku* = dark	L. *Limus* = slime

Table 17 continued

HORIZON Name and symbol	LITHON Lh	LUTON Ln	LUVON Lv	MARBLON Mb
Position	Middle or lower	Upper	Upper or middle	Middle or lower
Colour	Brown, yellowish brown.	Black, dark brown, brown, red, sometimes with faint mottling	Brownish grey, grey pale brown	Yellowish brown to brown with marbled pattern and faint mottling
Thickness (cm)	10–100	10–30	3–80	10–50
Texture	Sand or loam*	Loam to clay	Loam	Loam, silt
Stones and boulders (%)	> 80 wt > 60 v†	Absent to occasional	Occasional to absent	Absent to frequent
Structure	Single grain	Massive, blocky	Blocky, platy	Massive, prismatic, blocky
Consistence	Loose	Firm, hard	Loose, firm	Firm, hard.
OM (%)	< 2	1–10	0·5–5	< 2
C/N	8–15	8–15	10–20	8–12
pH	5·5–7·5	5·0–7·0	3·5–5·5	5·5–7·5
CEC (me%)	3–15	15–100	5–20	15–30
BS (%)	30–100	30–100	10–20	40–100
Carbonates (%)	0–100	0–80 from parent material	Absent	Absent
Salts (%)	Absent to rare	Absent	Absent	Absent
Clay minerals	—	Variable	Mica F – D Kaolinite F – D	Mica F – D Kaolinite F – D Montmorillonite < VF
Weatherable minerals (%)	0–100	Variable	0–60	5–60
Thin section morphology		Massive, isotropic	Single grain	

Table 17 continued

HORIZON Name and symbol	LITHON Lh	LUTON Ln	LUVON Lv	MARBLON Mb
Associated horizons Above: Below:	Hamadon, (HfLh) Cryon, parent material	Variable	Modon Argillon, pelon, krasnon, rosson, solon	Sesquon, alton, mullon, hydromoron Parent material
Intergrades and Variants	(LhSq) (LhAn) (LhAnc) (LhIn)	(LuMu)(LuCw)	(LvZo) (LvMi)	(MbGl) (MbAl) (MbPn)
Illustrations: Plates Figs.	41		Ic, Id, IIIb, IVc 62	
Other distinguishing features	* < 30% fine earth † Angular fragments Destruction of original rock structure.	Sharp lower boundary.	Contains less clay than the horizon below.	Ped surfaces often grey. Small black concretions absent to occasional.
Genesis	Fragmentation and accumulation of material in situ or at lower point on the slope.	Degradation of structure due to poor cultivation Decomposition and incorporation of organic matter in a moist soil.	Removal or destruction of clay.	Weak reducing conditions for a short period during each year.
Derivation	Gk. Lithos = stone	L. Lutum = mud	L. Luo = to wash	Gk. Marmaros = sparkle.

Table 17 continued

HORIZON Name and symbol	MINON Mi	MODON Mo	MULLON Mu	ORON Or
Position	Upper or middle	Upper	Upper	Middle or lower
Colour	Brownish grey with feint mottling	Black, very dark grey	Black, very dark grey, dark brown, brown.	Black, very dark brown.
Thickness (cm)	3–80	5–20	10–30	5–30
Texture	Loam	Loam or sand	Loam, silt,	Loam, sand
Stones and boulders (%)	Occasional to absent	Absent to abundant	Absent to abundant	Absent to frequent
Structure	Massive, blocky	Single grain crumb, granular, blocky.	Crumb, granular, subangular blocky	Single grain, massive nodular.
Consistence	Loose, firm	Loose, friable.	Friable, firm.	Loose, firm hard.
OM (%)	0·5–5	5–20	3–10	—
C/N	8–20	15–20	8–12	—
pH	3·5–5·5	3·5–6·0*	5·5–7·0*	4·5–6·0
CEC (me%)	5–20	10–40	20–40	—
BS (%)	10–20	10–30	40–100	—
Carbonates (%)	Absent	0–80 from parent material†	0–80 from parent material†	Absent
Salts (%)	Absent	Absent	Absent	Absent
Clay minerals	Mica F – D Kaolinite F – D Montmorillonite < F	Mica F – D Kaolinite F – D Vermiculite O – F	Mica F – D Kaolinite F – D Montmorillonite < F	Amorphous oxides of iron and manganese
Weatherable minerals (%)	0–60	0–50	20–100	0–50
Thin section morphology	Single grain	Single grain, frequent small opaque granules. Occasional faecal pellets.	Abundant worm casts or arthropod faecal pellets. Isotropic matrix.	

Table 17 continued

HORIZON Name and symbol	MINON Mi	MODON Mo	MULLON Mu	ORON Or
Associated Above: horizons Below:	Modon Argillon, pelon, planon, (PeGl)	Humifon, fermenton, hydromoron Zolon, sesquon, husesquon, hudepon, argillon.	Litter. Alton, argillon, pelon.	Marblon Gleyson, cerulon
Intergrades and variants	(MiLv)	(MoMu) (MoZo)	(MuMo) (MuTa) (MuVe)	(OrGl) (OrCu)
Illustrations: Plates Figs.		Va 77	Ia, IVa 56	
Other distinguishing features	Contains less clay than the horizon below. Sesquioxidic concretions may be present.	*Higher when carbonates present. †When present Moc Numerous strongly bleached sand grains Usually less clay and/or sesquioxides than horizon below.	*Higher when carbonates present. †When present Muc Many earthworms.	
Genesis	Removal or destruction of clay.	Translocation of fine particles of partly decomposed organic matter giving a crude mixture with the mineral material.	Decomposition and blending of organic and mineral material by soil fauna.	Accumulation of hydrous oxides of iron and manganese at the top of the water table.
Derivation	L. *Minor* = less	Kubiëna Moder	E. Mull	OE. *Ar* = brass

Table 17 continued

HORIZON Name and symbol	PALLON Pl	PELON Pe	PESSON Ps	PLACON Pk
Position	Lower	Middle or lower	Middle or lower	Middle or lower
Colour	Pinkish grey with red mottles.	Brown, reddish brown, red, black, grey*	Reddish brown, red, very dark brown	Very dark brown, reddish brown
Thickness (cm)	100–300	20–50	20–200	0·5–2
Texture	Clay loam to clay	Sandy clay to clay†	Sandy clay to clay	—
Stones and boulders (%)	Absent to rare	Absent to occasional	Absent to rare	Absent to abundant
Structure	Massive, prismatic.	Blocky, prismatic, massive.	Massive with abundant concretions	Massive
Consistence	Plastic, hard.	Firm, hard, plastic	Firm, hard, cemented	Hard
OM (%)	< 1	< 2	< 1	—
C/N	—	8–12		—
pH	4·5–5·5	5·0–7·5	4·5–5·5	—
CEC (me%)	10–20	20–40		—
BS (%)	< 50	30–100		—
Carbonates (%)	Absent	0–30‡	Absent	Absent
Salts (%)	Absent	Absent	Absent	Absent
Clay minerals	Kaolinite A – D Gibbsite < F Goethite < F Mica < O	Kaolinite F – D Mica < F Montmorillonite < F Vermiculite < F	Kaolinite F – D Gibbsite F – D Goethite F – D Hematite < F	—
Weatherable minerals (%)	≤ 5	5–50	< 10	—
Thin section morphology		Numerous medium or small random domains	Abundant concretions in a brown matrix containing macro-crystalline kaolinite.	Dense, isotropic; deep red in very thin sections.

Table 17 continued

HORIZON Name and symbol	PALLON Pl	PELON Pe	PESSON Ps	PLACON Pk
Associated horizons Above: Below:	Krasnon, veson Chemically weathered rock	Mullon, luton, tannon, candon Calcon, celon, halon, parent material	Krasnon, zhelton Flambon, chemically weathered rock	Candon, gleyson, husesquon. Sesquon, ison, parent material.
Intergrades and variants		(PeVe) (PeAt) Pec (PeAr)(PeGl)	(PsFb)	(PkHs)
Illustrations: Plates Figs.		60	95	IIIc 101
Other distinguishing features		*Colour usually determined by parent material †From parent material ‡From parent material		Irregular outline. Impermeable to water and roots. Continuous through stones.
Genesis	Progressive hydrolysis and clay mineral formation	Weak expansion and contraction. Weak weathering.	Segregation of iron to form concretions followed by crystallisation of kaolinite in the matrix.	Obscure.
Derivation	L. *Pallere* = pale	Gk. *Pelos* = mud	Gk. *Pessos* = oval stone	Gk. *Plax* = plate

Table 17 continued

HORIZON Name and symbol	PLAGGON Pg	PLANON Pn	PRIMON Pr	PROXON Px
Position	Upper or middle	Middle or lower	Upper	Middle or lower
Colour	Very dark brown, brown, greyish brown.	Greyish brown with yellow or brown mottles.	Yellowish brown* brown	Brown, yellowish brown, reddish brown, red
Thickness (cm)	50–100	30–100*	5–20	20–100
Texture	Loam	Clay loam to clay	Variable	Loam to clay
Stones and boulders (%)	Absent to occasional	Absent to frequent	Variable	Occasional to absent
Structure	Crumb, granular, blocky.	Massive, prismatic, angular, blocky	Variable	Blocky, granular
Consistence	Firm, friable.	Firm, hard, plastic	Variable	Firm, plastic
OM (%)	3–15	< 2	1–5	< 1
C/N	8–15	8–15	15–25	8–12
pH	5·5–7·0	5·5–7·0	Variable	4–5·5
CEC (me%)	20–40	20–50	Variable	5–20
BS (%)	60–100	20–100	Variable	10–25
Carbonates (%)	Absent	Absent	0–90	Absent
Salts (%)	Absent	Absent	< 0·5	Absent
Clay minerals	Variable	Mica F – D Montmorillonite F – D	Absent to variable	Kaolinite F – A Gibbsite F – A Mica < O
Weatherable minerals (%)	20–60	10–50	0–95	≥ 20
Thin section morphology	—	Birefringent matrix with frequent to abundant domains. Occasional to rare clay cutans. Frequent to common weathered mineral.	Discrete particles of mineral material, relatively un-humified organic matter	Abundant small domains

Table 17 continued

HORIZON Name and symbol	PLAGGON Pg	PLANON Pn	PRIMON Pr	PROXON Px
Associated horizons Above: Below:	— Variable	Minon, luvon, mullon, modon Pelon (PnGl), parent material	Litter Parent material	Tannon, mullon Sapron, weathered rock
Intergrades and variants	(PgMu)	(PnGl) (PnSl)	To all types of surface horizons	(PxSa) (PxRf)
Illustrations: Plates Figs.			141	
Other distinguishing features	Often crudely stratified.	*Often with sharp upper boundary > 10% clay than the horizon above or below	*Varies with the colour of the parent material	Low silt to clay ratio
Genesis	Continuous addition by man of organic and mineral material to the surface in order to maintain fertility.	Hydrolysis and clay mineral formation in a wet environment.	Little development being the initial stage in the formation of an upper horizon.	Profound weathering under humid tropical conditions
Derivation	G. *Plaggen* = meadow	L. *Planus* = level	L. *Primus* = first	L. *Proximus* = next

Table 17 continued

HORIZON Name and symbol	PSEUDOFIBRON Pd	ROSSON Ro	RUBON Ru	RUFON Rf
Position	Upper or middle	Middle	Upper or middle	Upper, middle
Colour	Black, very dark brown	Red, reddish brown*	Red, reddish brown*	Red, reddish brown*
Thickness (cm)	10–200	20–100	20–100	20–100
Texture Stones and boulders (%) Structure Consistence	— — Massive to fibrous. Spongy to plastic.	> 30% clay Absent to rare Blocky, granular. Firm, friable, plastic.	Loam to clay Absent to occasional Blocky, granular Firm, friable, plastic.	Loam to clay Absent to rare Blocky, granular Firm, friable, plastic.
OM (%) C/N pH CEC (me%) BS (%)	> 60 25–40 3–5.0* 80–120 10–30*	< 5 8–12 4.5–7.0 > 20 (clay) > 30	< 2 8–12 4.5–7.0 > 20 (clay) > 30	< 5 8–12 4.5–5.5 < 20 (clay) > 30
Carbonates (%)	Absent	Absent	Absent	Absent
Salts (%)	Absent	Absent	Absent	Absent
Clay minerals	—	Kaolinite F – D Mica O – F Montmorillonite O – F Gibbsite < R Goethite F – A Hematite < F	Kaolinite F – D Mica O – F Montmorillonite O – F Gibbsite < R Hematite < F	Kaolinite F – D Mica < O Gibbsite < F Goethite F – A Hematite < F
Weatherable minerals (%)	< 10	< 5	5–20	5–20
Thin section morphology	Isotropic, frequent cellular plant fragments.	Abundant small and medium domains.	Abundant small domains.	Abundant small domains.

Table 17 continued

HORIZON Name and symbol	PSEUDOFIBRON Pd	ROSSON Ro	RUBON Ru	RUFON Rf
Associated horizons Above: Below:	Litter, fibron Amorphon, cerulon, celon.	Tannon, mullon Chemically weathered rock	Tannon Weathered rock	Tannon, mullon Flambon, chemically weathered rock
Intergrades and variants	(PdAm) (PdFi) (PdCk) Pdh	(RoKs) (RoAr)	(RuAe)	(RfRo) (RfFb)
Illustrations: Plates Figs.		IIIb 141, 143	141	141
Other distinguishing features	*Higher when developed in base rich water. When wet < 30% hard fibrous material. Permanently saturated with water.	*10R and 2·5 YR	*10 R and 2·5 YR	10 R and 2·5 YR
Genesis	Accumulation and decomposition of organic matter under anaerobic conditions.	Hydrolysis and clay mineral formation in a warm or hot climate	Hydrolysis under hot, wet and dry tropical conditions	Hydrolysis and clay mineral formation in a humid tropical climate
Derivation	Gk. *Pseudo* = false L. *Fibra* = fibre	I. *Rossa* = red	L. *Ruber* = red	L. *Rufus* = brownish red

Table 17 continued

HORIZON Name and symbol	SAPRON Sa	SERON Sn	SESQUON Sq	SIDERON Sd
Position	Middle or lower	Upper	Middle	Middle or lower
Colour	Dark brown, yellowish brown, reddish brown	Yellowish brown, reddish brown, red.	Dark brown, brown	Very dark brown to black
Thickness (cm)	20–100	5–20*	20–60	20–100
Texture	Loam to clay	Loam to silt	Sand or loam	—
Stones and boulders (%)	Occasional to frequent	Absent to frequent	< 60 wt < 50 v	Absent
Structure	Massive*	Weak platy, alveolar massive with many pores.	Single grain, blocky massive.	Massive
Consistence	Loose, firm	Firm, crusty	Loose, firm friable*.	Plastic, hard when dry
OM (%)	< 1	< 1	3–10	50–90
C/N	8–12	8–12	12–20	15–40
pH	4–6	7·0–8·0	4·5–5·5	3·0–4·5
CEC (me%)	5–20	5–30	10–20	—
BS (%)	10–50	100	10–20	—
Carbonates (%)	Absent	1–5†	Absent	Absent
Salts (%)	Absent	< 0·5	Absent	Absent
Clay minerals	Kaolinite F – A Gibbsite F – A Mica F – A	Mica F – A Kaolinite O – F Montmorillonite < O	Mica F – D Kaolinite F – D Hydroxides > F	
Weatherable minerals (%)	> 5	10–60	5–60	
Thin section morphology	Partially preserved rock structure, pseudomorphs of oxides or clay minerals after original minerals	Isotropic matrix, thin layer of calcite crystals on ped surfaces and lining pores.	Small granules with isotropic matrix. Isotropic coatings on sand grains.	Partly decomposed plant fragments, small concretions and needle crystals of siderite

Table 17 continued

HORIZON Name and symbol	SAPRON Sa	SERON Sn	SESQUON Sq	SIDERON Sd
Associated horizons Above: Below:	Flambon, proxon, zhelton, krasnon Weathered rock	Hamadon Calcon, halon, argillon, clamon	Modon, zolon. Fragon, ison, parent material	Amorphon Parent material
Intergrades and variants	(SaPx) (SaZh) (SaKs)	(SnKa) (SnHa) (SnTn)	(SqAt)(SqAr)(SqFr)	(SdAp)
Illustrations: Plates Figs.	36, 37	IVb	108	
Other distinguishing features	Partially preserved rock structure	*Often less due to erosion. †May be higher from parent material then Sec. Contains less clay than horizon below	*When hard and cemented – Sqd. When loose the fine fraction has greasy feel. SiO_2/R_2O_3 less than horizon above by > 0.5	Permanently saturated with water
Genesis	Profound weathering under humid tropical conditions	Weak leaching, accumulation of some salts and carbonates. Removal of fine material by wind and surface washing.	Accumulation of oxides of aluminium and iron translocated from the horizons above.	Formation of siderite in amorphous peat
Derivation	Gk. *Sapros* = rotten	R. *Seri* = grey	L. *Sesqui* = one and a half	Siderite

Table 17 continued

HORIZON Name and symbol	SOLON Sl	SULPHON Su	TANNON Tn	THION To
Position	Middle	Upper middle or lower	Upper	Middle or lower
Colour	Brown, dark brown, black	Greyish brown, mottled, olive brown	Brown, yellowish brown, reddish brown.	Olive brown, greyish brown
Thickness (cm)	20–100	> 20	5–25	50–100
Texture	Loam to clay*	Variable	Variable*	Usually clay loam to clay
Stones and boulders (%)	Rare to absent	Variable	Absent to frequent	Absent to frequent
Structure	Columnar, prismatic	Massive prismatic	Crumb, granular, blocky, massive	Massive, prismatic, angular blocky.
Consistence	Firm, hard, plastic	Firm hard	Friable, firm, hard, plastic.	Soft, firm, hard, plastic
OM (%)	1–10	< 2	≤ 2	0–30
C/N	8–15	12–20	8–12	20–35
pH	8–9	8·0–9·0	4·5–7·0	3–4·5 or 6–8*
CEC (me%)	15–30	Variable	5–20	20–40
BS (%)	100*	100	10–100	30–50
Carbonates (%)	< 1	Variable	Variable from parent material†	Absent
Salts (%)	≤ 0·5	≥ 0·5	Absent	< 0·5
Clay minerals	Mica F – D Kaolinite F – D Montmorillonite < VF	Montmorillonite F – A Mica O – A	Variable	Mica F – A Montmorillonite < O Kaolinite F – A
Weatherable minerals (%)	5–50	Variable	0–100	5–30†
Thin section morphology	Clay cutans often present on ped surfaces, lining pores and within the matrix, variable domains usually medium and random.		Isotropic matrix.	Dense birefringent matrix. Abundant large random domains. Pyrite randomly distributed or associated with organic matter.

Table 17 continued

HORIZON Name and symbol	SOLON Sl	SULPHON Su	TANNON Tn	THION To
Associated horizons	Above: Luton, luvon Below: Parent material.		Litter Variable e.g. sesquon, alton.	Fibron, pseudofibron Parent material
Intergrades and variants	(SlBu) (SlKa) (SlLv)		(TnMu) (TnMo)	(ThCu) (ThJa)
Illustrations: Plates Figs.	IVc, IVd 61	69	IIc, IIIa	
Other distinguishing features	> 15% exchangeable sodium or sodium + magnesium exceeds calcium plus hydrogen. EC < 4 mmhos *Contains more clay than the horizon above	* > 60% SO_4^{--} in the saturation extract EC > 4 mmhos	*Similar to horizon below †When present Tac	*Very acid when dry, about neutral when wet. †Contains pyrite or elemental sulphur
Genesis	Formation of clay in situ and a little translocated from above.		Mixing of organic and mineral material by fauna or ploughing	Reduction of iron and formation of pyrite in an anaerobic environment containing organic matter and sulphur.
Derivation	R. Solonetz = salt	Sulphur	C. Tann = oak	Gk. Theion = sulphur

Table 17 continued

HORIZON Name and symbol	VERTON Ve	VESON Vs	ZHELTON Zh	ZOLON Zo
Position	Upper or middle	Upper, middle or lower	Middle or lower	Upper or lower
Colour	Black, very dark grey, very dark brown	Mottled red, brown and yellowish brown	Yellowish brown brown, yellowish red reddish brown*	Grey, brownish grey pale brown.
Thickness (cm)	50–200	100–300	25–200	1–100
Texture	Clay*	Clay loam to clay	> 30% clay	Sand, loamy sand
Stones and boulders (%)	Usually absent	Absent to rare	Usually absent	Absent to frequent
Structure	Blocky, wedge massive, prismatic†	Vesicular	Subangular blocky, granular, massive	Single grain, lenticular alveolar
Consistence	Firm, hard, plastic	Firm, plastic, hard*	Firm, hard, plastic	Loose, firm
OM (%)	< 2	< 1	< 5	0·5–5
C/N	8–12	8–15	8–12	10–20
pH	5·5–7·0	—	5·5–7·5	3·5–5·0
CEC (me%)	50–100	—	≤ 15 (clay)	5–20
BS (%)	50–100	—	10–50†	10–20
Carbonates (%)	Absent	Absent	Absent	Absent
Salts (%)	Absent	Absent	Absent	Absent
Clay minerals	Montmorillonite A – D Mica F – A Kaolinite < VF	Kaolinite F – D Goethite O – A Hematite R – F Gibbsite R – A	Kaolinite F – O Goethite F – VA Gibbsite < F Mica < O	Kaolinite F – D Mica F – D Montmorillonite < F
Weatherable minerals (%)	10–30	< 5	< 5	0–60
Thin section morphology			Massive, incomplete subangular blocky. Abundant large random domains.	Single grain, domains absent

Table 17 continued

HORIZON Name and symbol	VERTON Ve	VESON Vs	ZHELTON Zh	ZOLON Zo
Associated Above: horizons Below:	Crumon, dermon, luton Calcon, celon	Krasnon, flavon Flambon, chemically weathered rock	Tannon, mullon Chemically weathered material	Modon, sesquon Sesquon, husesquon, hudepon
Intergrades and variants	(VeCr) (VeDe) (VeCk) (VeTh) (VeHa) (VeHl) Vec (VePe) (VeMu)	(VsFb)	(ZhLv) (ZhZo)	(ZoMo) (ZoSq)
Illustrations: Plates Figs.	Vc 63, 116	94	Vd 67, 143	IIId 105
Other distinguishing features	* > 35% clay > 20% strongly expanding lattice minerals. †Often with slickensides. Cracks when dry	*Usually hardens irreversibly on exposure to the atmosphere, then Vsd	*5 YR, 7·5 YR, 10 YR †When higher Zhh Silt to clay ratio < 0·15 on crystalline rocks < 0·2 on sediments	Contains less ses- quioxides than horizon below.
Genesis	Clay formation or transformation in a base rich environment. Usually marked seasonal moisture variations causing churning, expansion, contraction and gilgai.	Precipitation of hydr- oxides, and formation of clay minerals at the top of the water table in a hot humid climate. Structure formed by termites.	Hydrolysis and clay mineral formation in a moist hot environment.	Removal of hydroxides in solution or suspension.
Derivation	L. *Vertere* = to turn	L. *Vesica* = cyst	R. *Zhelti* = yellow	R. *Zola* = ash

ISONS – In

General definition. Lower horizon which varies from friable to extremely hard. It has a characteristic lenticular or massive structure with cutans of silt on the upper surfaces of stones. This horizon is generally regarded as a relic cryon.

Synonyms. B_3, Bx, Cx, fragipan, indurated layer.

Position. Lower, with sharp upper boundary which may have a placon. Occurs at 45 to 60 cm from the surface may be shallower due to erosion or on north facing slopes, may be deeper on moderately steep south facing slopes.

Thickness. 10 – 200 + cm.

Colour. Often greyish brown (10YR 5/2, 2.5Y 5/2) or olive (5Y 5/3).
Usually has weak mottling, especially on ped surfaces. The colour is sometimes inherited from the parent material and may vary from strong brown to reddish brown to red.

Texture. Sand to loam with a high content of fine sand and coarse silt. The best developed isons have angular, faceted sand grains. The range for the < 2 mm fraction is: sand 60–75%, silt 20–40%, clay 3–10%. The clay content may be greater if it is dominantly kaolinite. The material is usually a till, solifluction deposit, or weathered rock.

Stones. Absent to dominant – some are usually present. When present they are often more frequent than in the horizon above. Almost invariably they have a strongly adherent, firm, cutan on their upper surface mainly of fine sand and silt. Frequently there is a nest of coarser particles beneath the stone. The particle size range for the silt cappings is, sand 15–55%, silt 25–75%, clay 10–25%.

On sloping situations the stones are often oriented with their long axis parallel to the slope and normal to the contour. On flat sites their orientation is more or less vertical.

Structure & Porosity. Medium to coarse lenticular or massive, in a number of cases a horizontal section shows a marked polygonal pattern of lighter coloured areas outlined by yellowish brown iron staining. In the more humid areas the lighter coloured areas are formed by the reduction and removal of iron since these areas are the most permeable and form the main conducting system for soil moisture. In semi-arid areas the lighter coloured areas are due to the deposition of calcium carbonate – such horizons are designated (InCk). Many subspherical or irregular pores are usually present.

Bulk Density. This is greater than that of the overlying horizon.

Handling Properties. This varies from friable to very hard and difficult to dig with a pick. Friable isons are designated In. Hard isons are designated Ind and in hand, specimens are fragile and easily broken producing an explosive rupture. This property does not vary with moisture content and therefore indicates physical compaction.

Permeability. Moderate to very slow for very hard isons; appears impermeable in some places and may severely restrict root development.

pH. 5.5–7 but higher when Inc (or InCk).

CEC. 3–15 me%.

BS. 30–100%.

OM. $< 2\%$.

C/N ratio. 8–12.

(Ca + Mg) CO$_3$. Absent to dominant.

Soluble salts. Absent.

Weatherable minerals. 0–100% of the sand fraction.

Clay. Kaolinite, mica, vermiculite are the main minerals; molecular ratios show high contents of aluminium.

Thin section morphology. Slightly porous, medium or coarse lenticular to massive, many subspherical pores. Single sand grains sometimes occur in the pore space and beneath the stones. When stones are present silt cappings are common (Fig. 103). There is only a small amount of matrix between the large number of angular and faceted sand grains, generally the matrix is isotropic but there may be a few small randomly distributed domains which can easily be confused with small flakes of mica. Clay cutans are rare or occasional on ped surfaces but more usually form concentric deposits on the inner surfaces of pores on ped surfaces or on the surfaces of the polygonal cracks. When isons are overlain by gleysons some of the ped faces and pores may have yellowish brown deposits of iron or black areas of manganese dioxide deposition.

Genesis. Most of these horizons were cryons (see page 186) during the Pleistocene period but with the amelioration of climate, the ice disappeared very gradually leaving behind the characteristic lenticular or massive structure. In soils that contain stones the disappearance of their sheath of ice caused the formation of large pore spaces above the stone which became filled with fine material. The polygonal structure is also inherited from the cryon many of which have thin veins of ice forming a polygonal structure extending vertically. The high bulk density is also inherited from the cryon in which the mineral material was strongly compacted during the development of ice lenses and sheaths. Occasionally isons are formed as a result of contemporary deep freezing and ice lens formation. This produces the characteristic structure but annual thawing does not usually allow the development of silt cappings. This may account for the formation of some isons in the Lake States of the U.S.A. It should be pointed out that in this area, throughout many soils, there is a platy structure which is apparently due to annual winter freezing.

Evidence from the clay data indicates that the preservation of the structure and a certain amount of cementation has resulted from the deposition of aluminium hydroxides, clay and possibly other substances during the Holocene period.

After the cryon thaws and the ison forms there may follow a number of evolutionary paths depending upon the chemical and mineralogical composition of the ison and the nature of the soil moisture regime as influenced by the climate. Isons – In form in acid or basic drift – ASP or BSP – and calcareous isons – Inc – form in calcareous drift – LSP. When these horizons form in areas where evapotranspiration greatly exceeds precipitation calcium carbonate will accumulate to form the compound horizons (InCk) or (IncCk). Such horizons occur in certain soils of the mid-western U.S.A.; the compound horizon (IncCk) also occurs in the drier parts of Eastern England.

If precipitation slightly exceeds evapotranspiration the original ison – In – or calcareous ison – Inc – may remain unchanged. Throughout a long period of time or when these two horizons occur in a more humid area the calcareous ison is steadily decalcified following decalcification of the horizons above and gradually the ison changes into an alton. On the other hand the ison formed in the acid or basic drift may accumulate aluminium hydroxide translocated from above becoming progressively more cemented to form the compound horizon (InFg). Since isons are usually impermeable, often there is a tendency for moisture to accumulate at their surface leading to reduction and the formation of a gleyson and the compound horizons (InGl) or (GlIn). In place of the gleyson there may be a candon or cerulon depending upon the degree of waterlogging.

It is conceivable that isons containing originally only a little calcium carbonate could become decalcified without losing their structure and changing into altons.

Variability. Isons are derived from cryons that developed in unconsolidated deposits but it is essential for the material to be of the correct texture otherwise an extreme degree of compaction will not be achieved. In some situations where a cryon formed in weathered material one may find compound horizons such as (InAKw) or (InBKw) which are usually not very hard because of the predominance of one particle size.

Associated horizons. The overlying horizons include altons, argillons, candons, gleysons, placons and sesquons.

Distribution. Isons are particularly widespread in areas that were subjected to periglacial conditions during the Pleistocene and more especially where there are superficial deposits derived from coarse grained rocks. Thus isons occur in central and southern Scandinavia, Belgium, British Isles, Canada, Alaska and the northern part of the U.S.A.

Discussion. These horizons appear to have received the greatest attention in Britain and the U.S.A. where they are often called fragipans. Generally most workers in Britain are agreed that the structural features are the result of freezing, presumably under periglacial conditions. In the U.S.A. there seems to be a lack of agreement and no definite proposals have been put forward for their formation in that country.

Because of the great thickness of the ison in some places and its occurrence in a number of soils it is most likely to be an inherited phenomenon rather than one developed under contemporary conditions.

Observations by the author and discussions with a number of other investigators have indicated that in the U.S.A. there are probably at least two different horizons that have been called fragipans hence the lack of agreement and confusion that exists. In the north central and north-eastern U.S.A. and New-foundland there are horizons which correlate exactly with the isons of Britain but elsewhere there are horizons that have some points of similarity but also differ in many important respects. The two most important are the marked absence of silt cappings and softening upon wetting which are found in many of the fragipans of the U.S.A. as well as in New Zealand. The isons in Britain and many of those in the northern parts of the U.S.A. remain hard at all times and have prominent silt cappings.

It should be pointed out that although the bulk density in all cases is higher than the horizons above, this property should be interpreted with caution since the bulk density of till and some other sediments is similar to that of isons. Therefore the change in bulk density could be due to a number of factors including settling, freezing or other methods of compaction.

Therefore it is necessary to recognise two separate horizons, isons and fragons. The former is due to freezing with the addition of a small amount of cementing material and the latter due entirely to the presence of cementing material.

The presence of polygonal cracks should also be interpreted with restraint since these occur in sediments of Holocene age then they are most likely to be due to shrinkage and cracking.

6

SOIL CLASSES OF THE WORLD

The intention in this chapter is to give information about a restricted number of classes of soils, concentrating on those which illustrate important pedological principles. Some classes are included only to give a balanced treatment and therefore are treated in a rather superficial manner.

The Primosols and Rankers are the first two classes to be discussed, therefore the presentation is based at the outset on the concept of soil evolution; Primosols forming the initial stages and Rankers the second stages of soil formation. As there is no continuing fixed path of evolution the other classes are set out alphabetically. A few distinctive evolutionary sequences are given in Chapter 8. It is hoped that the alphabetical presentation will facilitate the location of information and therefore is preferred to a system of grouping based on climate as used by many other authors. However, the frequency distribution and relationships between soils and climate are discussed on page 263 *et seq*.

The data for most classes are prefaced by a statement about the origin of the name of the class, followed where appropriate, by the common approximately equivalent names and in some cases pertinent references. The text commences with a brief statement about the general characteristics of the class followed by an account of the field and thin section morphology, analytical data and genesis of a common member of the class. As far as possible the generalised accounts are illustrated by colour plates but there are a few departures from this presentation. The discussions continue with information about the principal variations in the properties of the class, including statements about the parent material, climate, vegetation, topography, age and pedounit. Then a brief account is given of the distribution and utilisation of members of the class, stating the ameliorative methods and precautions that should be adopted during cultivation.

PRIMOSOLS

Derivation of Name: From the Latin word *primus* = first
Approximate Equivalent Names: Lithosols; Regosols; Skeletal soils; some Entisols.

General characteristics

These soils form the initial stages of soil development under all types of conditions, being composed of weakly differentiated horizons with little evidence of recent weathering of the parent material. They consist of a poorly developed upper horizon – the primon, overlying relatively unaltered parent material. The members of the class have extremely variable characteristics which are determined mainly by the composition and degree of consolidation of parent material. Possibly the greatest contrast and widest var-

iation is between primosol development on deep loose material such as sand dunes or alluvium, and those on bare, smooth, rock surfaces as found in glaciated areas. Although such widely divergent soils are not usually found together, it is possible, nevertheless, to find strongly contrasting primosols adjacent to each other, as encountered when loose, porous, volcanic ash lies next to hard consolidated lava or in areas where inselberge are surrounded by deep soils. The development of primosols is also determined, to a limited extent by other factors, particularly climate and vegetation.

Primosols developed on consolidated rocks are found under a wide range of conditions but more particularly in mountainous areas that were strongly glaciated. The pedounit has a thin mat of vegetation often dominated by lichens, mosses or grasses. Below is a very thin dark grey crude mixture of organic and mineral material which in turn rests upon hard unweathered rock and may have the formula Pr_2AK if the rock is granite. This group usually occurs as small discontinuous areas measuring not more than one metre in diameter, normally forming an intricate network with bare rock and rankers into which they intergrade.

Primosols developed in sand dunes usually have a thin primon composed of loose, pale, greyish-brown sand containing a few fragments of partially decomposed organic matter, overlying the relatively unaltered, stratified sand. This sequence of horizons may have the formula $Pr_{15}AS$. In many parts of Europe the pH of the primon developed in a sand dune is nearly neutral but decreases to about 4·5 on neighbouring sites that have been stabilised for a longer period and where rankers are developing. Further differentiation to podzols takes place quickly in humid environments such as under a marine climate (see page 272).

Many primosols have developed in alluvium as found in the great river valleys of the world. When these occur in arid and semi-arid areas the accumulation of salt is the main process. This contrasts sharply with the weathering by hydrolysis of granite or leaching of the dune sand in a humid environment.

These three examples show that the parent material controls the characteristics of the primon, particularly its thickness, for in the dune the primon is 15 cm thick but on the granite it is only 2 cm thick. It is likely that the middle horizons will develop fairly quickly in the dune or alluvium because of their high porosity and high surface area whereas development on the granite will not go beyond a primosol for some time. Primons intergrade in a number of directions as determined by the rate and type of pedogenesis which is itself controlled mainly by the type of climate and the nature of the underlying material, and will be discussed further under soil relationships.

Distribution

Primosols occur in all parts of the world, but their greatest extent is in areas where fresh parent material is either being constantly deposited or exposed. These include steep slopes and mountain ranges, areas of continuous alluvial deposition such as in the Nile Valley, areas of frequent volcanic activity as in Japan and areas of recent reclamation such as the Polders of Holland. However, their greatest total area is probably on coastal sand dunes.

Utilisation

Where these soils are developed in deep, loose, unconsolidated material or soft rock strata, they can be utilised, particularly if of medium texture. However, in many coastal or alluvial areas where they are forming in material reclaimed from the sea they require certain ameliorative procedures to be conducted such as drainage, construction of protective lévées or desalinisation. Afterwards they can then be adapted to the systems of land use in the immediately adjacent areas.

RANKERS

Derivation of Name: From the Austrian *rank* = steep slope.
Approximate Equivalent Names: Lithosols; Regosols; Skeletal soils; some Entisols.

General characteristics

These soils are regarded as the second stage in soil evolution, being characterised by the full development of the most easily formed horizons which are usually those in the upper position. Thus rankers usually have one or more well developed upper horizons overlying relatively unaltered material which may be unconsolidated with a similar mineralogy and particle size distribution or it may be hard rock. When it is the latter the material in which the soil is formed may be a thin cover of drift with a mineralogical composition which is different to that of the rock beneath.

Morphology and genesis

The morphology of the soils in this class is extremely variable. Since nearly every type of upper horizon can occur directly on unaltered material it is impossible to give modal characteristics for this class. Similarly they have few genetic processes in common with each other.

There would appear to be a few rankers that are more widespread than others and have formed in areas where fresh parent material has been exposed fairly recently. In large parts of North America and Eurasia the rocks were scraped bare by glaciers which left varying thicknesses of till. Where this is thin or absent rankers have often developed. One of the principal types has a sequence of upper horizons similar to those found in podzols and has the designation $Lt_2Fm_1Hf_2Mo_5Zo_5AK$. A second type of ranker found in moister parts of these areas has an accumulation of black, plastic, organic matter or hydromoron and may have the designation $Lt_2Hy_{10}Mo_5AK$. Usually, these develop at fairly high elevations where conditions are cool and humid.

In some parts of the world rankers have formed in volcanic ash, having as the principal upper horizon the very dark brown or black, humose, kuron containing allophane. Some of these soils may have the formula $Ku_{15}IZ$.

Rankers develop also in arid and semi arid areas where the accumulation of salts may be the principal process. Such soils can have the formula $Hl_{20}AZ$.

The rankers developed in fine textured sediments such as alluvium and marine clays exhibit a particularly interesting evolutionary sequence. Upon exposure to the atmosphere the material shrinks and cracks. This may be a cyclic process in the case of alluvium that is flooded every year during the rainy season and become exposed during the dry season. When these areas are protected by lévées or by dykes as in the case of the polders of Holland they are gradually colonised by vegetation and an upper horizon develops. This may be a massive luton if fairly moist conditions persist but often it is a granular mullon particularly if the material is calcareous. Because these materials are often of fine texture they continue to crack during the dry season in spite of the vegetative cover. The result is that organic and mineral material from the surface is washed into the crack to form coatings or organic-mineral cutans on the surfaces of the prismatic peds. Where this process has continued for a long period these cutans can exceed 1 cm in thickness (Fig. 74) with wide variations over short distances. This causes the prism faces to have a similar dark colour to the surface horizons but the interior of the prisms may be lighter in colour showing original stratification and very little sign of alteration.

Such rankers may have the formula $Mu_{13}(Mu/AC)_{30}AC$.

Principal variations in the properties of the class

Parent material, climate, vegetation, topography and age

Rankers develop in association with a wide range of parent materials, climate, vegetation and topography, therefore they have very diverse properties. It is the degree of consolidation of the parent material that has the most marked influence on their properties and development. Where there is a thin deposit of drift over hard rock, upper horizons develop quickly in the drift, but further differentiation involving the rock is very slow and the soil may remain in this state for thousands of years. On the other hand in a thick deposit of drift or in weathered material, differentiation of middle or lower horizons rapidly follows those at the surface and the ranker stage may last only for a few centuries; this is true particularly for material like volcanic ash which weathers rapidly.

Pedounit

Most upper horizons that develop quickly form the upper horizons of rankers. The precise type is usually related to the adjoining deep soils but modifications are sometimes introduced particularly in those developed on rock which often affects the drainage in such a way that the soils are very wet during the wet period of the year and very dry when precipitation is small.

Rankers with zolons occur in association with podzols, while those with mullons are frequently found with the altosols. Set out below are three common sub-classes and groups.

1. [Zo(ASSq)] $Lt_2Fm_1Hf_2Mo_5Zo_5(ASSq)_{20}AS$
2. [TaAZ] $Ta_{10}AZ$
3. [TaAKw] $Ta_{10}AKw$

While most of the above are common rankers the first example developed on an acid sand is beginning to intergrade towards a podzol. This is a common feature of many stabilised sand dunes in a marine environment.

The second example occurs naturally in a number of situations but its most interesting location is in those arid and semi-arid areas which are now being cultivated, with the result that the surface has been stabilised, some organic matter incorporated in the soil and any salts washed out to form a tannon.

The third example represents a new phase of pedogenesis starting on chemically weathered acid rock following the removal of the previous soil by erosion. This is found in many tropical areas where there is natural or man induced erosion but would have been more widespread at the end of a pluvial period during which large areas were severely eroded.

A further characteristic of many pedounits is a succession of buried soils caused by the repeated addition of fresh parent material to the surface. This is a feature of areas where loess, alluvium and volcanic ash are regularly deposited and also in mountainous regions where erosion of the steep slopes is common.

Distribution

Rankers occur in all parts of the world; the largest extent is either in mountainous areas or on valley floors. In the former position rock is normally near to the surface, while in the latter there are recent deposits of deep alluvium often with buried soils.

Utilisation

Where these soils are developed in deep, loose, unconsolidated material they can be cultivated, particularly if of medium texture. However, in many cases certain ameliorative procedures require to be conducted such as drainage, the erection of protective lévées or desalinisation. Following this they can be adapted to the systems of land-use practised in their immediate surroundings.

Shallow rankers in mountainous areas can be afforested but the forests seldom have economic value, such areas are best left to develop a natural vegetation and wild life and to be used for recreative pursuits.

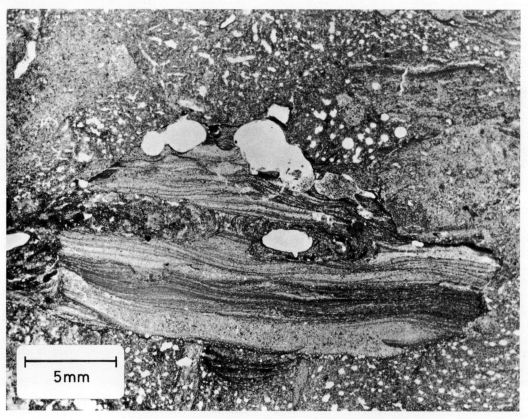

5mm

FIG. 74.   Horizontal thin section of cutans of organic and mineral material filling cracks.

ALTOSOLS

Derivation of Name: From the Latin word *alter* = other
Approximate Equivalent Names: Brown earths, Sols bruns acides, Brown calcareous
  soils, some Entisols.
References: Kubiëna 1953; Tavernier and Smith 1957

General characteristics

Altosols normally have a fairly uniform brown pedounit with a total thickness of about a metre and a half. There is usually a humose mullon at the surface grading into the granular alton. Altosols occur mainly in a marine or humid continental warm summer climate, beneath deciduous vegetation on a wide range of parent materials.

Morphology

Under natural conditions there is usually a loose leafy litter at the surface resting on a humose greyish brown or brown granular or crumb mullon which varies from about 5 to 20 cm in thickness. In thin section the granular peds have an isotropic matrix, being complete or fragments of faecal pellets and containing occasional fragments of partially decomposed plant fragments (Fig. 56). The mullon grades into the brown, angular or subangular blocky alton which also has an isotropic matrix or there may be rare, small, random domains (Fig. 75).

The lower horizons are variable and are determined largely by the evolutionary history of the soil, nature of the parent material and climate but often the alton grades into parent material which is usually basic or intermediate drift.

Analytical data – Fig. 76

These soils are usually of medium texture with the upper horizon having the maximum content of clay which decreases gradually with depth or it may remain fairly uniform throughout the pedounit. When there is a marked textural contrast between horizons it is because of original stratification of the parent material. The pH value in the mullon varies from 5·0–6·5 increasing with depth to about neutrality in the lower part of the alton. The organic matter at the surface varies from about 3–15% with a C/N ratio of 8–12, indicative of a high degree of humification. The cation exchange capacity is usually about 15–30 me% in the surface, decreasing with depth as the organic matter and clay contents decrease. Calcium is usually the principal exchangeable cation but magnesium may predominate when there is a high content of this element in the parent material. The percentage saturation is very variable as determined by many factors particularly the nature of the parent material, usually increasing as the percentage of ferromagnesian minerals in the parent material increases. When the base saturation of the alton is greater than 65% it is regarded as being high and the horizon is designated Ath.

The amount and distribution of free Fe_2O_3 are considered to be important criteria for differentiating between these soils and podzols. In altosols the content of free Fe_2O_3 should be uniformly distributed throughout the upper and middle horizons and in the alton it should not exceed 10% of the total clay fraction. In some cases this value is exceeded due to a high content of iron in the parent material.

Although many workers no longer use the $SiO_2:R_2O_3$ ratio of the clay fraction for categorising these soils it is probably a better criterion than free Fe_2O_3 because it takes into account the total composition of the clay fraction, not merely the rather indeterminate free iron content. In these soils this ratio follows a similar trend to that of the free iron by remaining uniform or being slightly less in the alton than in the mullon above.

Genesis

Like many soils of the middle latitudes these may have an evolutionary sequence going back to the Tertiary period but in many cases the Holocene processes predominate.

At the surface the litter is rapidly decomposed and partly incorporated into the soil by the microorganisms and mesofauna, the former are indicated by the frequent occurrence of fungal mycelia and the latter by the granular or crumb structure.

Hydrolysis is fairly active and most of the iron and aluminium released is precipitated fairly close to the point of release only a small amount being removed in the drainage or redistribution by leaching. This

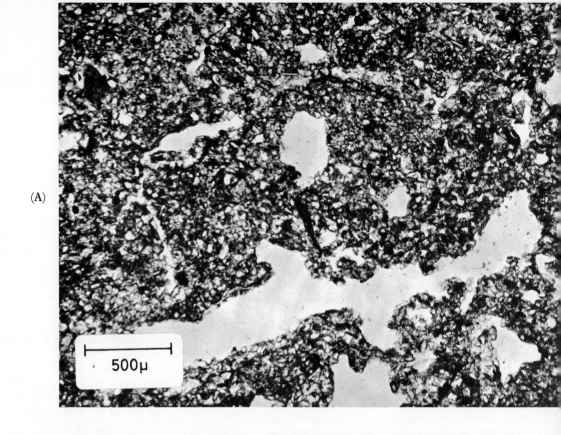

(A)

500μ

FIG. 75. Thin section of an alton. A, in plain transmitted light. B, in crossed polarised light showing the isotropic matrix and birefringent sand gains which are mainly quartz.

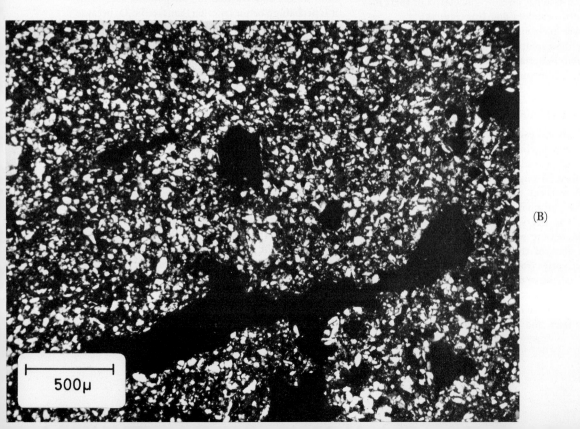

500μ

(B)

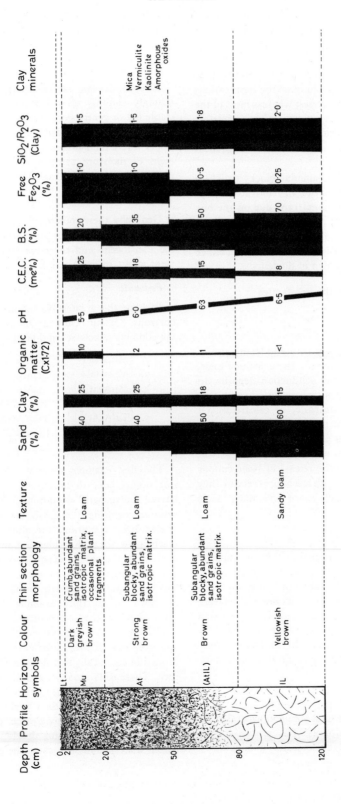

FIG. 76. Generalised data for altosols.

accounts for the fairly uniform distribution of the free Fe_2O_3 and uniform $SiO_2:R_2O_3$ ratios throughout the pedounit. The highest content of clay at the surface indicates that this is the position of maximum hydrolysis which is associated with the highest content of organic matter. Leaching of the basic cations is one of the principal processes; whereas most of the sodium, potassium and magnesium are lost completely from the soil into the drainage water, calcium carbonate may be deposited under certain circumstances to form a calcon.

The formation of the subangular blocky structure seems to be formed by expansion and contraction. The formation of the isotropic matrix is due on the one hand to the occurrence of organic matter and mixing by the organisms in the mullon. On the other hand in the alton it is due to amorphous oxides which either mask or inhibit the formation of domains.

Principal variations in the properties of the class

Parent material

Unconsolidated deposits of silty and loam textures are the usual parent materials, these include loess, glacial drift, alluvium, and solifluction deposits. In most cases the material is of intermediate composition but it is often basic, ultrabasic, or calcareous. Of interest, is their development on acid parent materials providing there is a high content of silt which gives a large surface area from which cations are released, since it is essential that the release of basic cations by weathering keeps pace with leaching otherwise they would become progressively more acid and podzols would form.

Climate

These soils develop most easily either in a marine or a humid continental warm summer climate. Variations in the character of these two climatic types particularly the former, lead to variations in the nature of these soils especially in the degree of leaching which is greater in a marine climate with high rainfall. These soils occur also in most other humid environments where they seem to represent an evolutionary stage of short duration.

Vegetation

Deciduous forest is the most common natural plant community found on altosols. The precise species vary from place to place; communities dominated by

oak (*Quercus* spp.), beech (*Fagus sylvatica*), hickory (*Carya* spp.) and hazel (*Corylus avellana*) are among the most common.

Topography

Altosols develop on sites that vary from flat to strongly sloping, their best development being on stable flat or gently sloping sites. In areas of moderate to high rainfall they usually occur on moderate to steep slopes where their presence is due to the lateral movement of moisture carrying dissolved cations which maintain a fairly high base status.

Age

Generally these soils occur within areas that were glaciated or subjected to periglacial conditions during the Pleistocene and therefore started their development about 10,000 years B.P.

Pedounit

Perhaps the greatest variations are in the lowest horizons. Set out below are the formulae for some of the common subclasses and groups.

1. [MuAtIn] $Lt_2Mu_{15}At_{30}In$
2. [MuAtGl] $Lt_2Mu_{12}At_{20}Mb_{15}Gl$
3. [MuAthCk] $Lt_2Mu_{15}Ath_{30}Ck_{15}IS$
4. [Ta(AtSq)] $Ta_{10}(AtSq)_{20}At_{15}AS$
5. [Ta(AtAr)] $Ta_{15}(AtAr)_{25}AS$
6. [MuAthBM] $Lt_3Mu_{15}Ath_{30}BMw_{200}BM$
7. [MucAtc] $Lt_1Muc_{10}Atc_{15}LFp$

The first example illustrates a soil in which an ison occurs beneath the alton, this is found in soils having polygenetic evolution from a cryosol, the ison representing the fossil cryon (see page 212). Under moist conditions the alton may grade into a marblon or through a marblon into a gleyson; this is given in the second example. At the dry end of the climatic range for this class, calcium carbonate sometimes accumulates beneath the alton to form a calcon as shown in the third example. Altosols intergrade towards many other soils principally podzols and argillosols. Two such intergrades are given in examples four and five. The sixth example is for an altosol that has formed in chemically weathered medium grained basic rock. These soils usually have altons of strong brown colour and high base status, often showing polygenetic evolution from the Tertiary period when the initial weathering of the rock took place (see page 212). The seventh example is a soil developed on calcareous parent material. In such cases the pH values are between 7·0 and 7·5 in all horizons, and the exchange complex is saturated mainly with calcium or magnesium depending upon the nature of the parent material. Usually the content of carbonate decreases sharply upwards due to leaching and in some cases where the original carbonate content was low it has been removed completely.

Sometimes the recognition of intergrades in the field is very difficult, greater reliance being given to laboratory data. The intergrades towards podzols have a smaller $SiO_2:R_2O_3$ ratio and more free Fe_2O_3 in the (AtSq) than in the tannon. At the same time the thin section morphology shows the development of areas of small granular structure such as is found in sesquons. In the intergrades towards argillosols the (AtAr) horizon tends to develop an angular blocky or massive structure with some of the ped faces and pores lined with thin clay cutans.

The colour of these soils varies somewhat from place to place, generally they tend to become yellower when the acidity increases and as they begin to intergrade towards podzols. On the other hand they become darker when there is an increase in the content of ferromagnesian minerals in the parent material which upon hydrolysis produces more basic cations as well as more ferric hydroxide.

Many of the natural chemical properties are altered by cultivation. This applies particularly to the pH and distribution of base cations, both of which are often higher in the surface due to liming and the application of fertilisers.

Distribution

These soils occur principally throughout central and western Europe, east and central North America, and eastern New Zealand. Often they do not form large continuous areas but small areas as determined by the occurrence of suitable parent material. They occur in many parts of the tropics in hilly areas where the slopes are too steep for deep soils to develop or as an early stage in an evolutionary sequence.

Utilisation

Altosols are highly prized because of their fairly high inherent fertility. When the natural deciduous vegetation is removed and the slopes are not too steep they can be adapted to a variety of systems of land use, more usually it is to mixed farming but large areas are used for dairying, orchards and other types of land use. When agriculture is practised annual applications of artificial fertilisers and periodic liming are necessary. Where possible these practices should be supplemented by the addition of organic matter either as dung or compost. Sometimes these soils are replanted with deciduous forests.

ANDOSOLS   [Plate Ib

Derivation of Name: From the Japanese word *an* = dark and *do* = soil.
Approximate Equivalent Names: Humic allophane soils; Acid brown forest soils; Kuro-
    boku; some Andepts.
Reference: F.A.O., 1964

General characteristics

These soils develop on volcanic ash and occur under a wide range of humid climates. They have dark brown pedounits about one metre thick, but probably their most important property is the presence of allophane as the dominant material in the clay fraction.

Morphology

Under natural conditions there is a loose leafy litter at the surface. This rests on the very humose, dark brown, crumb or granular kuron which may be up to 30 cm thick and is dense and isotropic in thin section. The kuron grades into the brown or yellowish brown andon which is 20–30 cm thick and has an angular or subangular blocky structure, also with an isotropic matrix. With depth the andon grades into the relatively unaltered volcanic ash.

Both horizons are fluffy and when rubbed in the wet state have a greasy as distinct from sticky consistence. In some cases the andon is thixotropic i.e. it becomes plastic when rubbed, yielding moisture but hardens when the rubbing is stopped. In most situations there may be several buried soils due to successive showers of volcanic ash.

Analytical data

The content of clay is low or very low, usually not

exceeding 20–25%. The greatest amount occurs in the kuron, decreasing with depth to less than 5% in the relatively unweathered parent material. An important and characteristic feature of all the horizons is fluffiness and high porosity which may exceed 70% in both the kuron and the andon. Both of these properties are attributed to the presence of allophane which is the main product of hydrolysis. These soils vary from moderately to strongly acid with pH values as low as 4·5 in the surface, however there is a steady increase with depth up to pH 6·0 or over in the relatively unaltered ash.

The content of organic matter is usually high and commonly there are values of over 20% in the kuron. Although the C/N ratio may be about 15 the material seems to be in a fairly advanced state of decomposition and appears to form a stable complex with allophane.

Due to the high content of humus the cation exchange capacity is high in the upper part of the kuron and may be over 35 me%, below in the andon it decreases sharply to about 10–15 me%. However these ion exchange data have to be interpreted with caution since allophane does not behave like other clay minerals. The base saturation is normally lowest in the upper part of the kuron, increasing with depth, but there may be high values at the surface due to cultivation. In some andosols manganese dioxide is a product of weathering and may be present in sufficient amounts to give a fairly vigorous reaction with hydrogen peroxide.

Genesis

The formation of andosols is a very rapid process resulting from the high surface areas afforded by volcanic ash which behaves in a unique manner under humid conditions. The principal process is hydrolysis of the volcanic ash which weathers initially to yellow, brown or orange palagonite which is presumed to be an amorphous alumino-silicate containing calcium, magnesium and potassium; however this changes quickly into allophane. Amorphous iron and aluminium oxides are also formed following hydrolysis and are uniformly distributed through the kuron and andon. The other main process is the partial humification of the organic matter and the formation of the stable complex with allophane.

Principal variations in the properties of the class

Parent material

This is mainly volcanic material in the form of vitric volcanic ash, cinders or other vitric pyroclastic material.

It may have accumulated from the atmosphere following a volcanic eruption, or it may be reworked material such as alluvium. There are some differences in the chemical and mineralogical composition of these materials, but the main variations are in the texture, however it is usually finely divided.

Climate

Because the development of these soils is determined mainly by the nature of the parent material, they are found under humid conditions ranging from the arctic to the tropics, but development is fastest under humid tropical conditions.

Vegetation

The nature of the vegetation is as variable as the climate with a tendency for natural plant communities to be more luxuriant than those on adjacent sites where the soils have developed from other parent materials. This is due to faster weathering of the fine grain volcanic material giving a more plentiful supply of plant nutrients in these soils.

Topography

The sites on which these soils develop vary from flat to steeply sloping, but their best development is on the more stable, flat or gently sloping situations.

Age

Most of the soils of this class occurring within areas that were glaciated or subjected to periglacial conditions, started to develop since the beginning of the Holocene period. In tropical areas the starting dates vary from middle Holocene to late Pleistocene.

Pedounit

The variations in the morphology of these soils do not seem to be as wide as encountered in some other classes. The formulae for the subclasses and groups for some of the principal variants are given below:

1. [AnFg] $Ku_{25}An_{30}Fg_{20}IS$
2. [AnDu] $Ku_{25}An_{30}Du_{25}IS$
3. [Ku(AnSq)] $Ku_{20}(AnSq)_{35}IS$

The first variant is similar to most normal andosols with the addition of a fragon which occasionally is a prominent feature in some of these soils. Similarly, durons may occur as shown in the second example. In a marine climate with high rainfall there may be translocation of material to form soils that have affinities with podzols as shown in the third example.

Probably the most common variants are those that have weakly developed andons – (AnIS) because the time interval since the last ash deposits has been too short for middle horizons to develop or because there is continual addition of material at the surface.

Some andosols tend to have a higher content of clay with some transformation of allophane to halloysite and kaolinite, therefore they are intergrading to other soils.

Distribution

Since these soils are associated primarily with volcanic ash their distribution is approximately the same as volcanic ash, occurring principally in Japan, New Zealand, north western U.S.A. East Indies, northern and southern extremities of the Andes and East Africa.

Utilisation

Andosols are reputed to be very infertile when cultivated, however they respond well to amelioration and can be highly productive. For example, in New Zealand grass-clover mixtures will provide adequate grazing for livestock if sufficient liming and fertilising is carried out. Phosphorus is particularly important because the soils have a very high capacity for absorbing and fixing this element. It is advantageous in warm climates to plant certain crops such as sweet potatoes (*Ipomea batatas*) which show little response to the addition of phosphorus. Rice will grow well because the soils have a high water holding capacity.

ARENOSOLS

Derivation of Name: From the Latin word *arena* = sand.
Approximate Equivalent Names: Red and yellow sands; Quartzipsamments.
References: Anderson 1957; Stace *et al.* 1968

General characteristics

These are very sandy soils, often composed mainly of quartz and other resistant minerals. They have very weak horizon differentiation and occur mainly in tropical and subtropical areas.

Morphology

The colours of these soils range from reddish yellow to yellowish brown to red as determined in part by the origin of the material and moisture status. The upper horizon is a tannon which is usually darker than the rest of the soil due to staining by organic matter, and typically it is a loose, coarse sand with single grain structure, clearly seen in thin sections. Faecal pellets and worm casts may also be present, but they are usually rare because very sandy soils are not suitable habitats for the mesofauna. Below the tannon is the arenon which may exceed 2 m in thickness. This is also a coarse sand but in spite of the low content of clay it is often massive and hard. The individual sand grains may have a thin isotropic coating which increases in thickness as the content of clay increases then there may be a few small random domains in the matrix (Fig. 55). In many arenosols some of the quartz grains may have red stainings in cracks (wustenquartz) indicating their origin from a previous phase of soil formation.

Analytical data

The outstanding physical property is the high content of coarse sand, which imparts a high permeability and causes the formation of the single grain or massive structure.

In spite of the high content of quartz these soils are often only weakly acid with pH values between 6·0 and 7·0. This is coupled with a high base saturation but the cation exchange capacity is very low because of the low clay content with kaolinite being the principal clay mineral. The thin tannon usually has < 2% organic matter which may have a C/N ratio of < 12 indicating a moderate level of humification.

Genesis

These soils illustrate a small part of the complexity that is found in tropical and subtropical areas. Since they contain a small amount of clay there is little horizon differentiation except for the formation of a tannon and in some cases a gleyson. Arenosols appear to form in a number of distinct phases the first often being a krasnozem, rossosol or a similar strongly weathered soil composed mainly of clay and the resistant residue. This soil was then exposed to insidious erosion so that the finest material was preferentially removed leaving a residue of coarse particles. This may have taken place at a specific period during the

Pleistocene, alternatively the accumulation of the sand may represent the normal relationship between soil formation and landscape development. In some cases the evidence for a period of erosion is clear when these soils are formed in alluvial terraces, sand dunes or in stratified colluvial deposits. Therefore they may occupy different parts of the landscape and have various relationships with other soils as discussed on page 270.

Termite activity has been evoked as playing a significant part in their formation. When termitaria are evacuated they collapse, then differential erosion takes place during which the fine material is washed into the valley leaving the coarse fraction on the slopes.

Principal variations in the properties of the class

Parent material

Unconsolidated deposits of aeolian, colluvial or alluvial origin are common parent materials and in many cases they seem to have developed from a previous soil.

Climate

The greatest extent of these soils seems to be in the wet and dry tropics and tropical semi-arid areas. This is probably because it is under these conditions that suitable parent material forms over large areas of the landscape.

Vegetation

The variation is from deciduous tropical forest to tropical and subtropical savanna.

Topography

This varies from flat to moderately sloping. On flat sites they are usually developed in Pleistocene or Holocene sediments, whereas on sloping sites they are common in cols at the heads of some small valleys.

Age

When these soils have formed through progressive weathering of the rock and differential erosion, they are very old and date back to the Tertiary period. On the other hand many of those on sediments may have started their development during the middle to late Pleistocene period.

Pedounit

The soils of this class vary in colour from yellow to red as determined by the origin of the material and the internal drainage status. There is also some variability in the texture through the soil; this may be due to pedogenic processes or to the original stratification of the material.

Arenosols grade gradually into krasnozems and zheltozems as the content of clay increases. There are also sequences to podzols or halosols depending upon the climate.

Distribution

These soils are found in most tropical and subtropical areas, throughout east and south-east Africa, central and western Australia and in various parts of equatorial South America.

Utilisation

In spite of their high content of sand these soils give good crop yields. Their reserve of nutrients is low as is their capacity to absorb and hold them therefore they have to be supplied constantly with fertilisers, supplemented wherever possible by the addition of organic matter. When the pedounits are thick, they have a large storage capacity for moisture which is readily available to plants but there may be a water shortage if there is a prolonged dry season.

ARGILLOSOLS   [Plate Ic

Derivation of Name: From the latin word argilla = clay.

Approximate Equivalent Names: Grey-brown podzolic soils; Parabraunerden; Grey wooden soils; some Alfisols.

References: Tavernier and Smith 1957; Thorp, Cady, and Gamble 1959; Walscher et al. 1960

General characteristics

The presence of a prominent middle horizon of clay accumulation – the argillon is the distinguishing characteristic of this class. The principal upper horizon may be a tannon, luvon, modon or mullon. These soils occur beneath deciduous forests in a variety of humid climates which have a dry season.

Morphology

At the surface there may be a loose leafy litter resting directly on the mineral soil or their may be a thin mat of humified plant remains – a humifon. This passes sharply into a dark grey or dark greyish brown, granular, modon which is a mixture of mineral grains and fragments of humified organic matter. In this horizon and the humifon above, there may be fruiting bodies of fungi and some mycelia (Fig. 77). The modon is about 10 cm in thickness and grades sharply into a leached grey or brownish grey luvon which may be loose with single grain structure or it may grade into a luvon-gelon intergrade with its characteristic lenticular structure (Fig. 62).

The luvon which may exceed 20 cm in thickness, usually changes fairly sharply into the underlying brown or yellowish brown argillon with a marked increase in the content of clay and angular blocky or subcuboidal structure. The surfaces of the peds and many pores in the argillon are normally shiny with waxy coatings of clay cutans which may occur also within the matrix where there are varying amounts of domains (Fig. 72), the majority of which are randomly oriented but some are oriented parallel to the surfaces of peds and sand grains.

With depth the argillon may grade into the relatively unaltered material with a lower content of clay and in which cutans are few or absent or there may be an intervening calcon characterised by concretions and cutans of calcium carbonate (Fig. 78).

Analytical data – Fig. 79

These soils usually develop in material of medium texture with a high proportion forming in loess, therefore they are composed mainly of material in the size range 2 to 50 μ. The content of clay ($< 2 \mu$)

100µ

FIG. 77. Fungal mycelia decomposing plant fragments in a modon.

is at a minimum in the modon and luvon increasing to a maximum in the middle or lower part of the argillon where there are prominent and often continuous cutans. The texture usually becomes finer by at least one texture class but there can be as much as 20% or more clay in the argillon than in the luvon.

The fine lenticular structure of the luvon and more particularly the subangular blocky structure of the argillon are characteristic features of many members of this class, but when these soils are developed in sandy material the subangular blocky structure of the argillon is not well developed.

The pH values have an interesting pattern, they tend to vary from about 5·5 to 6·5 in the modon decreasing to about 4·5 to 5·0 in the argillon. There is then a steady increase into the relatively unaltered material where the value may be over 7·5 if it is calcareous or where there is a calcon. The maximum amount of organic matter occurs in the modon where

it varies from 5–10% and has a moderately high C/N ratio of 12–18 reflecting the partially decomposed state of the incorporated organic matter. Throughout the rest of the soil the content of organic matter and C/N ratio are very low due to the advanced state of decomposition of the organic matter. Sometimes C/N ratios of less than 8 are found in the argillon and probably are due to ammonia fixed by the clay fraction. The cation exchange capacity normally has two maxima one is in the modon where it is due to the combined influences of the organic matter and clay and the second in the middle part of the argillon where the clay content is also at its maximum. In a similar way the individual exchangeable cations dominated by calcium often have two maxima, the first is in the modon and the second in the argillon or in the calcon when present.

The total amount of free Fe_2O_3 is usually small throughout the soil but it increases steadily with depth

500μ

FIG. 78.   Cutans of finely crystalline calcite in a calcon.

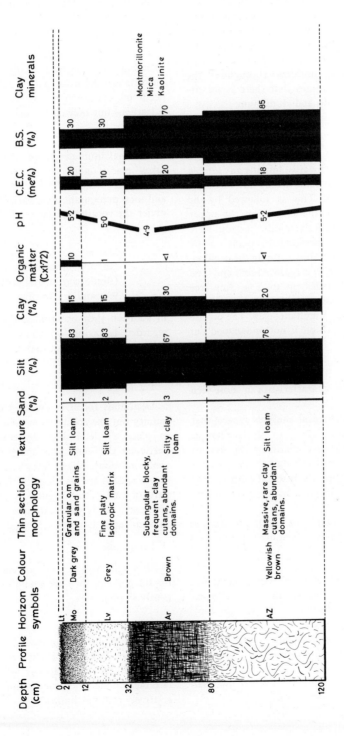

FIG. 79. Generalised data for argillosols.

from about 1% in the modon up to about 2% in the lower part of the argillon and remains constant thereafter. This shows that the free iron is not associated with the clay but appears to be lost from the upper part of the soil. The SiO_2/R_2O_3 ratio of the clay fraction often remains uniform throughout the pedounit indicating sometimes that there is no differential movement of iron and aluminium.

Genesis

These soils are formed in large part by progressive migration of material downwards. Initially any soluble salts and carbonates are removed by the moderate amount of precipitation. This is followed by the gradual translocation of the clay from the upper horizon or horizons to form the middle horizon. Here the clay is deposited as birefringent cutans on the surfaces of peds and pores thus forming the argillon. The very finest clay $< 0.5\ \mu$ is removed first and then follows the coarser particles up to $2\ \mu$. Any free oxides of iron, aluminium or silica associated with the clay are translocated simultaneously. The precise mechanism of clay deposition is not understood but some workers including Brewer and Haldane (1956) have shown that when a suspension of clay is passed through a column of soil it is deposited and in thin sections exhibits similar characteristics to those seen in argillons.

The data for the mechanical analyses suggest that the amount of cutanic material in the argillon is somewhat less than to be expected from the amount of clay removed from the overlying horizons. Therefore it seems probable that the textural contrast may be enhanced by clay destruction in the luvon and its removal from the system.

The precise time at which the argillon formed, is still open to question. It may be forming at present but it may have formed at an earlier phase in soil development.

One of the important climatic requirements for these soils is a distinct to marked dry season; hence their occurrence mainly in mildly to strongly continental climates. Under such conditions material is translocated in the wet season but during the dry period these fine particles on the surfaces of peds and pores become partly dehydrated and firmly attached. With annual repetition of this cycle layers of material gradually accumulate to form cutans.

Principal variations in the properties of the class

Parent material

Unconsolidated deposits of medium to fine texture including loess, glacial drift, alluvium and solifluction deposits are the usual parent materials, but they can occur on coarser materials. These deposits are often calcareous but they can be acid or intermediate.

Climate

The maximum development of this class invariably takes place under continental conditions with a marked dry season. The precise temperature regime and total precipitation are variable for they are found under the continental facies of the following climate types:

Marine
Humid continental warm summer
Humid continental cool summer
Wet and dry tropics.

In addition they occur in arid areas where they are regarded as having formed under more humid conditions and are now fossilised by a change to drier climatic conditions.

Vegetation

Deciduous forest is the most common plant community but the species composition varies from place to place as determined by environmental factors, particularly climate. Oak and beech are usual under cool conditions. Mixed forests of deciduous and coniferous species are associated with these soils in certain parts of the U.S.S.R. Argillosols can occur under grassland particularly when intergrading to other soils such as brunizems.

Topography

Although members of this class sometimes occur on steep slopes they are found most often on flat or gently sloping sites.

Age

The differentiation of most argillons seems to have taken place during the Holocene period but in certain soils of Europe they appear to be a feature of the late Pleistocene when they would have formed under relatively cold conditions.

Mu

1At

2At

AFp

cm
......0

......18

......40

......52

......60

Plate Ia   Altosol

Ku

1An

2An

AL

cm
......0

......12

......55

......150

......180

Plate Ib   Andosol

Lt
Mo

Lv

1Ar

2Ar

ALP

cm
......0
......2

......10

......20

......70

......120

......150

Plate Ic   Argillosol

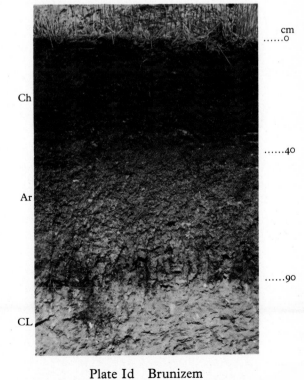

Ch

Ar

CL

cm
......0

......40

......90

Plate Id   Brunizem

Pedounit

Because of the relatively wide variation in the climatic condition under which these soils are formed there are wide differences in the nature of the pedounits. Set out below are a few of the common subclass and group formulae.

1. $[MuAr]$ $\quad$ $Lt_2Mu_{15}Ar_{35}AZ$
2. $[ZoHsAr]$ $\quad$ $Mo_{10}Zo_{20}Hs_{15}Ar_{30}AZ$
3. $[Mu(ArAt)]$ $\quad$ $Lt_2Mu_{20}(ArAt)_{30}IL$
4. $[LvArFg]$ $\quad$ $Mo_{10}Lv_{15}Ar_{50}Fg_{20}AZ$
5. $[Lv(ArRo)]$ $\quad$ $Lv_{10}(ArRo)_{35}AFw$

A relatively simple variant is given in the first example, which is found often under natural conditions but can result from cultivation during which the original upper horizon is changed into a mullon. The second example illustrates the polygenetic situation in which a thick luvon has formed, followed by the formation of a zolon and husesquon in the luvon resulting in the sequence — zolon, husesquon, argillon.

A common type of pedounit shows the intergrading

situation between an altosol and an argillosol; this is illustrated in the third example in which an alton has been changed partly into an argillon. Similarly argillons can intergrade to sesquons.

In certain places, particularly the central U.S.A. these soils have a prominent fragon beneath the argillon as shown in the fourth example.

In a number of areas polygenetic soil development has caused argillons to form in a previously developed soil. Some of the so-called red and brown mediterranean soils are good examples of this type of development and are illustrated in the fifth example.

With progressive development, the argillon becomes increasingly impermeable and can lead to the build up of moisture in the soil and the development of a minon above the argillon which itself may slowly change to an (ArGl). These soils are not regarded as belonging to this class and are classified with the supragleysols.

In some of these soils the luvon grades gradually into the argillon through the intergrades (LvAr) and (ArLv) and appears to result from the destruction of the argillon as the luvon extends downwards. In thin

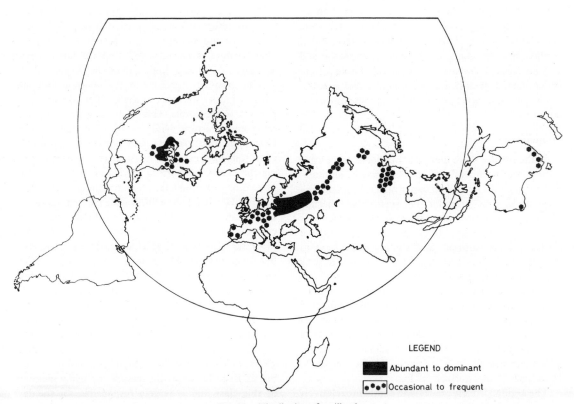

LEGEND

■ Abundant to dominant

●●● Occasional to frequent

FIG. 80. Distribution of argillosols.

section there are usually thin coatings of sand grains on the outer surface of the peds due to the destruction of the argillons.

Distribution – Fig. 80

Argillosols are of widespread distribution, being found more especially in west-central U.S.S.R., east-central and north-central U.S.A., the central belt of Europe, southern Australia and elsewhere.

Utilisation

The potential of these soils for agriculture varies from moderate to good. Soils with a thick luvon fall into the former category whereas those with a mullon or tannon are included amongst the world's most productive soils. Because they occur under moist conditions they are frequently used for mixed farming, dairying or horticulture but wheat, maize and oats can also be grown. Fertility is maintained by the normal procedure of liming and fertiliser application, only in exceptional circumstances is it necessary to adopt any special techniques. Erosion is a common feature and rigorous control methods must be maintained at all times.

BRUNIZEMS   [Plate Id

Derivation of Name: From the French word *brun* = brown and the Russian word
    zem = soil.
Approximate Equivalent Names: Grey forest soils; Argiudolls.
Reference: Walscher *et al.* 1960

General characteristics

The major upper horizon is a chernon with its characteristic very dark gray or black colour and granular structure. This merges into a dark brown argillon having cutans, a higher content of clay and angular blocky structure. These soils are found mainly under humid continental cool and warm summer conditions and occur beneath deciduous forests or tall grass.

Morphology

At the surface there may be a thin, loose, leafy litter resting on the mineral soil or there may be an intervening thin root-mat. The upper mineral horizon is a very dark grey granular or vermicular chernon which may be up to 50 cm thick. In thin section it is seen to contain many faecal pellets and to have a matrix that is isotropic and weakly translucent to opaque (cf Fig. 82). The chernon grades into the dark brown, angular or subangular blocky argillon with cutans on the surfaces of the peds. In thin section the isotropic matrix of the chernon gradually changes to birefringent in the argillon due to the presence of common, medium random domains. The surfaces of the peds and pores have varying amounts of cutans which are usually at a maximum in the middle or lower part of this horizon (cf Fig. 72).

The argillon usually grades with depth into relatively unaltered material.

Analytical data – Fig. 81

A high proportion of these soils are developed in loess; therefore they are dominated by particles $< 50\,\mu$. The maximum amount of clay occurs in the argillon where values of between 30% and 40% are common and are 10% to 20% higher than the chernon above.

The granular structure in the chernon and subangular blocky structure in the argillon impart a high porosity to these soils. This allows easy penetration of roots and moisture, while at the same time the peds have high moisture retaining powers. Consequently these soils have excellent moisture relationships allowing the excess moisture to percolate freely, and at the same time retaining a large amount in the porous peds.

The pH values have a fluctuating pattern which may be due in part to cultivation, to differences in the composition of the parent material or to pedogenic processes. The pH values at the surface are about 6·5 but decrease to about 5·0–6·0 in the lower part of the chernon and upper part of the argillon. Then follows a steady increase through the lowest part of the argillon and into the parent material.

The organic matter steadily decreases from about 5% in the chernon to 1–2% in the lower part of the argillon. The upper part of the argillon may contain 3% organic matter which is high for a middle horizon but it is a characteristic feature for many members of this class and is largely responsible for the dark colour.

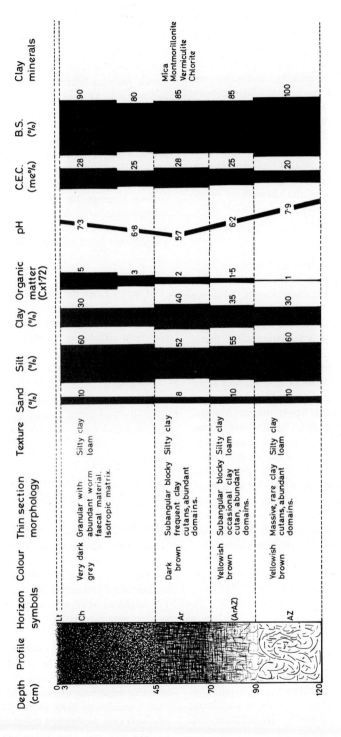

FIG. 81. Generalised data for brunizems.

The narrow C/N ratio of 10–12 in the chernon indicates that the organic matter is well humified while the thin section morphology shows that there is an intimate blend of the organic and mineral material. The cation exchange capacity is variable; in the U.S.A. it is about 25 me% and 30 me% in the chernon and argillon respectively due to montmorillonite being the principal clay mineral in that country. In the U.S.S.R. the values are lower because mica is the predominant clay mineral. The base saturation is normally high exceeding 80% in most situations with calcium being the predominant exchangeable ion. In some soils the lowest percentage base saturation occurs in the lower part of the chernon or in the argillon.

Genesis

These soils seem to have developed in two or more distinct stages and it appears that they may be out of phase with their present environment. Evidence from pollen analysis (West 1961) for the central U.S.A. indicates that successive waves of vegetation have colonised areas of these soils following the climatic trends of the Pleistocene. There was spruce, followed by hardwood and now prairie. Therefore, it seems likely that the first stage was the development of an argillosol superimposed upon which has been the formation of the highly organic upper horizon induced by the prairie vegetation and churning organisms. Further, it is suggested that the present vegetation should be deciduous woodland, the prairie communities being a relic from the climatic optimum but the deciduous forests now find it difficult to compete with the vigorous growth of the grasses (Walscher et al. 1960). However in the U.S.S.R. these soils carry deciduous forests so it would appear that the highly organic layers are due to the present conditions rather than to a grassy vegetation. In any case the argillon is regarded generally as partially if not totally a relic horizon.

Principal variations in the properties of the class

Parent material

Unconsolidated deposits including glacial drift, loess and alluvium are the usual parent materials. The texture is usually a silt, silty clay loam or clay loam with variable mineralogy, normally ranging from acid to strongly calcareous.

Climate

Brunizems are confined largely to continental conditions where the variation is from humid continental cool summer to humid continental warm summer conditions. Perhaps one of the widest ranges is found in North America where the soils extend from Alberta in Canada to Missouri in the U.S.A. Within this range there is a steady decrease in mean annual temperature from 18°C in Missouri to 2°C in Alberta, accompanied by a decrease in precipitation from 1200 mm to 400 mm. This is a good example of the same type of soil extending through contrasting atmospheric climate types. In all cases the amount of moisture passing through the soil is similar because evapotranspiration decreases with decreasing temperatures so that the increased precipitation does not produce a significant increase in percolation.

Vegetation

In many cases the natural communities are dominated by tall grasses such as big bluestem (*Andropogon furcatus*) in North America; in Europe there are often deciduous forest communities dominated by species such as oak (*Quercus* sp).

Topography

These soils are confined almost exclusively to flat or gently undulating situations and are almost completely absent from moderate or steep slopes.

Age

Most of these soils are developed in deposits of late Pleistocene age, therefore they are relatively young.

Pedounit

Generally the greatest variations occur in the character of the upper horizon which is sometimes a chernon while in other situations it may be a thick mullon or an intergrade between these two horizons. The argillons vary in colour and degree of development, commonly having more than 2% organic matter hence their dark colours.

The lower horizon is often a fragon or there may be a second sequence of horizons, composed of a luvon and a second argillon, and the intergrades form a complete sequences to argillosols and to chernozems.

In a number of cases these soils are developed in a relatively thin deposit of loess overlying till so that the argillon may rest directly on the till. Alternatively, where the till is impermeable an argillon-gleyson inter-grade develops over the till.

The type of clay minerals in this class varies fairly widely, whereas in the U.S.A. the principal component is montmorillonite, in the U.S.S.R. mica generally predominates. This suggests that the type of clay mineral is determined by the original composition of the parent material rather than by pedogenic processes.

Distribution

These soils have a high natural fertility and give good crop yields, but often these can be increased by the application of phosphorus, and when intensive cultivation is practised it is necessary to apply other fertilisers and lime. Traditionally these soils have been used for growing grain crops including maize, wheat and oats, other crops such as soya beans are now extensively grown. The utilisation of the crops varies from place to place, sometimes they are used exclusively for human consumption whereas in others they are fed to animals on the farms where they are grown. Erosion is a serious hazard and rigorous control methods have to be maintained.

BUROZEMS

Derivation of Name: From the Russian words *buryi* = brown and *zem* = soil
Approximate Equivalent Names: Brown soils, Orthic brown soils; some Ustolls.
Reference: Cline *et al.* 1960

General characteristics

These soils have a relatively simple morphology, being composed of a uniform brown or dark brown buron overlying a calcon or hardened calcon.

Morphology

Under natural conditions there is a very sparse leafy litter at the surface resting on the greyish brown or brown buron about 40 cm thick. The upper part of the buron is granular or subangular blocky and grades with depth into fine or medium prismatic which breaks down to angular blocky or subcuboidal. They have a fairly uniform texture and are often associated with clamosols which have a marked increase in clay in the middle horizon. The nature of this relationship varies from place to place, burozems predominate in the U.S.S.R. whereas clamosols with strong textural contrasts occur in the U.S.A. and Canada. Beneath the buron, calcium carbonate has accumulated to form a calcon or hardened calcon, which grades with depth into the relatively unaltered material.

Most of these soils have occasional crotovinas.

Analytical data

A high proportion of burozems are developed in loess therefore they are dominated by particles $< 50 \mu$ in diameter. Many burons are calcareous containing about 2–3% carbonate which increases to about 10–20% in the calcon, however the total amount is deter-mined by the initial content in the parent material. The pH value in the surface varies from 7·5–8·5 and increases to about 8·5 in the calcon, the pattern following in part the distribution of calcium carbonate, i.e. increasing as the carbonate content increases. The content of organic matter varies from 2–4% in the buron decreasing to $< 1\%$ in the calcon, but sometimes cultivation causes a reduction in the content of organic matter at the surface. The C/N ratio varies from about 8–10 on the surface indicating that humi-fication is rapid, and is a result of the warm summers and neutral soil conditions which together encourage vigorous microbiological activity.

The cation exchange capacity is generally less than 40 me% in the upper part of the buron and decreases to < 10 me% in the calcon. There is usually complete base saturation, the principal exchangeable cations being calcium and magnesium. Exchangeable sodium and magnesium are often relatively high as compared with soils of more humid areas.

Genesis

The principal processes taking place in these soils are the rapid decomposition of the plant remains and their incorporation into the mineral soil. Also taking place is the translocation and deposition of calcium carbonate, to form the calcon. Because of the lack of data on these soils it is not possible to be certain about the reason for the increase in clay with depth and their gradation to clamosols. Tentatively, *in situ* clay formation is considered to be the dominant process but clay translocation may also be taking place.

Principal variations in the properties of the class

Parent material

Unconsolidated deposits of silty and loamy textures are the usual parent materials. These are mainly loess and alluvium of Pleistocene or early Holocene age; in addition some denuded Tertiary or earlier deposits also appear to form the parent materials as in certain parts of the mid-western U.S.A. The most important variation in the chemical composition of the parent material is in the amount of carbonate which seems to have a strong influence on the development of horizons of carbonate deposition.

Climate

These soils are found predominantly in the mid-latitude, semi-arid and tropical conditions where the annual precipitation varies from about 250–500 mm.

Vegetation

Short grass is the principal natural plant community. At the dry end of the climatic range for these soils, succulents or cacti begin to appear as part of the plant communities.

Topography

These soils develop on sites that are usually flat to gently sloping. Moderately and steeply sloping sites lose a considerable proportion of the annual precipitation through run-off and therefore they are too dry to support a dense vegetation and may suffer progressive erosion.

Age

A high proportion of these soils occur just outside the limits of the Pleistocene glaciations and therefore started their development towards the end of the Pleistocene period. In tropical and subtropical areas they may be much older.

Pedounit

Variation within this class of soils is rather narrow but there are a few groups worthy of note. In a number of cases the calcon is hard but at the other extreme the calcon is very weakly developed. This range is probably due to variations in the character of the parent material, with more carbonate accumulating as the parent material becomes more basic or calcareous. There are a number of intergrading situations, chiefly to kastanozems, serozems, solonchaks and solonetzes and it is within the areas of burozems that solonetzes have their greatest distribution therefore the intergrades between these two soils are very common.

Distribution

Burozems occupy fairly large areas in semi-arid regions of the world particularly in the central and south western U.S.A., southern U.S.S.R.

Utilisation

Low rainfall imposes severe restrictions on the utilisation of these soils. At the wetter end of their range it is possible to grow grain crops such as wheat or sorghum but usually dry farming is practised. At the dry end of the range ranching is the normal practice. Where irrigation is possible these soils have proved to be extremely productive for a variety of crops such as maize and vegetables. Because the natural plant communities are open grass, wind erosion is a constant hazard.

CHERNOZEMS   [Plate IIa

Derivation of Name: From the Russian words *cherni* = black and *zem* = soil.
Approximate Equivlanet Names: Orthic black soils, some Mollisols.
Reference: Ministry of Agriculture of the U.S.S.R. 1964

General characteristics

The thick, dark, greyish-brown or black chernon is the differentiating horizon of this class. This often has an accumulation of carbonates at its base and grades into the relatively unaltered material; often krotovinas are abundant. These soils occur mainly in a mid-latitude, semi-arid environment where the natural vegetation is tall grass.

Morphology

At the surface there may be a thin loose leafy litter resting on a root mat up to 5 cm thick. Below is the dark coloured chernon which may exceed 2 m in thickness but is normally 50–100 cm thick. This horizon has a marked granular or vermicular structure which may be composed very largely of earthworm casts that are clearly seen in thin sections (Fig. 82). The chernon grades into a calcon in which calcium carbonate is deposited in the form of pseudomycelia or concretions, (Fig. 83) only in rare cases are there great thicknesses of carbonate. Underneath is the relatively unaltered material but this usually contains a few earthworm passages.

Krotovinas are common phenomena in most chernozems and where they are well developed there is a thick, mixed layer of chernon and relatively unaltered material from beneath the chernon. The organisms responsible for mixing the soil vary from place to place; in Europe the mole rat (*Spalax typhluon*) and gopher (*Citellus suslyka*) are mainly responsible.

Analytical data – Fig. 84

The thick chernon with its well developed vermicular structure is unique and has excellent moisture relationships for most of the growing season. It is capable of storing large quantities of moisture within the peds, at the same time excess water drains away freely and allows good aeration. Loess is the common parent material which imparts a fine texture and is largely responsible for the marked and well developed structure.

The distribution pattern of clay down the pedounit is either uniform or shows a small maximum about 50–200 cm from the surface, probably representing slight eluviation of fine material or weathering *in situ*. In view of the relatively low rainfall it is difficult to explain the slight accumulation of clay under contemporary conditions but there could have been some migration during the later part of the Pleistocene and early Holocene periods when the climate was cooler and probably more moist.

Perhaps the most important chemical property is the steady increase with depth in the amount of carbonate. Usually when the parent material contains only a small amount of carbonate most of the chernon has been decalcified and acid with pH values as low as 6·0, therefore it is not possible to estimate the rate of carbonate removal. In Chernozem Novi Sad in Yugoslavia the parent material contains over 50% carbonate but the surface only has 15% carbonate, therefore 35% carbonate has been lost from the uppermost part of the chernon during 10,000 years which gives a rate of removal of ·035 mg per gm per annum. This is a very crude estimate for there is some uncertainty about the time factor and further it is assumed that the climate has remained unchanged during that period. However it does give some indication of the rate of removal of carbonate under humid continental warm summer conditions.

The organic matter content in the surface of the chernon varies from about 3 to 15% with a C/N ratio of 8–12. With depth the organic matter content decreases steadily to < 1% in the calcon. Afanaseva (1927) has demonstrated that a chernozem may have 292 g/m^2 of plant material above ground and 1044 g/m^2 below i.e. there is three times more organic material in the soil than in the above ground foliage therefore a very high proportion of the organic matter in the soil is derived from decomposing roots.

The pH value varies from 5·5 to 8·0 in the upper part of the chernon as determined by the amount of decalcification and leaching that has taken place, increasing with depth to values up to 8·0 and over.

The cation exchange capacity tends to be rather low for many members of this class because the clay fraction is often composed mainly of mica but there is usually a significant amount of montmorillonite.

Genesis

The principal processes taking part in the formation of chernozems are the rapid incorporation of organic matter into the soil accompanied by its humification and the leaching of soluble salts and carbonates. The decomposition and incorporation of the organic matter is accomplished in some soils by earthworms which appear to have ingested at some time all of the material in the upper 50 cm. The great thickness of the chernon is due also to constant churning by small vertebrates. Leaching has removed the easily soluble salts completely from the soil but in some cases carbonate has been deposited as pseudomycelia or concretions in the lower part of the chernon and below to form a separate calcon.

Principal variation in the properties of the class

Parent material

Chernozems are developed almost exclusively in

180

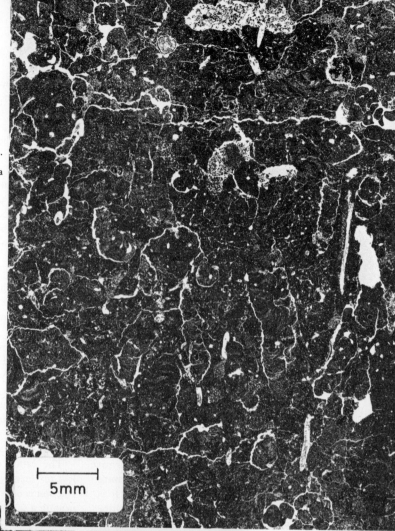

FIG. 82.

Thin section of vermicular structure in a chernon produced by earthworms.

FIG. 83.

Pseudomycelium of tightly packed acicular crystals of calcite.

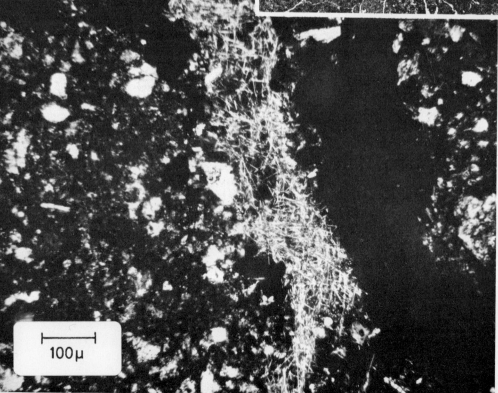

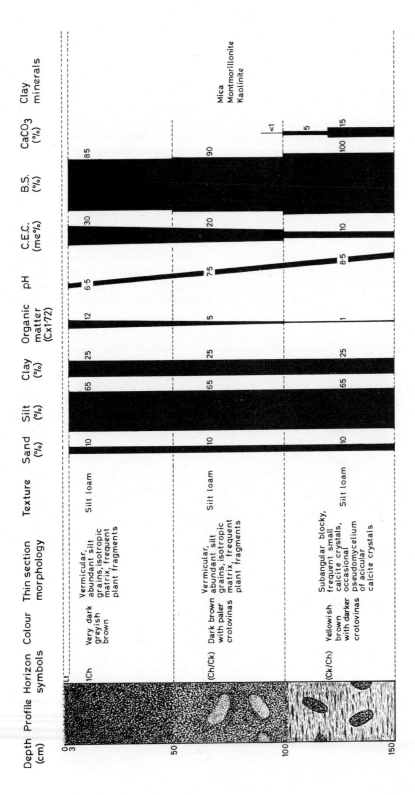

FIG. 84.   Generalised data for chernozems.

loess but they also occur on other sediments which are calcareous or otherwise contain easily weatherable minerals that release a high proportion of calcium. Where quartzose sands of coarse texture occur within areas of chernozems, podzols or argillosols may develop.

Climate

These soils are confined largely to continental conditions where the range is from humid continental warm summer to mid-latitude semi-arid but they can occur under humid continental cool summer conditions. The important features of the climate are the cold winters, hot summers and an excess evapotranspiration over precipitation.

Vegetation

Tall grass is almost the only plant community found on these soils but deciduous woodland dominated by species of oak is common in certain transition situations.

Topography

These soils are found on sites that vary from flat to moderately undulating. They are absent or only poorly developed on steeper slopes.

Age

Most chernozems started their development during the last 10,000 years. Some may have been forming throughout the whole of that period while others may only be about half that age.

Pedounit

There is a fairly narrow range of variability within this class of soils. This is in part due to the relatively short period during which they have been forming and secondly to the rather restricted environmental conditions. Chernozems grade gradually in at least eight principal directions into: altosols, argillosols, brunizems, clamosols, kastanozems, solonetzes, solonchaks and subgleysols. The spatial change to the

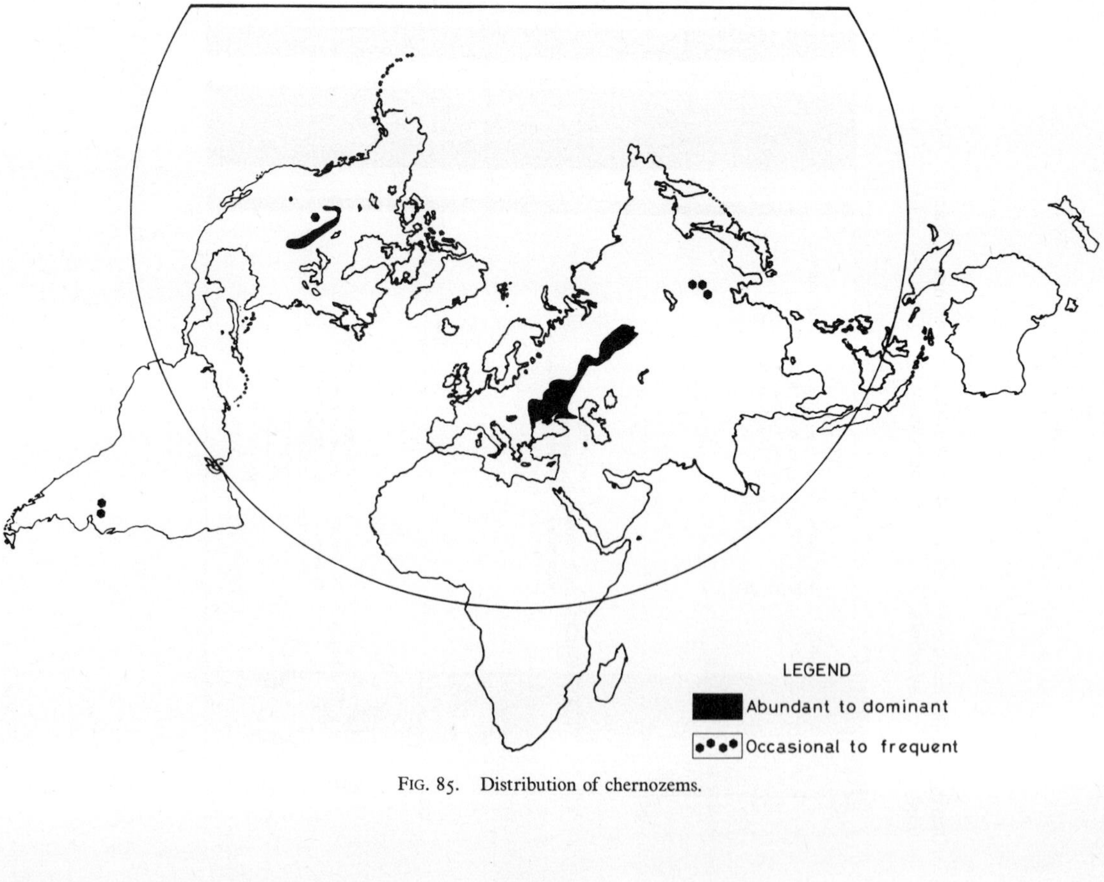
LEGEND
Abundant to dominant
Occasional to frequent

FIG. 85. Distribution of chernozems.

first five is due to a gradual change in climate, on the other hand the change to the latter three is due mainly to topography.

Within the class the principal variation is in the thickness and humus content of the chernon. The former varies from about 50 cm to over 2 m while the latter varies between 3% and 15%. Some workers have used the variability in the amounts of organic matter, thickness of chernon and colour as criteria to subdivide the class. The clay mineralogy also shows some variation whereas in the U.S.S.R. mica is predominant, montmorillonite is most common in the U.S.A.

Distribution – Fig. 85

The total areas occupied by these soils is relatively small. There is a belt in the south central part of western U.S.S.R., and one area in the central and north central U.S.A. Elsewhere in the middle latitudes they occupy very small areas.

Utilisation

These soils have a high nutrient status, excellent structure and high water-holding capacity thus imparting soils a natural high fertility that makes them eminently suitable for agriculture. Indeed it was once thought that they were inexhaustible. However overcropping can lead to deficiencies particularly in nitrogen and phosphorus which have to be added as fertilisers.

Since these soils are found in areas of relatively low rainfall summer drought is sometimes a hazard so that moisture conservation is extremely important. This can be achieved by retaining snow in furrows and through the use of shelter belts.

Wheat, barley and maize are the principal crops grown and where there is a plentiful supply of water, vegetables can be grown under irrigation. In recent years there has been a tendency to rear livestock and to practise dairying but the lack of water is again a limiting factor.

CLAMOSOLS [Plate IIb, IIc

Derivation of Name: From the Old English word *clam* = clay.
Approximate Equivalent Names: Red brown earths; Non-calcic brown soils; Cinnamonic soils; some Alfisols.
References: Stace et al. 1968

General characteristics

The middle horizon of these soils – the clamon has a markedly higher content of clay than those above and below, particularly the former. The upper and lower horizons are somewhat variable as determined by differences in climatic conditions which include arid to semi-arid of subtropical and mid-latitude areas.

Morphology

At the surface there is a sparse litter which rests on a greyish brown or reddish brown, sandy loam, tannon or mullon, 10–15 cm thick, with isotropic or weakly birefringent matrix. Below the tannon there is commonly a lighter coloured clay depleted minon about 15–20 cm thick with varying amounts of mottling and sesquioxidic concretions. The matrix is weakly birefringent with small and medium domains, occurring randomly, in clusters and surrounding pores and sand grains. Clay cutans are occasional to rare on pore and ped surfaces.

The tannon or minon changes sharply into the brown or reddish brown clay or clay loam clamon about 35–40 cm thick and with strong angular blocky or prismatic structure. The matrix is moderately birefringent with abundant small and medium domains occurring randomly, in clusters and surrounding pores and sand grains. Clay cutans are occasional to rare on pore and ped surfaces. The clamon grades with depth into a calcon or into a calcon-halon intergrade. In thin sections the calcon has calcite cutans on peds and surrounding stones, also present are carbonate concretions. The matrix is birefringent with many medium and large random domains, clusters and zones of domains and grain cutans.

Analytical data – Fig. 86

The most conspicuous property of these soils is the marked increase in clay in the middle horizons followed by a gradual decrease with depth. Carbonates are absent from the middle and upper parts of the soil but may exceed 15% in the lower part of the soil. Soluble salts follow a similar trend and are often

$>1\%$ in the lower part of the soil, forming the intergrade (CaHl). pH values are around neutrality in the surface but increase to over 8·0 in the lowest horizon. The content of organic matter varies from about $1-3\%$ in the upper horizon decreasing to $<1\%$ in the clamon, generally the C/N ratios are <10.

There is sometimes a small amount of exchange acidity in the surface horizon but from the clamon down there is complete base cation saturation. The proportion of the cations varies with depth, whereas calcium predominates at the surface, it is often less than 50% of the total exchangeable cations in the clamon where magnesium and sodium sometimes predominate.

The clay minerals are variable but are usually dominated by montmorillonite, mica and kaolinite.

Genesis

Oertel (1961) has shown clearly that the clamon is formed by weathering *in situ* with little addition of clay from the horizon above. The very much smaller amount of clay in the horizon above appears to be due to clay destruction and removal from the system. As shown by the thin section morphology there is some translocation of clay into the clamon but its contribution is minimal. The mottling and sesquioxide concretions in the minon may be fairly recent phenomena resulting from seasonal waterlogging caused by the relatively impermeable clamon.

The other main process is downward leaching of soluble salts and carbonates.

Principal variations in the properties of the class

Parent material

In most cases these soils develop in loose unconsolidated sediments of Pleistocene age including alluvium and loess. A great many clamosols seem to be polygenetic having formed in a truncated soil, therefore the lower horizon is often weathered rock.

Climate

These soils occur in semi-arid and arid subtropical areas with a tendency to occur in inland situations.

Vegetation

This varies from deciduous woodland to savanna or short grass of arid areas but the various plant communities do not seem to have a strong influence upon soil formation.

Topography

Clamosols are confined to flat or gently sloping situations. On moderate to steep slopes they are either poorly developed or absent.

Age

Most of these soils appear to have started to develop at the beginning of the Holocene period. However those occurring in the south western part of the U.S.A. and in some parts of Australia may date from the Pleistocene period.

Pedounit

Probably the most important variation is the thickness of the clamon which becomes progressively thinner, redder and nearer to the surface with increasing aridity, accompanied by a change from a minon to a luvon. In some cases the amount of soluble salts increases and comes nearer to the surface to form a (CnHl) or (HlCn). When it is the latter these soils are included with the halosols. A distinctive feature of the clamosols of the desert is the accumulation of gravel or stones at the surface to form a luvoncumulon intergrade or a hamadon. Here they may represent polygenetic evolution having formed when climatic conditions were slightly moister. When these soils have a very complex evolution which goes back to the Tertiary period the larger separates in the upper horizons may be composed of fragments of previous horizons such as vesons. At the very moist end of the range there is a mullon at the surface, and the whole pedounit is free of salts and carbonates.

Distribution

The greatest known occurrences of these soils are in Australia where they occupy large areas particularly in the eastern and western parts of the country. Clamosols are also common in the south western parts of the U.S.A. and in some of the arid and semi-arid parts of the U.S.S.R.

Utilisation

At the moist end of the range these soils are very productive, particularly for wheat and mixed farming. With increasing aridity the natural vegetation is grazed by sheep or cattle but some areas are irrigated and devoted to horticultural crops.

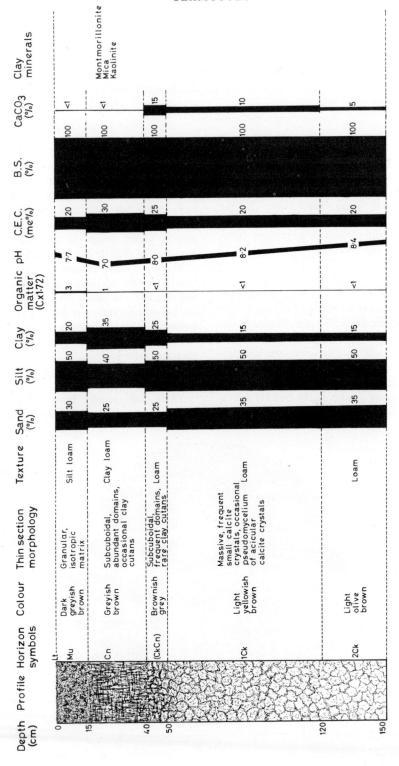

FIG. 86.   Generalised data for clamosols.

CRYOSOLS   [Plate IId

Derivation of Name: From the Greek word *kruos* = frost.
Approximate Equivalent Names: Tundra soils, Pergilic cryaquents.
Reference: Douglas and Tedrow 1960.

General characteristics

The distinctive feature of this class is the permanently frozen lower horizon – the cryon; hence their occurrence in polar areas particularly in the northern hemisphere. The cryon is impermeable to moisture, often causing periodic or permanent waterlogging of the overlying horizons.

Morphology

At the surface there is a thin litter which rests on a fermenton that may be up to 15 cm thick. The plant fragments in the fermenton are relatively fresh but are being decomposed by fungi whose mycelia are clearly seen in thin section. The fermenton is underlain by a massive, loamy, grey gleyson about 25 cm in thickness, with the characteristic ochreous or brown mottling. The gleyson grades fairly quickly into a mixed horizon of gleyson and fermenton – (Gl/Fm).

In thin section the features of the gleysons are typical for that horizon, it is massive with rare, circular, pores, abundant small random domains and diffusion cutans around sand grains. The ochreous mottled areas are more dense but they have a similar fabric. In the (Gl/Fm) the fabric of the gleyson is similar to the gleyson above while the fermenton component is composed of slightly decomposed plant fragments.

There is a sharp change to the cryon with its distinctive lenses of clear ice and laminae of frozen soil (Fig. 87). In many cases there are inclusions of frozen organic matter indicating that the cryon has formed in material similar to the horizon above. When stones are present in the cryon they have a complete sheath of clear ice.

Analytical data

These soils usually have a loamy texture with uniform particle size distribution throughout the

FIG. 87.

The polished surface of a cryon with the characteristic laminar structure. The dark areas are clear ice and the light areas are frozen soil.

5 cm

pedounit, however the very low temperature and the impermeable cryon are the two most important physical properties. The former severely restricts the type of vegetation to a few grasses, sedges, lichens and mosses, while the latter inhibits drainage causing wetness.

The surface horizon is usually moderately acid but the pH values increase with depth. The cation exchange capacity is low when the content of organic matter is low; similarly the base saturation is low but this is very variable as determined by the nature of the material. The organic matter is usually at a maximum in the surface, but there is often a second maximum just above or just within the cryon. The C/N ratio is usually wide due to the low degree of humification. When stones are present in the upper or middle horizons they have a marked tendency to be oriented vertically.

Genesis

These soils have formed either as a result of a progressive change from warm to cold climatic conditions or they have developed following the deposition of unconsolidated sediments in a cold environment. At the onset of a climatic change from warm to cool conditions the heat lost from the soil during the winter is balanced by the incoming heat received during the summer period. Gradually the stage is reached when the soil freezes during the winter period and thaws during the summer. With continued change to cooler conditions the depth of freezing is greater than the depth of thawing because the heat lost during the winter is greater than the incoming summer heat. The result is that a lower layer of frozen soil – the cryon extends from one winter to the other. If conditions become even colder the cryon becomes thicker and thicker and maybe > 300 m as encountered in certain parts of Siberia. Thus the cyclic condition is attained in which the upper part of the soil freezes every winter and thaws every summer. When this process is continuous for long periods, the attendant churning and wetness eliminates the previous horizons that were present at the initial stage. On a sloping site solifluction takes place and any previous horizontation is also destroyed. Ultimately the cryosol is formed in the previously underlying material (see page 273).

When unconsolidated materials are deposited in a cold environment they usually rest on material which itself has a cryon so that a single winter period is sufficient for a cryon to form in the new deposit but if the new deposit is thick it may require more than one winter. When a thick deposit is laid down on a previous fully developed crysol, it may be preserved. This happened periodically during the Pleistocene period so that buried soils of various ages and types are found in Pleistocene deposits.

After the formation of the cryon, annual freezing and thawing causes a considerable amount of churning of the soil so that various surface patterns develop and there is little tendency for horizons to develop except those produced by homogenisation (Figs. 38, 39, 40, 43). The presence of large lumps of fermenton just above and within the cryon suggests that churning in some of these soils was more active in the past when the freeze-thaw horizons were thicker.

The wet or moist conditions above the cryon have caused the reduction of iron to the ferrous state and the development of a mottled pattern.

Principal variations in the properties of the class

Parent material

Unconsolidated deposits including glacial drift, alluvium, marine deposits and solifluction deposits, are the usual parent materials. The mineralogy and particle size distribution vary considerably from place to place, but they do not seem to play a major role in the formation of these soils. However the mineralogy does exercise a little influence on the base status, particularly if carbonates are present.

Climate

The range of climate under which these soils are found is fairly narrow, always having a mean annual temperature below $0°C$. They are found under the taiga conditions of central Siberia and Alaska. They also occur under cold oceanic conditions as on the west coast of Alaska.

Vegetation

Grasses, sedges, lichens and mosses are the principal members of most plant communities, but in many places dwarf *Salix* spp. are frequent. Where these soils grade into those of warmer areas the tree species increase in frequency and the plant communities are dominated by various conifers.

Topography

Flat or gently sloping sites are the normal situations for these soils, but they are found also on moderate slopes particularly those with a cold northern aspect. Although they are associated usually with low elevations, they can occur on flat or gently sloping sites up to 1,000 m in the taiga.

Age

Most of these soils started to develop towards the end of the Pleistocene as shown by the C^{14} date of 11,000 years B.P. for the organic matter in the (Gl/Fm) horizon of Cryosol Barrow in Alaska. However, some of the areas that have these soils, like central and northern Alaska, were not glaciated during the Pleistocene so it is possible that their development may have started about 125,000 years B.P. i.e. at the beginning of the last major cold period or glaciation.

Pedounit

The following subclasses and groups are of common occurrence:

1. [At(Mb/Fm)Cy] $Lt_3At_{40}(Mb/Fm)_{10}Cy$
2. [GnCy] $Lt_4Gn_{40}Cy$
3. [LhCy] $Lh_{80}Cy$
4. [CuCy] $Fi_{15}Cu_{40}Cy$

The variations in the soils of this class occur mainly in the middle and upper horizons. Usually there is some wetness but on sloping situations this may cause only the formation of a marblon or a (Mb/Fm) above which an alton may develop as shown in the first example given above. The second example is found in the high arctic (Svalbard) where despite large amounts of moisture in the soil reduction of iron to the ferrous state is not as extensive as further to the south, therefore a fairly thick gelon with its characteristic lenticular structure often overlies the cryon. In the absence of a covering of superficial deposits frost action produces a thick covering of angular rock material or a lithon which may be the only horizon over the cryon as shown in the third example (Fig. 41). The fourth example occurs in very wet situations where a thin cover of peat has formed over a blue or grey cerulon.

In nearly every case where there is a cover of vegetation some organic matter is incorporated into the mineral soil below. This can vary from a few

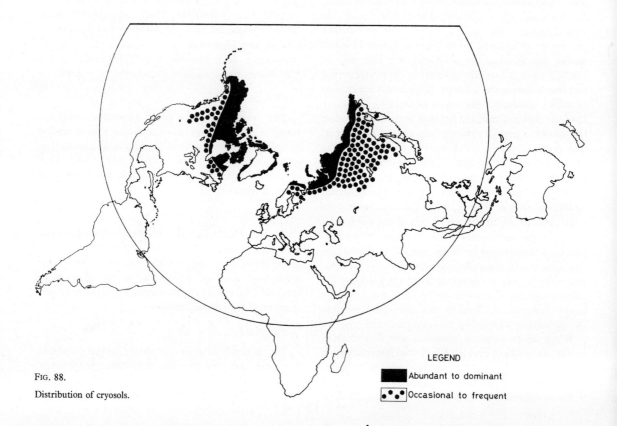

FIG. 88.

Distribution of cryosols.

LEGEND

■ Abundant to dominant

•°•° Occasional to frequent

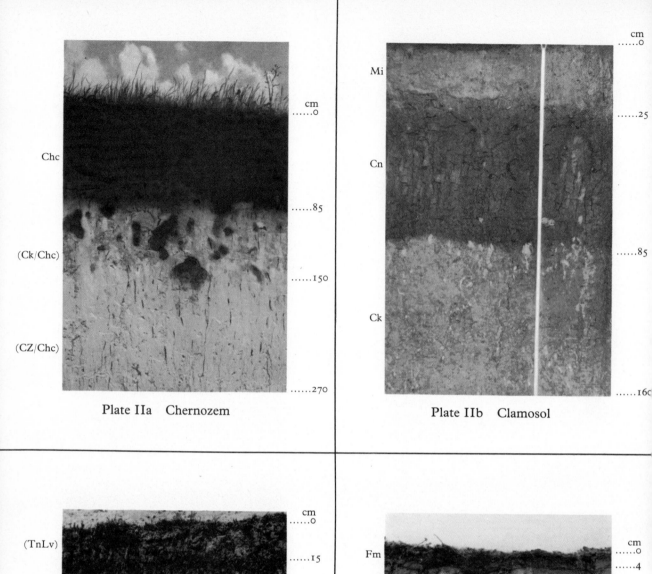

Plate IIa   Chernozem

Plate IIb   Clamosol

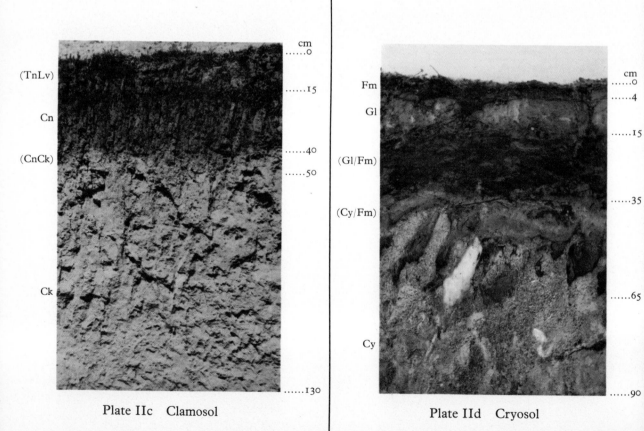

Plate IIc   Clamosol

Plate IId   Cryosol

leaves scattered throughout the pedounit to a fairly large amount.

Distribution – Fig. 88

The greatest extent of these soils is in the northern parts of the U.S.S.R., Canada and Alaska and Antarctica.

Utilisation

The low temperatures of these soils, their wetness and the presence of the cryon prevent their utilisation for agriculture or forestry. In the Northern hemisphere they produce herbage for the natural herds of reindeer and caribou, but in some places the herds are crudely managed.

FLAVOSOLS

Derivation of Name: From the Latin word *flavus* = yellow

Approximate Equivalent Names: Yellow earths; Xanthozems; Terra fuscas; Sols ferrugineux tropicaux.

References: Kubiëna 1953; Stace *et al.* 1968

General characteristics

This is a class of yellow or yellowish brown or brown soils that occur predominantly in humid subtropical and tropical areas. These soils are formed by progressive hydrolysis and solution during which mica and kaolinite are the principal weathering products, together with small but variable amounts of goethite and vermiculite.

Flavosols form on a variety of rock types and sediments both of which are often calcareous. The designations of a few subclasses and groups are as follows:

1. $[FvFb]$ $Mu_5Fv_{50}Fb_{200}AFw_{100}AF$
2. $[MuFv]$ $Lt_2Mu_{10}Fv_{50}LF$

The first example developed on slate is of fairly common occurrence in tropical and subtropical areas. This is a deep soil in which there is gradual transformation of the rock into soft weathered rock followed by the formation of a flambon which in turn grades into the flavon above.

The second example is of a flavosol formed on limestone. In these soils there is usually a fairly sharp break between the flavon and the rock, or the flavon may form tongues down solution cracks in the rock.

Many areas where flavosols developed extensively during the Tertiary period were subjected to climatic changes during the Pleistocene period. Consequently they often show polygenetic evolution, the recent phase being translocation of clay to form a luvon at the surface or a flavon-argillon intergrade as the middle horizon. The leached terra fuscas of Kubiëna (1953) fall into this category.

GELOSOLS

Derivation of Name: From the Latin *gelare* = to freeze.

Approximate Equivalent Names: Alpine soils, Sub-arctic brown soils.

General characteristics

This class of soils occurs principally in cold mountainous areas but is also common on the fringes of the arctic. In the former situation vegetation may be sparse or absent but in the arctic there may be dense ground flora and even forests of spruce or aspen.

Commonly the pedounit is developed in stony material which leads to the formation of a hamadon at the surface. This horizon rests on a gelon with its marked lenticular structure and brown or yellowish brown colour. The gelon grades with depth into unconsolidated drift or there may be a sharp change to rock. In these situations the amount of organic matter is small due to the sparse or absent vegetation.

During the winter the soil freezes to a metre or more but thaws during the summer. This cycle causes frost heaving of stones to the surface and together with deflation results in the formation of the hamadon. The lenticular structure of the gelon develops when freezing produces ice lenses which isolate the lenticular peds. In thin sections the upper part of the peds have a much finer texture due to the migration of fine particles mainly of silt size through the large pore spaces left after the ice has thawed.

When the material is of fine texture the gelon has a cuboidal structure but these occurrences are rare.

Although the distribution of gelosols is somewhat restricted there are many horizons occurring outside the very cold areas that display the lenticular structure of gelons as well as properties of other horizons and therefore they are regarded as intergrades. The most important include modon-gelon, zolon-gelon and luvon-gelon intergrades. The first intergrade occurs in many situations but is the principal horizon in some soils of the arctic that may have the designation $Lt_2(MoGn)_5Gn_{10}Ar_1Gn_{100}AZ$ and have been called sub-arctic brown soils. Zolon-gelon and luvon-gelon are two of the most distinctive intergrades occurring in many soils of the higher latitudes and have the usual properties of zolons or luvons with the addition of the laminar peds which have finer textures on their upper surface (Fig. 62). In some situations the gelon characteristics penetrate into the middle horizons.

In addition to a hamadon, the surface of the ground may have tundra phenomena such as polygons or terraces.

HALOSOLS

Derivation of Name: From the Greek word *hals* = salt.

Approximate Equivalent Names: Solonchaks; Desert loams; Saline soils; Salorthids.

General characteristics

This class of soils is categorised by the presence of a thick halon with its high content of soluble salts. This horizon may extend from the surface to over 2 m in depth or there may be a thin seron or tannon forming the surface horizon. When the halon occurs at the surface there may be an efflorescence of salts and a marked absence of vegetation.

Halosols form where the soil received seepage or flood water containing a large amount of soluble salts which are left when the moisture evaporates.

Generally these soils have pH values < 8.5 and electrical conductivity values of > 4 mmhos throughout the pedounit.

Halosols are confined largely to arid areas and intergrade to solonchaks in which salt accumulation is associated primarily with a high water table. Halons intergrade to gypsons and to calcons and it is common to find thick calcons or (HlCa) horizons in these soils. The amelioration and utilisation of halosols is similar to that for solonchaks as described on page 230.

KASTANOZEMS

Derivation of Name: From the Russian words *kastano* = chestnut and *zem* = soil.

Approximate Equivalent Names: Chestnut soils; Orthic dark brown soils; Xerolls.

Reference: Ministry of Agriculture of the U.S.S.R. 1964.

General characteristics

These soils have a dark brown, granular upper horizon – the kastanon which grades into a calcon or hardened calcon. Kastanozems occur in semi-arid areas beneath short grass communities.

Morphology

Under natural conditions at the surface there may be a loose leafy litter resting on a dark brown, granular kastanon which is up to 50 cm in thickness. The upper part of the kastanon contains abundant fine roots which decrease sharply in frequency with depth and are almost absent in the lower part. Thin sections show that the kastanon is composed largely of granular faecal material and that there are frequent fragments of organic matter (cf Fig. 83). With depth the structure in the upper part of the kastanon usually changes from granular to fine prismatic before grading into a firm, massive or prismatic calcon with pseudomycelia and concretions of carbonate. In thin section the secondary formation of carbonate is very evident in the form of microcrystalline calcite on pores and tightly packed acicular crystals in the pseudomycelia (cf Fig. 84). Between 1–2 m there is usually marked accumulation of gypsum to form a gypson (see Fig. 113). Krotovinas occur in kastanozems but their frequency is much less than in chernozems.

Analytical data

A high proportion of kastanozems are of uniform texture throughout the pedounit and are dominated by particles $< 50 \ \mu$ indicating the loessic nature of the parent material. There are a few of these soils in which the clay content is at a maximum in the lower part of the kastanon.

The content of organic matter in kastanons varies from 3–6% with a C/N ratio of 8–12 thus indicating the high degree of humification. The pH values are usually above neutrality increasing from about 7·0 in the surface to over 8·0 in the calcon. The cation exchange capacity in the kastanon varies from about 20–30 me% and is usually saturated with basic cations of which calcium is dominant followed in amount by magnesium, potassium and sodium. Carbonates are usually absent or present in only small amounts in the upper 60 cm of the soil but they can be present in large amounts if the content in the parent material is high while in the calcon the content of carbonate is 10–20% greater than in the parent material. Soluble salts are present in very small amounts and never exceed 0·1%.

Genesis

One of the principal processes taking place in these soils is the decomposition and incorporation of organic matter into the mineral soil. This process is fairly complete because the supply of organic matter is small while the activity of the soil fauna is high due to the basic nature of the soil and high summer temperatures. The second major process is leaching of the more soluble constituents, ions such as chloride, sulphate and sodium are either removed completely into underground waters or are deposited at great depths. Calcium is translocated as the bicarbonate which is deposited as calcium carbonate to form a calcon. Where calcium and sulphate ions are translocated they often combine to form calcium sulphate below the layer of calcium carbonate accumulation.

The reason for the increase in clay in the middle of some kastanozems is not clear and may result from at least three different mechanisms. It may be due to the translocation of clay since it appears that clay cutans are present in some cases. Weathering *in situ* in the middle part of the soil might be responsible or destruction of clay at the surface could be a third possibility. Until more evidence is available the increase in clay will be regarded as due in part to weathering *in situ* and the horizon designation (KtCn). In certain places kastanozems have developed in previous soils of early Pleistocene or Tertiary age, the so-called reddish chestnut soils of the south central U.S.A. probably represent polygenetic development extending back to the Sangamonian interglacial or to an earlier period, since the underlying material is sometimes strongly weathered.

Principal variations in the properties of the class

Parent material

Unconsolidated deposits including glacial drift, alluvium, and solifluction deposits, are the usual parent materials. The mineralogy and particle size distribution vary considerably from place to place but they do not seem to play a major role in the formation of these soils. However the composition of the parent material does exercise a little influence on the base status particularly if carbonates are present.

Climate

These soils are found predominantly in the mid-latitude semi-arid conditions.

Vegetation

Grassland of medium height is the principal plant community. In Europe the principal plants are *Stipa*, *Festuca* and *Poa*, spp. In the U.S.A. mixed short and tall grass form the main members of the plant communities.

Topography

These soils are found on sites that vary from flat to moderately undulating and are absent or only poorly developed on steeper slopes. They tend to be absent from depressions which are usually more saline.

Age

Most of these soils started their development during the last 10,000 years, some have been forming throughout the whole of that period while others may be only about half of that age.

Pedounit

There is a fairly wide range of variability within this class as determined by differences in parent material, age and topographic situation. Thus it is possible to find kastanozems that vary from high in clay to those containing a high content of boulders. Similarly the carbonate content of the parent material largely determines the amount that will be deposited in the calcon.

Usually these soils are found on elevated situations but they intergrade to solonetzes or solonchaks in depressions. They also intergrade to chernozems, burozems and clamosols.

Distribution

Kastanozems occupy fairly large areas in the semi-arid regions of the world particularly in the mid-western part of the U.S.A. and Canada, and southern U.S.S.R.

Utilisation

The low rainfall imposes severe restrictions on the utilisation of these soils. Generally they are used for growing grain on an extensive scale but drought is frequent at the drier end of the climatic range where dry farming is fairly common. When irrigation is possible, these soils have proved to be extremely fertile for a variety of crops such as maize and vegetables. Wind erosion is a constant hazard, so that preventative methods should always be in operation.

KRASNOZEMS

[Plate IIIa

Derivation of Name: From the Russian words *krasni* = red and *zem* = soil
Approximate Equivalent Names: Ferrallitic soils, Latosols, some Oxisols
References: D'Hoore 1963; Maignien 1966

General characteristics

The principal horizon of this class is the very thick, red or reddish brown krasnon which represents a very advanced state of hydrolysis and soil formation. The krasnon is composed of kaolinite, goethite, gibbsite and the resistant residue and has been forming from the Tertiary period in a humid tropical environment under a fairly wide range of forest and grassland communities.

Morphology

Under natural conditions the surface of the ground may be bare or there may be a very sparse litter. In some cases there are termitaria particularly when the plant communities are open forest or grassland. The uppermost mineral horizon is a red or greyish red tannon changing with depth into the red or brownish red krasnon, which may be over 3 m thick. With depth the krasnon may change into a number of horizons as discussed below, commonly it grades into a flambon which usually grades into chemically weathered rock with core stones which in turn grades into the solid rock.

When considering this sequence in detail it is better to start with the rock and work upwards through the progressively more weathered material and finally into the krasnon. A development sequence on granitic gneiss will be described because this seems to provide good illustrations of the important features. Within the first metre, disaggregation of the rock takes place along the surfaces of the joint blocks but the greater part of the rock is hard. Within the next stage which occupies 1–2 m or more, disaggregation penetrates deeper into the joint block to form core stones accompanied by marked exfoliation of biotite (see Fig. 34). In hand specimens many of the felspars appear fresh but in thin section they are in a fairly advanced stage of weathering being replaced by secondary products, mainly kaolinite, gibbsite and a little mica. Generally the degree of hydrolysis increases upwards with some decomposition products occurring as cutans around grains and in pores, and the whole mass is stained yellow by hydrated goethite. This stage grades into the flambon with its characteristic yellow and red mottled pattern which is usually reticulate or laminar. In the field some of the rock structure is evident but this is more obvious in thin section particularly when the parent material is a coarse grained, acid, metamorphic rock such as a granitic gneiss with a high content of quartz and a marked crystal orientation (Fig. 89). In thin section the flambon exhibits some of the most fascinating and complex phenomena to be found in soils since it represents an advanced but not complete stage of hydrolysis. The less resistant minerals have either been completely decomposed or are in an advanced stage of decomposition. In some cases the yellow material outlining the cleavages of the felspars becomes impregnated with a very dark brown substance which remains as a cellular outline when the remainder of the felspars has weathered away. This in turn becomes encapsulated to form a concretion (Fig. 90). The formation of macrocrystalline kaolinite (Fig. 35) seems to be at its maximum in this horizon, occurring as pseudomorphs after felspar but usually it has grown in spaces left after the decomposition of more than one mineral. Perhaps the most important and dominant features found in some flambons are the thick and

continuous cutans that cover nearly every surface and partly fill all the pore space (Fig. 91). On sedimentary rocks with a lower content of hydrolysable silicates the frequency of cutans is lower.

The flambon grades gradually upwards into the krasnon through a zone that appears in the field to be of uniform red or reddish brown colour but in thin section it is seen to be composed of some uniform areas but a considerable proportion is almost identical to the flambon (Fig. 92).

The krasnon has a clay loam to clay texture and massive to incomplete subangular blocky structure. In thin section it is usually composed of rare to occasional sand grains set in a red or reddish brown matrix. In sections of normal thickness the matrix appears dense and isotropic but in very thin sections $(5-10 \mu)$ it is seen to be dominantly birefringent and composed of

5mm

FIG. 89. Partial retention of the rock structure in a flambon as shown by the rectangular grains of quartz.

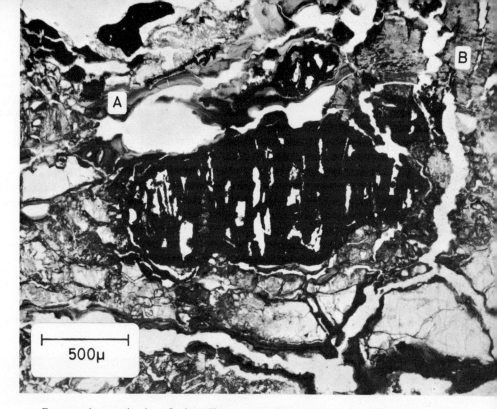

FIG. 90. A concretion in a flambon. The concretion has a characteristic cellular structure which apparently is determined by the original cleavages of a felspar crystal. The spaces between the cell walls are empty. Also present are A, cutans and B, macrocrystalline kaolinite.

FIG. 91. Clay cutans between grains of quartz in a flambon. The grains of quartz partly retain the original rock structure.

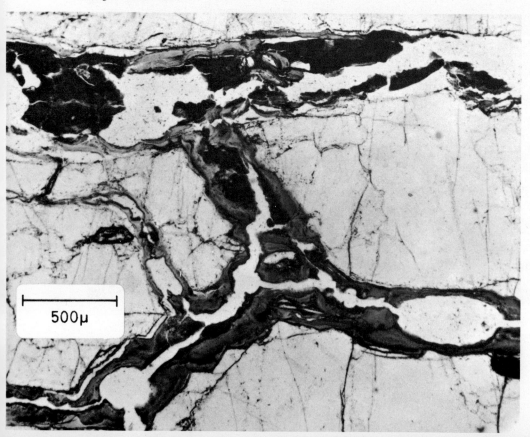

abundant small domains arranged randomly or in thin zones. In some krasnons cutans are rare, occurring on the surfaces of peds or as concentric deposits in pores.

At the surface is the reddish brown tannon with granular or subangular blocky structure but often has a complex structure of subangular blocky and granular, the latter component being due to earthworms or termites. In thin section it is isotropic or there may be rare to frequent small random domains.

Analytical data – Fig. 93

Krasnons have a clayey texture and in a number of cases they have well developed aggregates which may inhibit dispersion for mechanical analysis therefore it is not possible to determine accurately their true clay content. When the aggregates are well developed, the term pseudosand is applied to this type of structure. It is commonly assumed that iron oxides are the cementing agents but DesPande *et al.* (1964) have shown that aluminium oxide is responsible. Where dispersion is achieved it is the fine clay ($<\cdot 5\ \mu$) that is the dominant fraction $< 2\ \mu$. This shows a gradual increase from the surface downwards to a maximum in the middle of the krasnon followed by a considerable decrease to very low percentages in the chemically weathered rock.

In most cases kaolinite accounts for more than 80% of the clay fraction, the remainder being goethite and gibbsite but the latter may be absent. Mica sometimes occurs in small amounts.

A characteristic feature of the krasnon is the very low silt: clay ratio, due to the profound weathering that has formed these soils. This ratio varies from about 0·15 for soils developed on igneous rocks to 0·2 for soils on sedimentary rocks; the latter higher ratio being due to the higher initial content of resistant minerals in sedimentary rocks.

FIG. 92.   Decomposition of cutans in a flambon to form a krasnon.

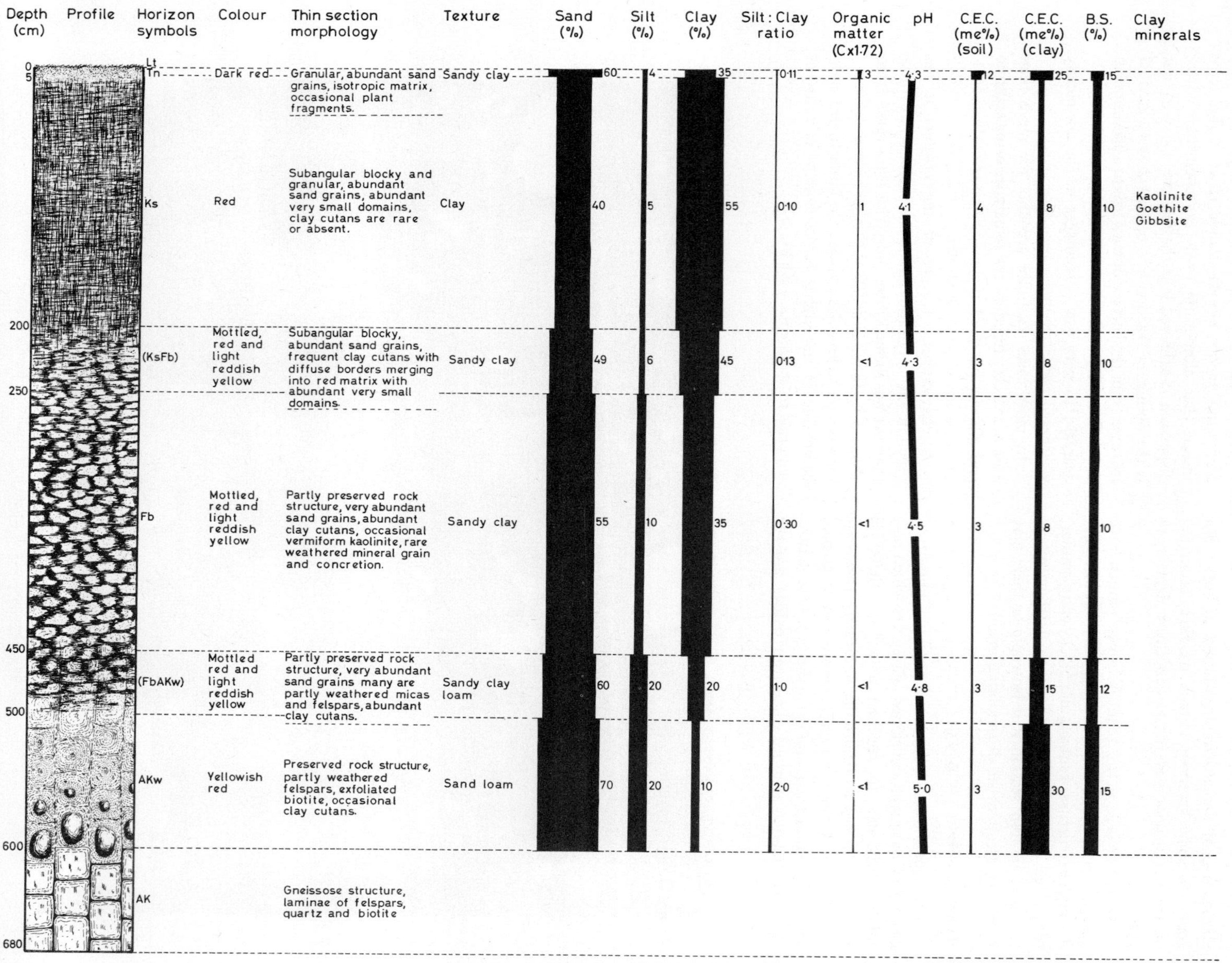

FIG. 93. Generalised data for krasnozems.

The acidity is remarkably uniform with pH values about 5·5 at the surface in the tannon, decreasing to about 4·5 in the middle of the krasnon then increasing to 5·0 in the flambon.

Because kaolinite is the principal clay mineral the cation exchange capacity is very low with values of <15 me% per 100 gm of clay, which give total soil values of about 12 me%. The amount of exchangeable cations and the base cation saturation are also low but the latter can be high in some situations. Calcium is the principal exchangeable cation with smaller amounts of magnesium, potassium and sodium.

The surface horizon seldom contains more than about 5% organic matter, usually there is only 2–3% decreasing to <1% with depth. The C/N ratio is 8–12 which indicates an advanced stage of humification.

Partial ultimate analysis reveals that from the flambon upwards the main mineral constituents are silicon, aluminium, and iron with lesser amounts of titanium and practically no basic cations thus the reserve of basic cations is extremely low, see Table 12.

Genesis

These soils are formed by the progressive hydrolysis and complete transformation of the rock into clay minerals, oxides, concentrations of the resistant residue and the loss in the drainage of much material, particularly basic cations and silica. The particular course of development seems to follow a number of different pathways as determined by the nature of the parent material and drainage characteristics. On very basic parent material the stages may be similar to those given on page 59 but in the example given above the stages of decomposition are more gradual and decomposition products are dominated by goethite and kaolinite. The latter may form pseudomorphs but more commonly appears to be redistributed within the weathering zone to form cutans which become progressively larger and more numerous as decomposition proceeds, attaining their maximum development in the flambon. A satisfactory explanation has not been offered for the formation of the mottling in the flambon, but presumably it is due to the segregation of compounds of iron to form the red areas and aluminium in the pale coloured areas. The change in colour from yellowish brown in the initial stages of weathering to red or reddish brown in the flambon suggests that the latter contains a less hydrated form of goethite. The development of the krasnon seems to result from the homogenisation of the flambon by a mechanism that is not understood. Termites and other organisms may be responsible but expansion and contraction may be the controlling process mechanism since the cutans of the flambon seem to merge and disappear into the matrix to form the krasnon (Fig. 92).

Principal variations in the properties of the class

Parent material

All types of consolidated rocks are the predominant parent materials, but these soils also form on unconsolidated deposits such as Tertiary sediments and old volcanic ash.

Climate

Characteristically these are soils of the rainy tropics, wet and dry tropics, humid subtropics and monsoon tropics. The optimum conditions for their development appear to be in areas with a mean annual temperature of > 25°C and mean annual precipitation of over 1000–1200 mm. Where precipitation is much higher or temperatures slightly lower they tend to be confined to the more basic parent materials such as basalt. Pleistocene and Holocene climatic changes have caused some of these soils to occur outside their normal climatic range; particularly good examples of this are seen in the area bordering the southern Sahara desert and in western Australia.

Vegetation

Rain forest and semi-deciduous tropical forests are the principal plant communities. When these soils occur outside their normal climatic range they may carry communities such as thorn woodland, deciduous tropical forests and savanna communities.

Topography

Krasnozems are soils of the lowlands, they are not found above about 1200–1500 m where they give way to soils of cooler, wetter conditions. They occur on sites that vary from flat to steeply sloping but they tend to be more common on moderate slopes.

Age

The majority of these soils are found in materials that are at least of mid-Tertiary age and therefore they are very old. Some occur on late Tertiary or Pleistocene materials but these usually have a low content of weatherable minerals.

Pedounit

The variations in the properties of this class are

FIG. 94.   A block of veson, (top) the outer weathered surface shows the characteristic pitted structure; (bottom) a cross section through the same block showing the distinctive vesicular structure.

wide and as a consequence offer a number of problems with regard to their designation and classification. Probably the most important variable is the thickness of the krasnon which may range from < 1 m to > 10 m. This range can be due to time, the thicker ones being older but the thin ones are often due to erosion during the pluvial periods. The formulae for a few of the common subclasses and groups are given below.

1. $[KsVsFb]$ $Tn_5Ks_{300}Vs_{200}Fb_{200}AKw_{300}AK$
2. $[KsPsAKw]$ $Tn_{10}Ks_{200}Ps_{100}AKw_{100}AK$
3. $[(KsCm)KsFb]$ $Tn_5(KsCm)_{25}Ks_{100}Fb_{400}AKw_{200}AK$
4. $[KsLF]$ $Tn_5Ks_{100}LF$
5. $[Zv(KsAr)Ks]$ $Lt_2Mo_5Zv_{15}(KsAr)_{30}Ks_{100}AF2w_{100}AF$

It is common to find vesons (Fig. 94) or pessons in krasnozems; these two horizons may occur at many positions in the pedounit but usually at the upper part of the flambon when it is present, or they may occur between the krasnon and the weathered rock; these variants are given in the first two examples.

The chemical composition of the pesson or veson can vary widely in the proportions of goethite and gibbsite. The latter is usually abundant to dominant in upland freely draining situations but goethite is the major constituent in less well drained sites.

The thin section morphology of pessons is somewhat variable; the example shown in Fig. 95 is composed of concretions embedded in material containing abundant thin veins of kaolinite which form the distinctive threadlike ramifications.

In many places, particularly in west Africa there has been differential erosion of the soils so that gravel and concretions from the krasnon have accumulated in the upper part of the soil. Sometimes the accumulation has been sufficient to form a cumulon (Fig. 96) but often there is the intergrade (KsCm). This is given in the third example. A krasnozem formed on limestone is given in the fourth example which shows

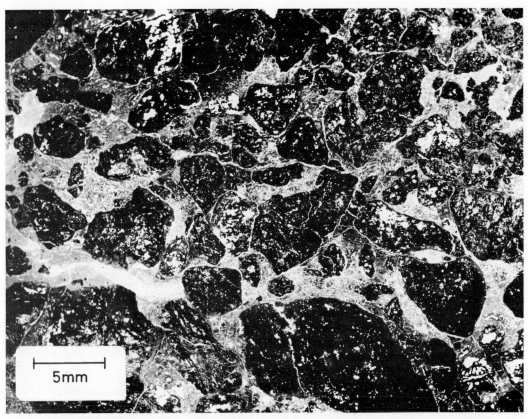

5mm

FIG. 95. Thin section of a hardened pesson composed of concretions set in a matrix containing thread like growths of finely crystalline kaolinite.

that the krasnon is relatively thin. This is the normal feature because the residue left after the weathering of the limestone is usually small.

Owing to a change in climate or some other factor, clay translocation may have been superimposed upon a krasnon; then the krasnon has the additional characteristic of frequent cutans and is a krasnon-argillon intergrade such as given in the fifth example.

The colour of krasnons vary from hues of 2·5YR to 10R as determined in large measure by the nature of the parent material, the redder colours occurring on material richer in ferromagnesian minerals.

Distribution

These soils are of widespread occurrence in tropical and subtropical areas. They occupy large areas in the north of Australia and throughout much of India. In Africa they occupy large areas on both sides of the equator and have a similar distribution in South America particularly in Brazil. They are found also in a number of subtropical countries (Fig. 97).

Utilisation

The low nutrient status and low content of organic matter give these soils a very low fertility but they often support high forest. The relationship between krasnozems and their natural forest vegetation is a good example of the delicate balance of nature in which nutrients are constantly recycled to maintain the natural forest community. When the forest is removed and agriculture is practised, fertility is quickly exhausted and crop failure is the normal result. This accounts for the practice of shifting cultivation so highly developed in parts of Africa. This system of land use involves felling the forest, cultivating the ground for a few years during which the nutrients are exhausted and then moving on to a fresh site leaving

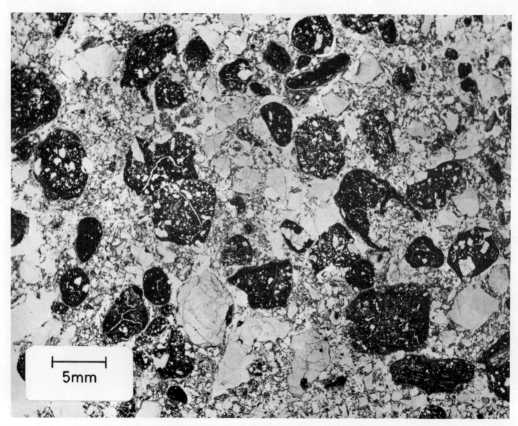

FIG. 96.   Thin section of a cumulon composed of residual concretions and quartz gravel.

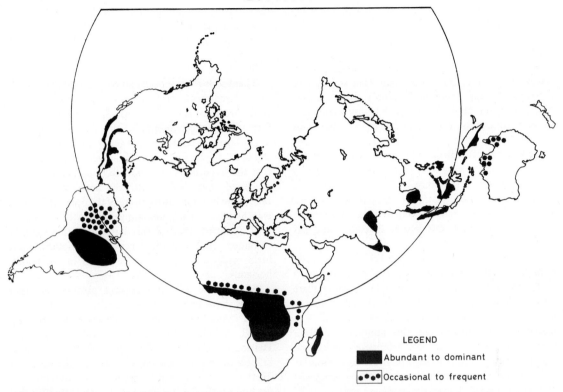

FIG. 97.   Distribution of krasnozems.

the area to develop a new forest community and re-build the fertility for another period of cultivation. In recent years, modern practices including liming and the applications of fertilisers have made possible a more stable system of agriculture but as yet a complete-ly satisfactory system of land use has not been de-veloped for many of these soils. In many cases the greatest success has been achieved with tree crops such as, cocoa, oil palm, tea, coffee and rubber. Very hardy plantation crops such as sugar cane have also been extremely successful, however these soils seldom sustain a high standard of living except when their utilisation is on an extensive scale because the profit per hectare from carbohydrate crops is small.

LUVOSOLS [Plate IIIb

Derivation of Name: From the Latin word *luo* = to wash

Approximate Equivalent Names: Red yellow podzolic soils; some Ultisols; some Alfisols.

References: Racz 1964; Racz et al. 1967

General characteristics

This class is established on the basis of unique associations of horizons rather than on the presence of a specific horizon. The principal upper horizon is a strongly leached modon or luvon which overlies a variety of horizons including krasnons, zheltons, rossons, flavons and pelons. The change from the upper to the middle horizon is usually fairly sharp and accompanied by a marked increase in the amount of clay but there is little evidence to show that it is due to translocation.

Morphology

There may be a loose leafy litter at the surface but this is usually very thin so that the bare soil may be exposed. The upper mineral horizon is usually a dark greyish brown granular modon with loamy texture and varies in thickness from < 10 cm to > 30 cm. This horizon grades sharply into the grey or brownish grey luvon which is often firm and massive except when the content of clay is very low, then it is loose and may have a single grain structure.

In thin sections the matrix of the modon is isotropic and contains many partly decomposed plant fragments. The luvon is similar but lighter in colour and has fewer plant fragments.

The lower horizon is very variable but is usually strongly weathered material *in situ* or transported over a short distance. In the example shown in Plate IIIb the lower horizon is a rosson that is *in situ* or may have been transported a short distance. The matrix of this horizon contains abundant medium domains some in zones having oblique orientation. There are abundant papules which are small circular birefringent areas of redder colour, being composed of abundant small domains. Cutans are rare.

Analytical data

The smallest amount of clay occurs in the modon followed by a slight increase in the luvon and a further increase in the rosson which remains fairly uniform with depth.

In the modon there may be up to 10% organic matter with a C/N ratio of 15, therefore it is not in a very advanced stage of humification. The cation exchange capacity has two maxima, one of about 35 me% occurs in the modon due to the presence of organic matter while the higher of about 45 me% occurs in the rosson due to the greater amount of clay. The amount of exchangeable cations is low throughout the soil increasing from about 15% base saturation in the modon to a maximum of about 35–50% in the rosson, with calcium as the dominant cation. The pH values are about 5.5 and are relatively uniform throughout the soil, but there is usually a slight increase with depth. The silica:sesquioxide ratios for the clay fraction are uniform and show clearly that there has been no differential leaching of any of these three constituents. However the same ratios for the total soil decreases with depth, therefore there has been a considerable loss of iron and aluminium from the primary minerals in the modon and the luvon.

Genesis

These soils form most commonly in acid or extremely acid sediments of Pleistocene age or they evolve from previous soils. The principal process in their formation is the reduction of the clay content in the upper horizons but there is no general agreement about the precise processes involved, however it seems likely that the hypothesis of Simonson (1949) is correct. He suggested that clay is destroyed and removed from the system but some authors are of the opinion that clay translocation is the dominant process (McCaleb 1959). It is doubtful whether this latter hypothesis can be sustained since the thin section morphology shows little evidence for clay accumulation in the middle horizons.

When these soils develop in a previous soil they usually show an interesting sequence of evolutionary stages that often can be traced back to the Tertiary period. A high proportion lay outside the limits of Pleistocene glaciation but occur within the areas influenced by periglacial conditions therefore the Tertiary soil was in part eroded by solifluction. On flat and gently sloping sites the effect was small so that a considerable part of the Tertiary soil remained. On sloping sites most of the old soils were removed and deposited in the valleys below, but where the movement was slow the material deposited in the valleys differed little from the original soil as in the example given above. Also, the amount of material remaining in place gets progressively thicker away from the influence of periglacial activity. This is very well demonstrated by a transect through France or through the eastern U.S.A. from New Jersey to West Virginia. Luvosols also occur in areas not strongly influenced by glacial or periglacial processes but some of these areas were also subjected to climatic change. This is seen in Australia where conditions for these soils are cooler at present than when original soil formation took place. This new set of conditions seems to induce greater acidity and more vigorous hydrolysis causing decomposition of the clay minerals and oxides in the modon and luvon and their removal from the system. This accounts for the sharp change in texture between the upper and middle horizons and the uniform texture through much of the remaining part of the soil except when there is a change to the lower horizon.

Principal variations in the properties of the class

Parent material

Luvosols usually show some measure of polygenesis therefore the relatively unaltered underlying material may have little direct influence upon their formation. They are often found on fine or medium textured drift deposits derived from a previous soil which was transported only a short distance as mentioned above or they may be developing directly in a previous soil; this gives rise to the wide variety of middle horizons.

Climate

The climate ranges from humid continental warm summer to rainy tropical; in all cases precipitation is heavy and fairly uniformly distributed through the year so that water is constantly moving through these soils. There is a tendency for the texture to become coarser with increasing temperature.

Vegetation

The natural vegetation varies with the climate; under humid continental warm summer conditions there is usually a poor deciduous forest. In contrast coniferous forests dominated by pine, form one of the principal communities in the humid subtropics of south-eastern U.S.A. and in the tropics there are usually rain forest communities. When the natural vegetation is removed regeneration is difficult because of their low fertility; thus a plagioclimax often results.

Topography

These soils develop on stable sites that range from flat to steeply sloping, providing that water can percolate freely, but their most common occurrence is on flat or undulating landscape.

Age

The development of luvosols has extended over most

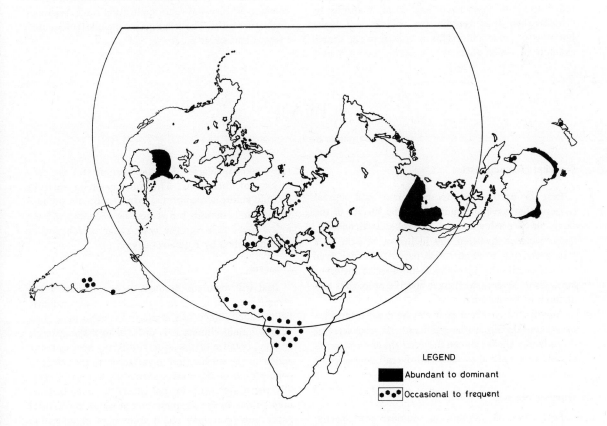

LEGEND

■ Abundant to dominant

•●•• Occasional to frequent

FIG. 98.   Distribution of luvosols.

of the Holocene period but in some areas these soils seem to be very much older, for example, in part of the southern U.S.A. they can be traced from the present surface to beneath late Pleistocene loess.

Pedounit

As a result of polygenetic evolution the variability in this class is very wide particularly the nature of the middle horizons. Set out below are a few of the principal subclasses and groups.

1. $[LvZh]$ $Lt_2Mo_5Lv_{20}Zh$
2. $[LvKs]$ $Lt_2Lv_{40}Ks$
3. $[MoKs]$ $Lt_5Mo_{20}Ks$
4. $[LvKsFb]$ $Lt_2Lv_{25}Ks_{150}Fb$

As seen from the variation given above the common feature linking the various groups is the strongly leached surface horizons which are usually luvons or modons. The horizons in the middle and lower positions are very varied depending upon the evolutionary history. Where they have developed from old soils that may have lost some material by erosion then they contain the middle and lower horizons of these soils. This is shown in the fourth example in which there is a krasnon followed by a flambon and a few groups contain vesons. Also, there appears to be a continuous gradation between luvosols and argillosols.

Distribution

The principal areas where these soils occur are in south-eastern U.S.A., some of the Mediterranean countries, south-eastern Australia and extensively throughout the wet and dry tropics, rainy and monsoon tropics (Fig. 98).

Utilisation

The thick, acid and strongly leached upper horizons create many problems for utilisation. After the removal of the natural vegetation the soils must be limed heavily, followed by the addition of large amounts of fertilizers particularly nitrogen and phosphorus. Even so their productivity is low and small areas are unable to provide a livelihood, therefore they are often devoted to plantation agriculture.

Deficiencies in a number of micro-elements are a common feature and their application can bring about dramatic results. The crops grown include fruit, maize, tobacco and sweet potatoes.

PEAT

Reference: Farnham and Finney 1965; Fraser 1933

General characteristics

Peat is partly decomposed, plant and animal remains that have accumulated at the surface of the earth in a predominantly anaerobic environment. Some workers question the inclusion of peat as a soil, preferring to regard it as parent material, even so it is possible to recognise a number of different horizons or layers which have specific properties and methods of formation.

Any discussion about peat can be divided into two parts, considering on the one hand, the character of the individual layers and on the other the environmental condition, rate of growth and overall character of the deposit.

Horizons in peat

Peat deposits contain a number of specific horizons as determined by the original character of the material, degree of decomposition and the conditions of formation. The most common horizons include amorphons, fibrons, gyttjons, limons, pseudofibrons and siderons but it may be necessary to have a few more to cover the wide range of subaqueous soils described by Kubiëna (1953).

Genesis

Peat will form continuously on those surfaces that remain wet and to which organic matter is constantly being added. Since such surfaces occur in most areas with a humid climate it is difficult to make generalisations because of the widely differing species being added to the surface and to variations in the chemical composition of the water causing the wetness. Generally there are two principal causes of anaerobism. It may be due to the accumulation of water in a natural depression or it may result from high precipitation and high humidity. The former conditions lead to the

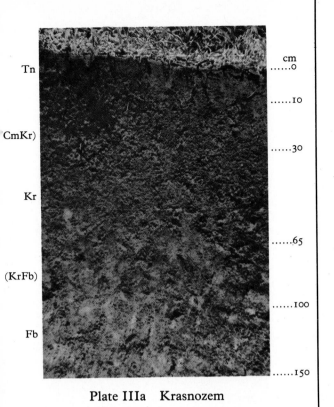

Plate IIIa   Krasnozem

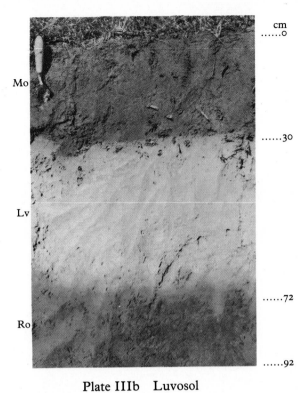

Plate IIIb   Luvosol

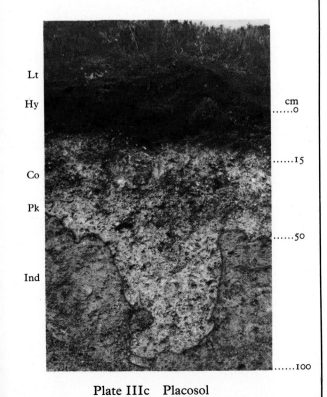

Plate IIIc   Placosol

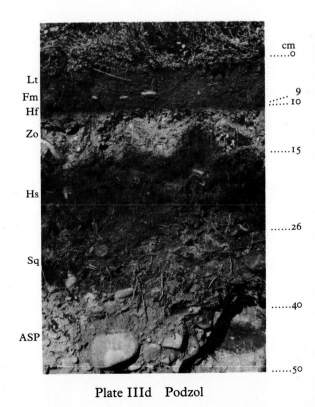

Plate IIId   Podzol

formation of *basin peat* and the latter to *blanket peat*. Sometimes an impermeable layer such as a placon or cryon may cause water to accumulate at the surface causing peat to form, such formations are included with the blanket peats.

Basin Peat

This form of peat commonly occurs on a very flat, low-lying landscape, on valley floors, between moraines and in lagoons often situated in deltas. These latter two may be closed depressions or they may be fed and drained by very slow flowing streams as are the former two. Most basin peats started to accumulate at some period during the last 10,000 years and usually have a distinctive layered structure produced by the successive plant communities that have colonised the area. There may have been a straight accumulation under uniform climatic conditions resulting in little variation throughout the deposit. On the other hand the various layers may be composed of contrasting species caused by climatic changes. An example of this in Europe is the characteristic *grenze* layer containing tree species, and resulting from the warmer and drier sub-boreal period which extended from 5,000 to 2,500 years BP. Subsequently conditions became cooler which increased the wetness of these peats resulting in their recolonization by wet habitat species. As the thickness of basin peat increases it steadily rises above the level of the ground water but development continues to take place because of very slow drainage and the presence of plants such as sphagnum that have a high water retaining capacity. Eventually the surface rises to form a domed outline and it is then dry enough to support certain tree species such as the birch forests of the U.S.S.R. This is the hochmoor or highmoor stage as compared with the low moor stage when the surface is flat. Examples of basin peat include the Everglades of Florida, the peat bogs of northern Europe and the peat deposits of the many coastal situations in the humid tropics such as in Guyana and northern Sarawak. Fig. 99 illustrates the developmental stages in the formation of a peat deposit.

Basin peats sometimes exceed 10 m in thickness and vary widely in their mineral content. The lower part of these deposits usually contains a fairly high content of mineral material being either an amorphon or gyttjon and above are the various layers of material, some more decomposed than others. Gradually the fibrous peat formed by recent additions of material—

the fibrons change into pseudofibrons which in turn change into amorphons. Thus the general trend is for basin peat to become more decomposed with depth i.e. with age. Amorphons may not be the final stage of transformation for under certain circumstances the organic material is slowly and steadily replaced by siderite to form a sideron. In a hand specimen amorphons and siderons are almost indistinguishable but in thin sections the siderons seem to be composed of a few irregular concretions and large masses of needle crystals of siderite.

The content of cations and pH values of peat vary widely as determined by the composition of the water causing the anaerobism. When the water flows from areas of acid crystalline rocks the peat is acid but if it flows from an area containing limestone or chalk then the pH of the peat may be around neutrality and calcareous skeletons of aquatic organisms may accumulate to form limons. Occasionally some peat deposits show interstratified layers of limons and amorphons for under alkaline condition amorphons appear to form more readily. The fens of eastern England are among the better known examples of a base rich peat.

Blanket peat

This form of peat occurs in areas where precipitation is high but more particularly where the humidity is high producing a constantly moist or wet surface, thus prohibiting decomposition of plant remains by aerobic soil flora and fauna. Such conditions are common throughout a considerable part of a wide belt that runs through northern Europe and central Canada where the muskeg of Canada is probably the largest single area in the world. Similar conditions occur at the higher elevations in the humid tropics.

Whereas the mineral content of basin peat is very variable, depending upon the nature of the water entering the deposit, blanket peat is invariably acid because the water is derived directly from precipitation. There is also a greater tendency for this form of peat to be more decomposed and to contain pseudofibrons and amorphons because the peat may dry to a limited extent during a very dry year. In addition there is a constant movement through the peat of moisture which will contain some dissolved oxygen thereby allowing a limited amount of decomposition to take place.

These formations are usually not more than 3 m thick and rest upon rock but more usually upon an old soil especially a placosol or supragleysol. In large parts of north western Europe the major phase of blanket peat formation started about 7,500 years BP

so that it may rest on soils that developed during the early post glacial period.

Since climatic conditions are the principal reasons

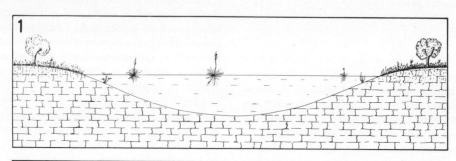

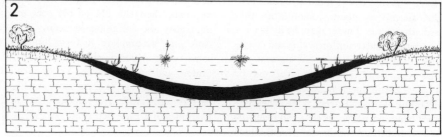

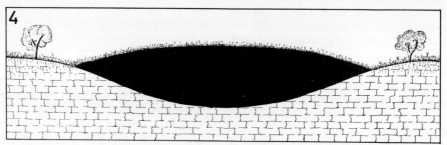

FIG. 99.   Stages in the formation of peat.

1. A small pond or lake which could be situated between two moraines.
2. At the bottom of the pond there is a thin accumulation of organic matter from the plants growing in the pond and the surrounding soil.
3. Considerable increase in the thickness of the organic matter and spread of vegetation on to its surface.
4. Continued thickening of the peat to develop the characteristic domed form of the final stage.

for the formation of blanket peat, it is found in a number of topographic situations including slopes up to 45°. In such situations it can be eroded easily and tends to flow *en masse* down the slope particularly in the cooler areas where it may form solifluction lobes.

Botanical composition of peat

The study of the composition of the plant remains in peats reveals the character of the communities that once grew in the area. The principal studies are conducted on the amount and distribution of pollen in the individual layers. This provides a fuller set of data because the pollen grains have an outer coating of cutin which protects them from decomposition except under neutral or alkaline conditions so that they remain preserved in the deposit very often long after other plant tissue has been altered beyond recognition. The pollen present in the peat is derived from the plants at a distance of over one kilometre as well as those in the immediate neighbourhood thus the information which is obtained refers to the area in general.

In contrast to mineral soils it is possible to determine the age of peat deposits fairly accurately by using the technique of radio carbon dating. Thus by combining the C^{14} date and pollen data for a peat deposit it is possible to reconstruct the vegetative history of the area and indirectly the soil history.

Distribution – Fig. 100

The total area of peat in the world is about 150,000,000 hectares of which the greatest occurrences are in the cooler humid parts of the world. Of the total area, countries having more than 5,000,000 hectares each, are the British Isles, Canada, Finland, Germany, U.S.A. and the U.S.S.R.

Utilisation

Peat has been the traditional fuel in many parts of the world but it is really too valuable for horticultural purposes to be burned. However its use as a fuel is likely to continue in certain areas, indeed mechanised peat harvesting followed by its use to generate steam for large turbines is practised in parts of the U.S.S.R., Germany and Ireland.

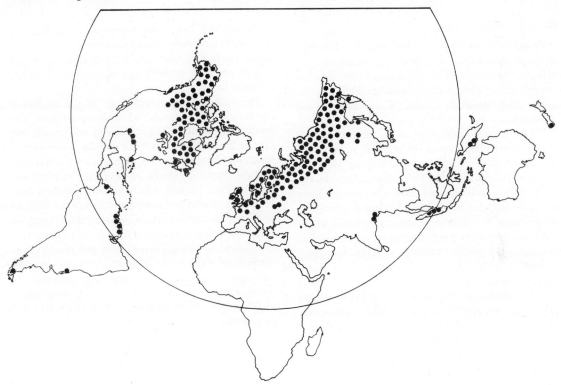

FIG. 100.   Areas of major peat deposits.

The use of peat for agriculture offers a number of problems but when these are overcome consistently high crop yields can be obtained. If a forest is present the trees and tree stumps must be removed followed by drainage which is probably the most critical operation. Over-draining peat can produce excess drying resulting in the formation of hard granular structural units which do not rehydrate when wet. This is particularly characteristic of those horizons containing much silt or appreciable amounts of iron. The result is rapid percolation of water and drought for the plants. In most cases the application of lime and fertilisers is essential and it is usually necessary to add microelements mainly, copper, cobalt, magnesium and boron. After these operations arable cultivation or dairying can proceed normally, the only limitation being those common to all agricultural practices, such as climate.

A common and sometimes serious problem following prolonged cultivation is subsidence which may lower the area below the possibility of drainage by gravity.

The growth of trees on peat is probably a more tenuous operation particularly in the case of blanket peat which is often in highly exposed situations making wind-throw an additional potentially serious hazard. However many areas have been drained, ploughed, fertilised and planted to trees in recent years with encouraging results. Peat is widely used as a potting substance in horticultural operations and it finds a few applications in the chemical industry.

PELOSOLS

[Fig. 60

Derivation of Name: From the Greek word *pelos* = clay or mud.
Approximate Equivalent Name: Heavy clay soils.
Reference: Mückenhausen 1962

General characteristics

These soils are formed in fine textured sediments often containing a high proportion of primary minerals with mica and kaolinite as the principal clay minerals. Generally they have poor horizon differentiation due to the relatively small amount of translocation and redistribution that has taken place.

Under natural conditions at the surface there is usually a granular mullon up to about 25 cm thick, this grades fairly sharply into the angular blocky or prismatic pelon which may have a few slickensides and has a similar texture to the mullon. The pelon grades in turn into the relatively unaltered material which may be stratified. The colour of these soils varies widely and is determined to a very large extent by the original colour of the parent material, usually the upper horizon is darker than the pelon because it contains organic matter.

When these soils occur in a cool humid environment they are always present on freely draining sites otherwise they are quickly transformed into supragleysols because of their low permeability (see page 261).

On slightly calcareous parent material leaching in a humid environment may have removed some or all of the calcium carbonate to a depth of three to four metres where it is deposited as crystals of calcite or forms concretions.

The thin section morphology of the mullon is characteristically granular with isotropic matrix. The pelon is prismatic or angular blocky and usually has rare to frequent small ovoid or circular pores some of which may have cutans. The matrix of the pelon is birefringent being composed of abundant medium and small random domains, which may occur singly or form zones. When sand grains are present they are outlined by thin diffusion clay cutans resulting from expansion and contraction. These soils intergrade to altosols as the content of clay decreases, they also intergrade to vertisols as the content of montmorillonite and base saturation increase. In a semi-arid environment intergrades to halosols and solonetzes are common.

Pelosols occur in many parts of the world but they seldom occupy large areas.

PLACOSOLS

[Plate IIIc

Derivation of Name: From the Greek word *plax* = plate
Approximate Equivalent Names: Peaty gley podzols; Thin iron pan soils; Placaquods;
 Placorthods.
References: Muir 1934; Glentworth and Muir 1963

General characteristics

A thin hard continuous iron pan or placon is the principal feature of this class. It is relatively impermeable causing moisture to accumulate at the surface and in some instances peat may form as a consequence. These soils develop mainly in marine areas with high atmospheric humidity and carry a heathy vegetation.

Morphology

Commonly there is a thin litter at the surface, changing with depth into a black, plastic hydromoron up to about 30 cm thick. In thin section it contains a few fragments of decomposing organic matter but the greater part is composed of closely packed granular aggregates of organic matter. These are interpreted as faecal pellets but they could be due to the flocculation of organic matter. The hydromoron usually changes sharply into the greyish brown or olive grey, loamy candon which varies from 10–50 cm in thickness and may have fine ochreous mottles visible in the field but they are more apparent in thin section which also shows the presence of frequent, weathered sand grains, very frequently randomly oriented domains and rare reddish brown, isotropic cutans.

Next comes the thin placon which is dark brown at the top becoming yellowish brown below, and in some cases there is a thin root mat on its upper surface. This horizon usually follows an irregular path through soil and of special interest is its continuity through stones and boulders (Fig. 101). Thin sections show that the placon is composed of thin, alternating bands of reddish brown and dark reddish brown material which is dense and isotropic. Some soils have multiple placons there being as many as three or four within about 10 cm or they may be 10–20 cm apart. The examination of a number of thin sections of placons has revealed that the frequency of sand grains is less as compared with the horizons above and below suggesting that there has been some form of *in situ* growth, displacing the sand grains.

The horizons occurring beneath the placon are very varied and appear to belong to a preceding phase of pedogenesis. There is often a sesquon or there may be an ison, the latter is very common in the British Isles and eastern Canada where these soils are extensive.

Analytical data – Fig. 102

The most important physical property of placosols is the impermeability of the placon which inhibits the penetration of roots and moisture. This severely restricts root development and their potential for cultivation.

These soils usually have medium to coarse texture which is fairly uniform through the pedounit and usually the content of stones varies from frequent to abundant. Placosols are very acid particularly in the surface organic horizons where the pH values may be as low as 3·5 but they increase with depth to about 5·0. The exchangeable cations, dominated by calcium and the percentage base saturation, are very low, further indicating low potential fertility. The organic matter is high in the surface with a wide C/N ratio indicating a low degree of humification.

Partial ultimate analysis of the total soil as well as the clay fraction show that the placon has the greatest amount of iron whereas aluminium has accumulated further down in the ison. These trends are shown better by the clay analyses, the placon may contain 10% more Fe_2O_3 than the ison, whereas the ison may contain 15% more Al_2O_3 than the placon.

Genesis

These soils display interesting evolutionary sequences in which the formation of the placon is probably the most significant development. Under marine conditions where these soils are most widespread the evolutionary sequence is often from cryosols of the Pleistocene period to podzols of the early part of the Holocene and then to placosols of the middle to late Holocene period, but in some cases the evolutionary sequence can be traced back to the Tertiary period as discussed on page 272 *et seq.* In all cases the placon has formed in a fully developed soil which is usually a podzol.

The precise reason for the formation of the placon is still obscure but it probably formed at the interface between the moister upper part of the soil and the relatively dry lower part, the upper part being kept wet by the higher rainfall and humidity of the Atlantic and subsequent periods. Such interfaces can probably be produced by a simple moisture gradient or occur at the upper surface of the ison which may be the reason in this case. Although the precipitation of iron and aluminium hydroxides from solution might account for the formation of a placon in the porous soil this mechanism cannot account for its occurrence within stones and boulders. Therefore it is necessary to speculate about bacteria as possible agents. If these organisms are responsible, the placon can be regarded as a growth phenomenon which might account for the reduced frequency of sand grains within the placon itself as compared with the material on either side. The formation of the placon has prevented vertical drainage thus causing anaerobism leading to the formation of a candon and hydromoron. The anaerobism has caused the reduction of the compounds of iron in the upper mineral horizons to give olive and grey colours as well as the formation of mottles. Where the formation of the placon is a recent phenomenon, traces of the previous horizons can be seen in the candon above, but where anaerobism has extended over a long period all evidence of previous horizons appear to be destroyed; even sesquons have had their iron reduced to the ferrous state. The occurrence of weathered sand grains, formation of cutans, and many domains seems to be due to wet conditions because similar features are seen in gleysons. Since the formation of the placon very little alteration has taken place in the underlying horizon.

25mm

FIG. 101.   A placon within a granite boulder.

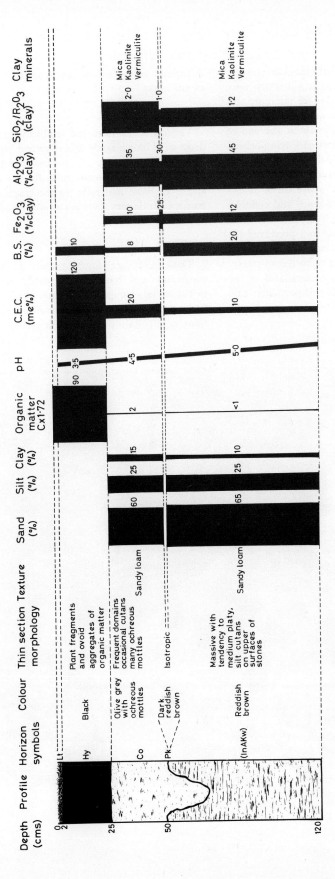

Fig. 102. Generalised data for placosols.

Initially the ison was a cryon with the characteristic laminae of ice and lenses of frozen soil which were compacted by the growth of the ice laminae; also present were sheaths of ice around stones and boulders. With the amelioration of the climate at the end of the Pleistocene period the ice melted leaving behind compacted mineral material. As the ice disappeared from around the stones they would gradually settle thus creating a large pore space above them. Subsequently silt and clay filled this space and colloidal material particularly aluminium hydroxide washed in from above has partly cemented and preserved the lenticular structure (Fig. 103). The result is material with a high bulk density and firm to hard consistence. In some cases fine material has accumulated in the spaces occupied by the ice as in this particular soil.

Principal variations in the properties of the class

Parent material

Various coarse textured sediments such as glacial drift, wind blown sands, alluvium and highly siliceous Tertiary deposits are the normal parent materials. However these soils occasionally form in intermediate and even basic materials. In the tropics they are a recent evolutionary stage in a number of soils including zheltozems.

Climate

These soils are found mainly under maritime conditions and have their greatest known extent in the areas of a marine climate. However they also occur under oceanic tropical conditions at high elevations as in certain parts of south eastern Asia.

2mm

FIG. 103.   Silt cutan on the top of a stone in an ison.

Organisms

These soils usually have a heathy vegetation dominated by species of the *Ericaceae* or sedges such as *Trichophorum caespitosum*. Under tropical conditions the principal plant community is evergreen upper montane forest containing coniferous species, members of the *Ericaceae* and mosses, particularly *Sphagnum*.

Topography

The slope is usually gentle or moderate but the soils occur also on slopes up to 20°.

Age

It would appear that the placon can form in about 1–2 hundred years and that many formed about 7500 years BP. Others developed at various times since then right up to the present.

Pedounit

These soils vary mainly in the thickness and colour of the candon. When it is very well developed due to a large amount of surface moisture it is olive grey and there is usually peat at the surface.

In a number of cases the placon occurs above a sesquon or husesquon or it may rest on the relatively unaltered material, and is characterised by its irregularity in depth from the surface. Placons sometimes occur in the most unexpected places such as within the plough layer of old cultivated fields that have been abandoned for several centuries.

Distribution

These soils occur in the cooler oceanic areas of north-western Europe including Scandinavia, northern Germany, northern France, eastern Canada and the British Isles. They are also found in Alaska, Malaysia and the Solomon Islands.

Utilisation

The low nutrient status, sandy texture, thick surface organic matter, candon and placon make these soils particularly unadaptable to utilisation. Until recently little use was made of them but with the introduction of powerful tractors it has been possible to rupture the placon by deep ploughing. This improves drainage and aeration thus rendering some of these soils suitable for forestry but even when trees are planted the application of fertilisers is essential, particularly phosphorus. The role of phosphorus in these soils is not fully understood, it seems to act directly as a plant nutrient and also as a nutrient to the soil microorganisms causing them to proliferate and break down the organic matter thereby releasing nitrogen and other plant nutrients.

PLANOSOLS

Derivation of Name: From the Latin word *planus* = flat.
Approximate Equivalent Names: Surface water gleys; Stagnogleys; some Aqualfs.

General characteristics

The most conspicuous feature of this class is the marked and often abrupt increase in the clay content on passing from the upper to the middle horizons. The former may be a mullon but there is often a mullon followed by a minon and then the sharp increase in clay to the mottled brown grey or olive planon. This horizon has a coarse angular blocky, prismatic or massive structure and in some cases cutans are present but the change in texture appears to be due in part to strong weathering *in situ* under wet conditions for there are usually more weathered minerals in these soils than in the adjacent freely drained soils. The texture differential is probably enhanced by clay destruction and removal from the minon and other upper horizons containing organic matter or receiving its acid decomposition products.

The initial wetness in the planon can result from a number of reasons, the soils may be in a depression or on a poorly drained flat or gently sloping site where water will accumulate. Alternatively there may be an impermeable horizon or initially a high content of clay in the parent material causing slow permeability.

Planosols appear to evolve from supragleysols and have their maximum distribution under humid continental warm summer conditions, they occur sometimes in the dry summer subtropics and marine climate where the soils are warm and moist during the summer.

PODZOLS

Derivation of Name: From the Russian words *pod* = under and *zola* = ash.

Approximate Equivalent Name: Iron podzols; Humus iron podzols; Humus podzols; Spodosols.

References: Kubiëna 1953; Muir 1961

General characteristics

The fully developed podzol pedounit has an upper, pale grey, strongly leached horizon – the zolon which underlies a surface organic horizon and overlies a brown to very dark brown horizon where iron and/ or humus has accumulated. These soils are associated mainly with coniferous vegetation and a cool humid climate, but they do occur under other circumstances.

Morphology

One of the more common types of podzols contains a husesquon in which iron, humus and aluminium have been deposited.

In this type, organic matter usually accumulates at the surface and shows progressive decomposition with depth. Generally there is a surface layer of litter 1–5 cm in depth or more; this is loose and spongy and

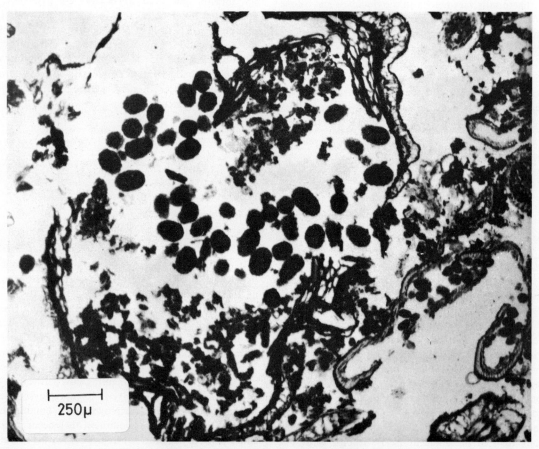

250μ

FIG. 104. Faecal pellets of an arthropod replacing the cellular plant tissue.

grades with depth into the partly humified horizon — the fermenton which is usually of similar thickness (Fig. 65). Here the plant fragments are still recognisable but they have been attacked by many organisms and in thin section decomposition is seen to be progressing rapidly. Many of the plant fragments have growths of fungal mycelia and the softer inner parts of stems and needles are partly replaced by the faecal pellets of arthropods (Fig. 104). Beneath the fermenton is the humifon in which the material is in a very advanced state of decomposition with few plant fragments. The black organic material tends to form loose aggregates or become massive. In thin sections it is isotropic except for the occasional mineral grain.

The humifon changes fairly sharply into a dark grey mixture of organic and mineral material, the former is similar to the horizon above, the latter is mainly quartz with many grains having dark brown or black isotropic coatings. This is the modon which grades into the lighter coloured zolon with its characteristic alveolar or single grain structure and pale colour, contrasting with the horizons above and below (Fig. 105). In the cooler latitudes the zolon may have a marked lenticular structure due to freezing (cf Fig. 62). Beneath the zolon there may be one of four different horizons, commonly it is a brown or dark brown husesquon containing an accumulation of humus, iron and aluminium. This varies from loose, through subangular blocky, to massive. In thin section the matrix is isotropic and arranged in individual granules or clusters of granules. Around most of the sand grains there is an isotropic coating of brown or dark brown material (Fig. 106). With depth the husesquon grades into relatively unaltered material or there may be a fragon or an ison.

The two most common parent materials for these

500μ

FIG. 105. A zolon with characteristic alveolar structure.

soils are alluvium and glacial drift therefore they have varying degrees of stratification and varying proportions of stones and boulders.

Analytical data – Fig. 107

The content of clay is usually low or very low and seldom exceeds about 10% in the zolon but often shows an increase of this size fraction in the middle horizon due to leaching and deposition of aluminium and iron hydroxides rather than to translocation of discrete clay particles.

The organic matter distribution has two maxima the greater of >70% occurs at the surface and the lesser, below in the husesquon where it has accumulated through leaching from above. The C/N ratios have a similar trend, with the maximum value of 25–30 in the fermenton, below in the zolon the value normally decreases to 10–15 and then increases to 15–25 in the husesquon. The cation exchange capacity also

has two maxima coincident with the distribution of organic matter which must be responsible because of the small amount of clay. The content of exchangeable cations is low throughout, the greatest amount occurs in the fermenton where humification is vigorously releasing basic cations. The base cation saturation is very low in all horizons, the maximum being in the lowest horizon. The surface is very acid at pH 3·5 to 4·5 which is followed by a steady increase with depth up to a maximum of about 5·5 in the underlying alluvium. These low figures are due on the one hand to the acid parent material and on the other to the acid litter.

The distribution of silica, alumina, iron oxide in the $< 2\mu$ fraction is the important distinguishing characteristics and have characteristic trends. The maximum free Fe_2O_3 occurs in the husesquon, the modon and zolon have the smallest amount, but perhaps the clay ratios are more revealing. The

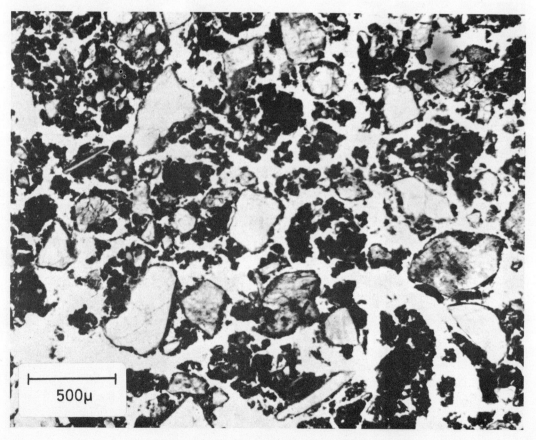

FIG. 106.   Husesquon with granular structure and coatings on sand grains.

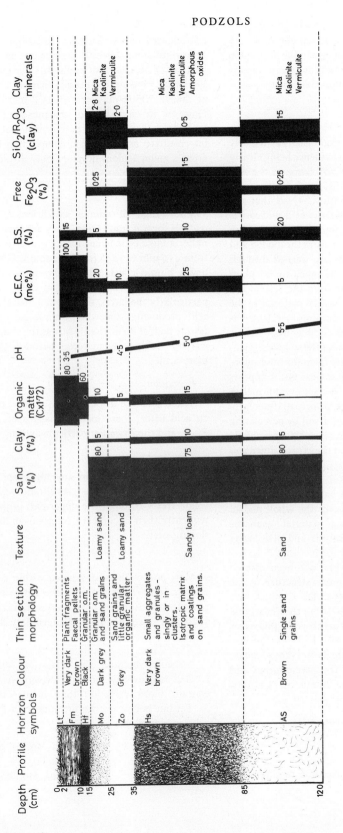

FIG. 107. Generalised data for podzols.

SiO_2/R_2O_3 ratio of the clay fraction shows a very great increase of iron and aluminium in the husesquon while the Al_2O_3/Fe_2O_3 ratio shows that there is really more aluminium in the husesquon than iron.

Genesis

The fermenton is formed by the progressive decomposition of the litter by organisms particularly fungi and small arthropods, the evidence being clearly seen in thin sections (see page 70). Some of the acid decomposition products are dissolved in the percolating rainwater charged with CO_2 so that the solution entering the mineral soils is acid and causes profound hydrolysis of many primary silicates, release of various cations, and leaching to form the modon and zolon. Most of the basic cations are washed through the soil system but some may be taken up by plant roots. Some of the silica is lost from the system but a little is deposited as white powdery material in the zolon. Some of the iron and aluminium released is also lost by leaching but a considerable proportion is deposited as oxides in the middle of the soil accompanied by the deposition of humus to form the husesquon. The details of this process are not clear but it appears that polyphenols formed in the surface organic matter by microbial decomposition are principally involved. They are carried into the mineral soil by percolating water where they chelate inorganic substances chiefly iron and aluminium and together they travel down and are deposited to form the husesquon. Alternatively, oxides are precipitated as the soil solution percolates into a zone with a higher pH value. Probably some silica is also translocated. However the whole podzol soil system is constantly losing material, even the husesquon is undergoing hydrolysis and also losing material.

A satisfactory explanation has not been offered for the formation of the granular structure of the husesquon seen in thin sections. De Coninck and Laruelle (1964) have suggested that the individual granules are the faecal pellets of arthropods; this seems unlikely, since it would require a high level of biological activity for which there is not an adequate supply of food. Alternatively the precipitation and flocculation of the oxides and humus in a porous medium may be responsible.

Principal variations in the properties of the class

Parent material

This is always a medium to coarse textured un-consolidated deposit such as, alluvium, glacial drift or a solifluction deposit, often containing a high proportion of stones and boulders. The mineralogy is somewhat variable but usually there is a high content of quartz which may exceed 95% in some deposits. However in cool, humid, marine conditions where leaching is intense the parent material may be of intermediate or even basic composition but in these areas a zolon seldom develops only a thick modon is found.

Climate

The range of climate conditions under which podzols form is quite wide. They are most widespread under a taiga or marine climate with rainfall variation from 450 mm to 1250 mm per annum. They occur also under humid continental cool summer climate, and tropical climates but in both of these environments particularly the latter, the parent material is extremely acid producing somewhat unique conditions which favour podzol formation. Thus as the climate becomes warmer the parent material becomes progressively more acid.

Vegetation

These soils are associated with plants that produce an acid litter, these include many species of *Pinus*, *Picea*, and the *Ericaceae* particularly *Calluna vulgaris*.

Topography

These soils develop in any topographic situation where aerobic conditions prevail and water can percolate freely through the upper part of the soil. Consequently their occurrence ranges from flat to steeply sloping sites.

Age

Most podzols have developed within the Holocene period, some have required the entire span of this period whereas others developed within the first 5000 years and since then have changed only slightly.

Pedounit

Probably, because this class of soil has been studied extensively and intensively it appears to contain more variation than many other classes.

Set out below are the formulae of a few common subclasses and groups.

1. $[ZoSq]$ $Lt_2Fm_3Hf_2Mo_5Zo_{10}Sq_{30}AS$
2. $[ZoFr]$ $Lt_1Fm_1Zo_5Fr_{50}AS$
3. $[ZoHd]$ $Lt_2Mo_3Zo_{110}Hd_{60}ES$

4. $\lceil$ZoSqIn$\rceil$ $Lt_2Fm_4Zo_6Sq_{40}In$
5. $\lceil$MoSqIn$\rceil$ $Lt_2Mo_{15}Sq_{25}In$
6. $\lceil$MuSq$\rceil$ $Mu_{30}Sq_{20}AST$
7. $\lceil$Mo(SqAt)$\rceil$ $Lt_1Mo_{15}(SqAt)_{30}BSG$
8. $\lfloor$ZoHsd$\rfloor$ $Lt_2Fm_2Hf_1Mo_5Zo_{20}Hsd_{30}ES$

The foremost variation is in the nature of the middle horizon which might be occupied by a ferron, sesquon, husesquon or hudepon as determined by local factors. At the drier end of the climatic range a ferron or a sesquon is more common as given in the first two examples and it is in sesquons that there is the best development of the small granular structure (Fig. 108). At the wet end there is often a hudepon which might be underlain by a husesquon or a sesquon or both. Under conditions of intense hydrolysis and leaching the zolon may be over a metre in thickness as seen in some tropical podzols typified by the third example.

Often the lower horizon is occupied by an ison as given in the fourth and fifth examples. The sixth example illustrates one type of cultivated podzol in which the surface horizons have been homogenised by cultivation and transformed into a mullon.

Podzols intergrade into a number of other soils with intergrades to altosols being among the most common. One type of intergrade is given in the seventh example. Hardening of the sesquon, husesquon or hudepon is a fairly common feature of podzols, this is illustrated in the eighth example. In some zolons there may be considerable amounts of chalcedony (secondary quartz) formed by the deposition of material released by hydrolysis, this is the characteristic feature of the so-called Kauri podzols in New Zealand.

An interesting type of variant is one that develops in the luvon of an argillosol so that there is an argillon beneath the husesquon. These are regarded as argillosols and discussed on page 173.

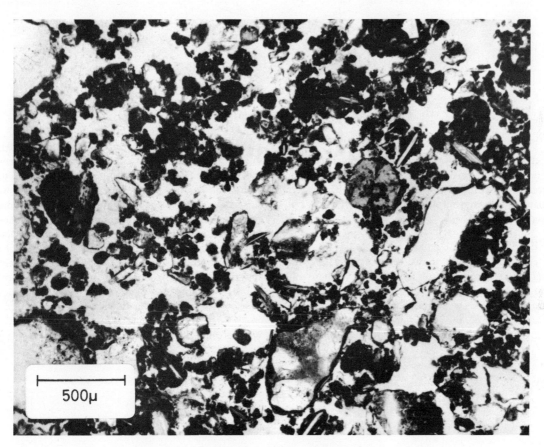

500μ

FIG. 108.   Sesquon with well developed granular structure.

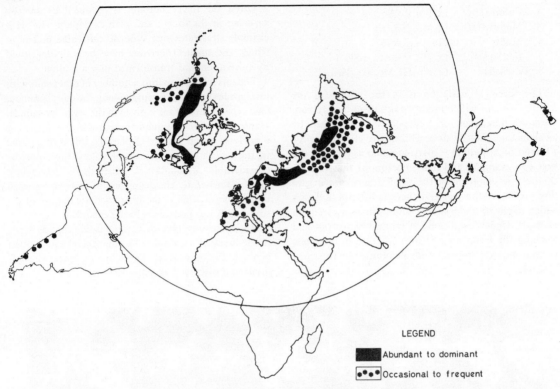

LEGEND

▮ Abundant to dominant

●●●● Occasional to frequent

FIG. 109.   Distribution of podzols.

Distribution – Fig. 109

Podzols occur over a large part of northern Asia, Europe, North America and New Zealand but they also occur elsewhere with a suitable climate-parent material relationship as in parts of Borneo, Brazil, Malaysia and Ghana.

Utilisation

The low nutrient status, sandy texture, and low pH values usually make the utilisation of these soils for agriculture difficult or impossible; often they are used for forestry or rough grazing. Where agriculture is practised, the vegetation has to be removed and the soil ploughed, followed by the addition of an adequate amount of liming material to raise the pH. In addition fertilisers containing nitrogen, phosphorus and potassium and possibly other elements have to be added. For the first few years the area may be devoted to a grass-legume mixture which allows the large amount of surface organic matter to be decomposed as well as for the grass roots to develop a good structure. Subsequently these areas can be adapted to the system of land use on the adjacent better soils. In the cool temperate areas, some form of mixed farming may be practised but in the tropics amelioration is often very difficult and special hardy crops such as cashew-nuts and coconuts are grown. Wind erosion may be a hazard in some areas.

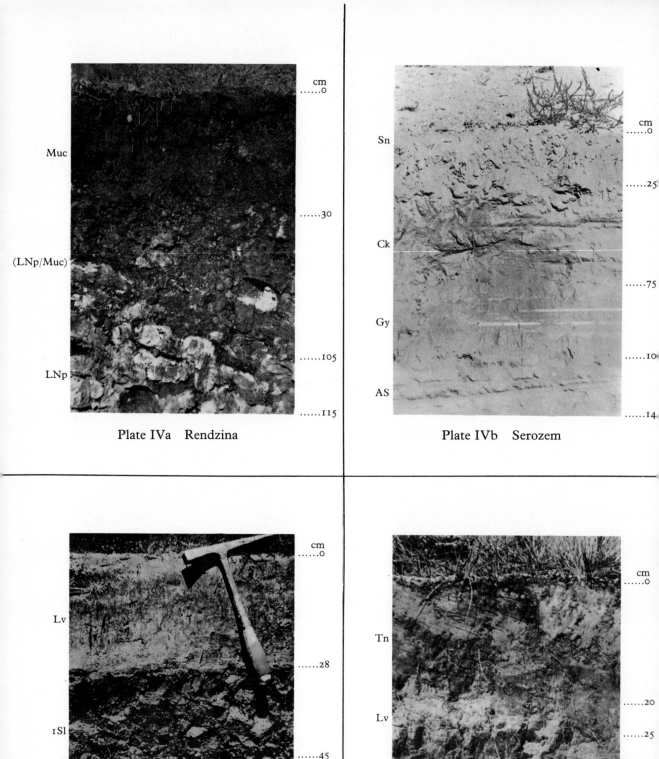

Muc

(LNp/Muc)

LNp

cm
......0

......30

......105

......115

Plate IVa   Rendzina

Sn

Ck

Gy

AS

cm
......0

......25

......75

......10

......14

Plate IVb   Serozem

Lv

1Sl

2Sl

cm
......0

......28

......45

......70

Plate IVc   Solod

Tn

Lv

Sl

(CkHl)

cm
......0

......20

......25

......40

......48

Plate IVd   Solonetz

RENDZINAS

Derivation of Name: From the Polish word *rzedzic* = to tremble, due to the penetration of implements through the shallow soil and their impact on the underlying rock.

Approximate Equivalent Names: Humus carbonate soils; some Rendolls

Reference: Avery *et al.* 1959.

General characteristics

These are very shallow soils composed of a thin dark brown or black calcareous mullon or calcareous modon lying on pale coloured limestone or limestone drift. These soils form due to the presence of limestone therefore they are distributed throughout a wide range of climate and carry many different plant communities.

Morphology

The characteristics of the surface horizons vary somewhat with climate and vegetation. A common type has a loose leafy litter resting on the dark brown or black calcareous mullon-Muc, speckled with white fragments of parent rock. The mullon has a well developed crumb or granular structure with abundant earthworm casts; therefore it is moderately to highly porous. In thin section each aggregate is seen to be outlined by a thin layer of calcite crystals. Fragments or complete mollusc shells are rare to occasional.

The calcareous mullon either rests directly upon the parent rock or there may be a narrow transitional horizon containing a mixture of these two horizons forming the intergrade (LFp/Muc).

Analytical data

Perhaps the most important physical properties of these soils are their shallowness, medium to fine texture and well developed granular to fine subangular blocky structure. Together these features allow rapid percolation of moisture which causes drying out and in some years there may be a period of drought.

The calcareous mullon usually contains up to 80% calcium carbonate giving pH values over 8·0. The organic matter content varies from 5−15% and is usually in an advanced state of humification as indicated by the C/N ratios of 8−12. The fine texture and high organic matter contents have resulted in cation exchange capacity values of up to 50 me%. The main exchangeable cation is calcium or magnesium on dolomites and there is complete base cation saturation.

Genesis

The principal process taking place in rendzinas is the solution and removal of the carbonate in the drainage water, leaving behind a relatively small residue. This is mixed with the humifying organic matter and small fragments of undissolved material by the very active mesofauna to give a granular structure but expansion and contraction probably accounts for the blocky structure when it is present. The characteristic dark colour is due to a calcium-humus complex and is very similar to that found in chernozems. When the residue accumulates to a very great thickness and the carbonate is below the depth of root or faunal penetration, decalcification proceeds. This is followed by the formation of soils similar to those developed on fine grained igneous rocks or sediments viz. altosols, argillosols and luvosols.

Principal variations in the properties of the class

Parent material

This is an example of a class of soils whose characteristics are determined almost entirely by their parent materials which are always composed of material containing a high proportion of calcium and/or magnesium carbonate. They are usually consolidated rock such as limestone and chalk but they can be unconsolidated sediments or drift deposits.

Climate

Since the parent material is the controlling factor in the formation of these soils they are found under nearly every type of climate outside very cold and very arid areas.

Vegetation

The natural vegetation varies with the climate but usually it contrasts strongly with adjacent plant communities on soils developed from different materials. Generally these soils have a richer flora except where they are very shallow, then drought may become a limiting factor to plant growth. Earthworms, arthropods and bacteria are the chief members of the soil organisms but again limitations may be imposed if there is a shortage of moisture.

Topography

The areas where rendzinas occur are usually diversified by karst phenomena and vary from flat to steeply sloping (Fig. 30). The old landscapes in many tropical and Mediterranean areas with marked polygenetic development often have rendzinas on the recently eroded sloped while older soils such as krasnozems and rossosols occupy the flatter sites.

Age

When these soils occur in areas that were subjected to considerable erosion during the Pleistocene period they are about 10,000 years old but in some tropical areas they appear to be much older.

Pedounit

The members of this class have a relatively small degree of variation which is confined largely to the thickness and regularity of the upper horizon. When they develop under fairly dry conditions they are lighter in colour and lower in organic matter; in addition some carbonate is deposited in the upper part of the underlying material to form a calcon. In

contrast the surface may develop a humifon, fermenton and calcareous modon under cool humid conditions. In moist situations montmorillonite develops and gradually these soils intergrade to vertisols. Where the parent material is low in carbonates they tend to be browner in colour and intergrade to altosols.

Distribution

Rendzinas occur in small areas throughout much of the world where environmental conditions are suitable. In some places, such as in the countries bordering the Mediterranean they occupy large areas.

Utilisation

Since rendzinas develop under a wide variety of conditions their utilisation is governed by local practices, therefore it is possible to find them producing a wide variety of crops including grapes, grain, sugar cane and cocoa. Some of the largest areas are devoted to vineyards.

Severe limitations are imposed by their shallowness and high permeability. The former often prevents the use of large implements whereas the latter causes drought even in a humid environment. The high content of calcium can induce microelement deficiencies by replacing them on the exchange sites. In spite of these deficiencies these soils have a high natural fertility and are highly prized by farmers but when they are shallow or occur on steep slopes they are devoted to forestry but the quality of the trees is sometimes poor.

ROSSOSOLS

Derivation of Name: From the Italian word *rossa* = red.
Approximate Equivalent Names: Terra rossas; Red Mediterranean soils; Eutrophic brown soils; some Inceptisols.

General characteristics

This is a class of red or reddish brown soils that is widely distributed in the wet and dry tropics, semi-arid tropics, and dry summer sub-tropics. The principal horizon is the rosson which occurs in the middle position and in addition to its distinctive colour the clay fraction contains some 2:1 minerals, mainly mica but there are often large amounts of montmorillonite with subordinate kaolinite. Goethite and hematite are the principal compounds of iron, gibbsite is absent or present only in very small amounts.

These soils develop from a variety of rocks and sediments both of which are often calcareous.

Rossosols form through progressive weathering by hydrolysis or solution in a moderately humid environment therefore leaching is not rapid, thus they remain about neutral with moderate to high base cation saturation. These conditions result in the formation of clay mica, rather than kaolinite and gibbsite which form in the wetter equatorial regions. Thus the cation exchange capacity of rossons is above 20 me per 100 gm clay.

Many of the areas where these soils occur, have been strongly affected by Pleistocene climatic changes consequently they show polygenetic evolution and there are many intergrades. The designations of a few subclasses and groups are as follows:

1. [TnRo] $Tn_{10}Ro_{10}LF$
2. [TnRo] $Tn_3Ro_{70}BFw_{30}BF$
3. [TnRo] $Tn_{10}Ro_{15}Ckd$
4. [Tn(RoAr)] $Tn_{20}(RoAr)_{50}AFw$

The first example developed on limestone includes some of those soils generally known as terra rossas. Previously all red soils developed from highly calcareous materials were placed in the same class but the range is sufficiently wide to justify placing them into two different classes; some with the rossosols and some with the krasnozems. The second example includes some of the eutrophic brown soils of the wet and dry tropics as well as some soils of tropical semi-arid areas. The third example includes those soils that occur mainly in a semi-arid environment where leaching is very restricted thus a calcon or hardened calcon has formed.

It is in the countries bordering the Mediterranean that rossosols developed extensively during the later part of the Tertiary period and subsequently have been subjected to polygenetic evolution, varying in kind from place to place. The principal phase has been translocation of clay to form a rosson-argillon intergrade and in some cases an argillon-rosson intergrade; some of the red Mediterranean soils fall into this category as shown in the fourth example.

SEROZEMS

[Plate IVb

Derivation of Name: From the Russian words *seri* = grey and *zem* = soil
Approximate Equivalent Names: Desert soils; some Aridisols
Reference: Jackson 1962

General characteristics

These soils develop in arid areas and are characterised by a thin greyish brown seron which often grades sharply into a calcon, hardened calcon or gypson. In many situations the seron rests on an horizon formed in a previous climatic phase.

Morphology

The upper horizon is often a loamy calcareous seron which varies in colour from yellowish red to greyish brown and is about 15–20 cm in thickness but can be much thinner. It has a thin crust at the surface and is usually firm or friable with a laminar or massive structure and grades quickly into a calcon up to a metre in thickness followed by a gypson or halon which may be over 3 m in thickness. In many cases there is a slight increase in clay beneath the seron to form a clamon or calcon-clamon intergrade.

In thin section the seron has frequent small and medium circular pores and there are usually numerous faunal passages and faecal pellets. The thin section morphology of the calcon the gypson is similar to the descriptions given on pages 179 and 230.

Analytical data

The seron usually contains a smaller amount of clay than the underlying horizon and may often contain less silt and fine sand but greater amounts of coarse sand and gravel. The content of organic matter seldom exceeds 2% and is usually less than 1%, the C/N ratio is distinctive being less than 10 and often less than 8. When the parent material contains carbonates the content may exceed 10% in the seron increasing to a maximum of over 12–15% in the calcon then decreasing in the parent material. Consequently the pH values in the seron vary from 7·0 to 8·0 and increase to a maximum of about 8·5 in the calcon.

The cation exchange capacity varies between 10 to 25 me% in the seron decreasing with depth to about 8–10 me% and there is complete saturation of the exchange complex with calcium being the dominant ion.

Genesis

Most of the processes taking place in these soils are operating very slowly because there is only a small amount of water passing through the system. However the small amount of organic matter contributed by the sparse vegetation is quickly humified. The presence of many faunal passages and faecal pellets do not indicate a high level of biological activity for features once formed tend to become preserved in the hot, dry environment.

The easily soluble salts are removed from the upper part of the soil and deposited between 1 to 5 m from the surface and small amounts of carbonate are removed from the seron and deposited below to form a calcon.

Because of the lack of a vegetative cover the surface of the soil is subject to deflation and rapid run off, both processes causing a removal of fine material resulting in concentration of gravel at the surface. However some of the textural difference may be enhanced by weathering *in situ*, thus forming a clamon.

Principal variations in the properties of the class

Parent material

The most common parent materials are sediments of Pleistocene or Holocene age. In Europe and the Americas it is often loess or alluvium but in old landscapes it may be a pedi-sediment.

In some cases serozems evolve from a previous soil, the most dramatic examples of this occur in the semi-arid areas of Africa and Australia where they are formed in Tertiary soils.

Climate

These soils are restricted to the arid areas of the world occurring in both mid-latitude and tropical areas.

Vegetation

Although most of the plant species growing in these soils have adaptions which permit them to grow in arid areas, the character of the communities is very different. In Eurasia, annuals including grasses and succulents are the principal species but in the Americas various cacti are usually dominant. In tropical and subtropical areas of Africa and Australia there are scrub woodlands dominated by species of *Acacia* and *Eucalyptus*.

Topography

These soils tend to be restricted to flat or gently undulating situations. On moderate or steep slopes erosion reduces the development of soils so that primosols and rankers are the main formations.

Age

Many of these soils started their development at the end of the Pleistocene period but those of the tropical and subtropical areas may have started to form at considerably earlier phases of the Pleistocene or even in the Tertiary period.

Pedounit

There is a considerable variation in the properties of this class as determined by differences in parent material, climate and polygenesis. Where these soils are formed in late Pleistocene sediment, variations in the pedounits are determined largely by the presence or absence of carbonates, for when carbonates are present in large amounts the soils are calcareous throughout and a calcon or hardened calcon is formed. In the absence of carbonate the soils are usually mildly to moderately acid at the surface. This is most marked in areas of old soils where the seron may rest on a rosson, pallon, flambon or hardened veson.

The colour of serozems can vary from brown to red, the former colour occurs mainly in the mid-latitude arid areas whereas in tropical and subtropical areas they are red and may have hues of 2·5 YR or 10 R. Although goethite is the principal iron oxide in these redder soils, hematite is usually present in small amounts. In some cases the red colouration is inherited from a previous soil cover but reddening seems to be a characteristic feature of these warmer areas and in spite of the dry conditions some weathering does take place in serozems which usually results in the formation of a thin clamon. Serozems have continuous gradational sequences to other soils including clamosols, burozems, solonetzes and solonchaks.

Where deflation is a vigorous process the gravel and stones increase in frequency to form a hamadon which varies very widely in character depending upon local conditions. It may be composed of stones and boulders if the parent material was a sediment or there may be fragments of weathered rock when the soil is developed in an older soil. However in many tropical and subtropical areas fragments of hardened vesons form the hamadon.

Distribution

Serozems are confined to the arid parts of the world. They are common in southern Russia, in a belt on either side of the Sahara Desert, in parts of Persia, west Pakistan, Afghanistan and Australia.

Utilisation

Under natural conditions rough grazing for cattle is the only form of land use that is practised and even this is a somewhat tenuous system because of the uncertainty of an adequate supply of water for the animals. These soils usually prove to be highly fertile if irrigated but this can be made difficult or impossible through the lack of water. Usually there are very few rivers in areas of serozems and the artesian water often has a high content of salts making it unsuitable for irrigation or domestic purposes. However there are a number of places where rivers flowing through these areas supply suitable water. In some other areas very large installations are used to convey water over long distances. Perhaps the most complex is found in the state of Colorado in the U.S.A. where water is pumped over the mountains to irrigate the soils in the valleys on the other side.

SOLODS   [Plate IVc

Derivation of Name: From the Russian word *sol* = salt
Approximate Equivalent Names: Some Aridisols, some Alfisols.
References: Stace *et al.* 1968; Reeder and Odynsky 1965.

General characteristics

Solods can be regarded as strongly leached solonetzes. Their morphology shows a thin litter and humifon at the surface resting on a modon 5–10 cm thick containing up to 20% organic matter and having a C/N ratio of about 15. Below the modon is the clay-depleted luvon which may be 20–50 cm thick but in extreme cases it may be over a metre in thickness and has thickened through the removal of clay from the upper part of the underlying solon with its columnar or prismatic structure. Frequently the rounded outline of the columns of the solon are still visible within the luvon which grades sharply into the finer textured solon or solon-argillon intergrade. With depth there is a change to a gypson, halon or gradation to the underlying material. The thin section characteristics are similar to those given on other pages for modons, luvons and solons.

Since solods are formed by continued leaching of solonetzes they become progressively more acid causing the luvon to have pH values as low as 5·0 and less than 30% base saturation. Similarly, the solon which originally had a pH value of 8·5 or over may now have a lower pH value and have lost some of the original high content of exchangeable sodium and magnesium.

Many solods occur in shallow depressions which receive run-off water, consequently they are moister for longer periods of the year than the adjacent soils, the result is greater leaching and often a denser plant community. For example in the U.S.S.R. it is common to find communities dominated by birch or aspen on solods whereas grass communities occur on the adjacent slightly elevated and drier sites.

The acid strongly leached luvon imparts an extremely low fertility to these soils, therefore it is necessary to add large amounts of liming materials and fertilisers, and where possible organic material followed by deep ploughing. Since solods occur mainly in arid and semi-arid areas, improved utilisation can be achieved by irrigation which will also remove salts and help to reduce the pH values in the solon thereby producing more root-room for plants.

SOLONCHAKS

Derivation of Name: From the Russian word *sol* = salt
Approximate Equivalent Names: Saline soils; Salorthids.

General characteristics

These soils contain more than 0·5% of soluble salts which have accumulated because saline ground water comes to the surface and evaporates leaving the salts behind. The principal horizon is often a halon-gleyson intergrade formed by the simultaneous accumulation of salts particularly chlorides and reduction of iron to the ferrous state due to the high ground-water-table. The greatest extent of these soils is in arid and semi-arid areas.

Morphology

Generally these soils show only weak contrasts

between horizons. The whole soil is normally grey or greyish brown with varying amounts of mottling, the greatest being in the middle of the soil forming the halon-gleyson intergrade. Usually the halon at the surface is slightly darker due to staining by organic matter and in many cases it has a thin crust of salt at the surface (Fig. 110) or there may be takyrs which are small domed shaped areas with polygonal outlines. In other cases the upper horizon may be massive, coarse platy, puffy or crusty.

In thin sections the (HlGl) horizon is seen to contain varying amounts of carbonate concretions while all the ped and pore surfaces are outlined by fine calcite crystals which usually form incrustations around small objects such as root fragments.

Analytical data – Fig. 111

The most important property of these soils is their high content of salts which are usually highest at the surface decreasing with depth. The most common ions are chloride, sulphate, carbonate, bicarbonate sodium, calcium, magnesium and small amounts of potassium but the variability in the proportions of these ions is extremely wide however the data given in Fig. 111 is representative of large areas of these soils.

As stated on page 93 the principal criteria used in the classification of solonchaks are pH values and the electrical conductivity of the soil or saturation extract. The distribution of the individual ions is also taken into account and further subdivision of the class is often made on the basis of ratios of the various ions present. This is useful in some places where there are regional variations in the proportions of ions as found in central U.S.S.R.

The structure throughout the pedounit is usually

FIG. 110.   Salt efflorescence on the surface of the soil and a partly exposed solonchak profile.

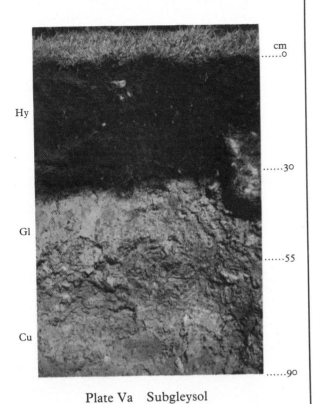

Hy

Gl

Cu

cm
......0

......30

......55

......90

Plate Va   Subgleysol

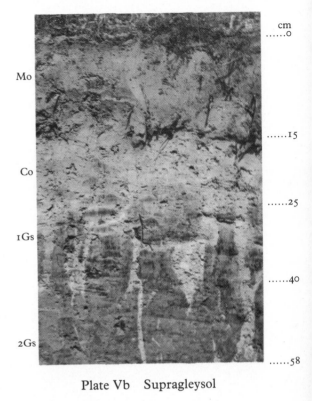

Mo

Co

1Gs

2Gs

cm
......0

......15

......25

......40

......58

Plate Vb   Supragleysol

Cw

1Ve

2Ve

Ck

cm
......0

......15

......60

......115

......135

Plate Vc   Vertisol

Lt
Fm
Mo

1Zh

2Zh

cm
......0
......5
......10

......60

......100

Plate Vd   Zheltozem

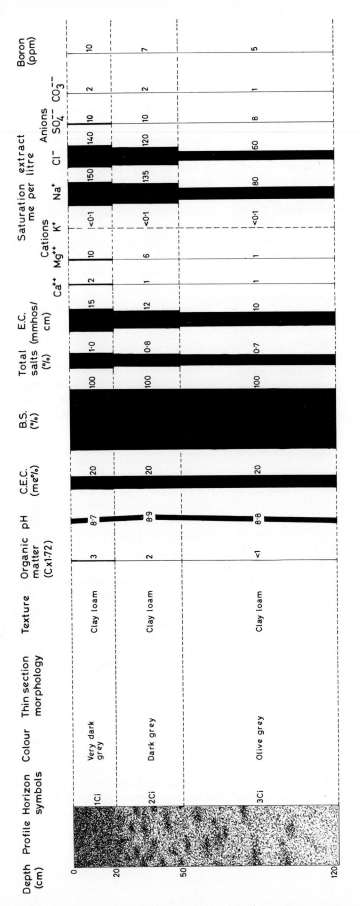

Fig. 111. Generalised data for solonchaks.

massive except on very sandy materials and it is interesting to compare this type of structure with the granular structure of a kastanozem which may develop in the same environment on similar parent material.

Genesis

These soils commonly develop in an arid or semi-arid environment on flat situations, or in depressions where the ground-water-table during the dry season is usually less than about 3 m from the surface. During the wet season the water-table rises often coming to the surface causing some reduction of iron and the development of a mottled pattern in the anaerobic environment. At the same time evaporation or evapotranspiration results in the loss of water causing some of the salts dissolved in the ground-water to be deposited on the surface and within the upper part of the soil. When the ground-water-table recedes during the dry period of the year the capillary water left in the upper part of the soil is lost by evaporation and any dissolved salts are deposited. Annual repetition of this cycle of wetting and drying causes a considerable amount of salts to accumulate within the zone of moisture fluctuation but there is no accumulation within the zone of permanent saturation. Therefore the pattern of salt distribution tends to form a maximum at the surface.

If the water-table never comes to the surface the maximum content of salt may occur at some depth within the soil. On the other hand the absence of a salt maximum at the surface may be due to secondary leaching. Kovda (1946) has demonstrated that the amount of soluble salts in some soils of the U.S.S.R. can vary from season to season. During the autumn, following the dry summer, the salt concentration at the surface is several times that of the spring period when melting snow causes some downward leaching thereby reducing the salt content.

The texture of the soil seems to influence the rate of salt accumulation; fine textured soils have a higher retentivity therefore they hold more saline water which upon evaporation leaves a higher amount of salt. Fine textured soils also have a lower permeability which decreases as the content of sodium increases because sodium disperses the clays. Sometimes it is not clear how the salts in the ground-water have originated, usually they are derived from the weathering of rocks but it is difficult to account for large amounts of chloride and carbonate. The former is not a normal constituent of rocks except for small amounts in some sediments, it is usually

regarded as coming from the sea either as spray or due to a previous inundation. The presence of carbonate particularly sodium carbonate is more difficult to explain particularly when calcium or magnesium carbonate was either absent or present in only small amounts in the original material. The greatest amount of sodium comes from the weathering of rock either by hydrolysis as in the case of the orthoclase felspars or by solution from sedimentary rocks. It seems that most of the sodium carbonate is formed from the CO_2 of the atmosphere in a number of stages. At first the CO_2 is dissolved in the soil solution to form H_2CO_3 which reacts with sodium to form sodium bicarbonate which is relatively unstable and is readily transformed into sodium carbonate.

Principal variations in the properties of the class

Parent material

Unconsolidated deposits including loess, alluvium and pedi-sediments are the principal parent materials. On the older landscapes colluvium derived from old soils may form the parent material.

Climate

The most common occurrence of solonchaks is in mid-latitude and tropical arid and semi-arid areas. They are found at the margins of humid continental warm summer conditions where evapotranspiration is much greater than precipitation. Solonchaks have also been reported from the drier polar areas.

Vegetation

The natural plant cover varies from quite dense to absent, depending upon the degree of salinity. Where the salt content is fairly low the species differ little from the adjacent non-saline areas which usually carry a grassy plant community. As the salt content increases to over 0·5% only halophytic species can grow (see Fig. 110). The plants growing on solonchaks have a high content of ash containing larger amounts of sodium, chloride and sulphate.

Topography

The flat or depression areas where these soils develop are often alluvial terraces, beds of old lakes or else they are basins surrounded by mountains which shed large amounts of moisture during the wet periods of the year so that they are temporarily waterlogged. The water from the mountains brings with it varying amounts of salts which are left behind when the water is lost by evaporation.

In many areas salt efflorescence occurs in shallow isolated depressions which are known as slickspots, easily recognised by the absence of vegetation and their characteristic pale grey colour (Fig. 110). As salinity increases the slickspots grow larger and eventually coalesce to form an almost unbroken salt crust.

Age

These soils form rapidly and most groups seem to have developed during the Holocene period. Since their main property is the presence of salts they can be changed rapidly by man's influence or by slight changes in elevation or climate, therefore it is only under somewhat exceptional circumstances that they would be preserved from an earlier period.

Pedounit

The principal variations in the properties of these soils lies in the amount and type of ions which can exist in various combinations creating many difficulties with regard to classification and delimiting boundaries because variations can take place rapidly over short distances. This problem can probably be solved by creating four horizons on the basis of the ternary diagram given in Fig. 69. The amount of mottling also varies from place to place but probably the greatest variations are found in the upper horizon which can be saline intergrades, since solonchaks intergrade into a number of directions principally to subgleysols, chernozems, kastanozems, burozems, serozems and halosols.

Distribution – Fig. 112

Every major land mass has its area of solonchaks which occur mainly in the drier central parts of continents. However they are common in many coastal situations where the main wind currents blow from off the land. This is marked on the N.W. coast of Africa and South America and the west coast of Australia.

FIG. 112. Distribution of solonchaks.

Utilisation

This class of soils probably offers the greatest amelioration problems primarily because of the difficulties encountered when trying to remove the salts. Since water conveyed the salts into the soil it is necessary to use water to remove them, but in an arid or semi-arid environment there is usually a shortage of water or the available supply may have such a high content of salts as to make it unsuitable for leaching. There are a few notable exceptions such as the waters of the Nile and Colorado rivers. Even when there is an adequate supply of water there still exist many problems. The soils have a poor structure which makes them slowly permeable so that much of the water that is applied may be lost by evaporation or run off. Secondly, drains have to be installed so as to remove the saline leachate as well as to reduce the height of the ground-water-table but since these soils occur in flat or depression situations there are difficulties with regard to removing the drainage water. Even when the drainage water can be removed there remains the problem of disposal for it cannot be put into the natural drainage of the area such as rivers because it would increase their salinity. Where the area is very low-lying the ground-water-table can be lowered by digging wells and pumping out the water.

In addition to adding water it is normal to add gypsum (calcium sulphate), this dissolves gradually and the calcium is slowly absorbed on to the exchange site thereby replacing sodium which is lost in the drainage. As the clays become more saturated with calcium they flocculate and gradually the structure improves causing an increase in permeability. In some soils the initial content of calcium sulphate is very high so that it is necessary only to irrigate. After these operations have been completed these soils can then be adapted to the system of land use of their surroundings providing there are no other hazards. Sometimes however microelements such as boron may be present in toxic proportions.

In a number of arid areas the saline water-table may be several metres from the surface and without any capillary rise to the surface therefore the soils are not saline. Often when these soils are given an excess of irrigation water, it percolates down through the soil and makes contact with the underground water. This causes capillary rise of moisture to take place and the saline waters are drawn to the surface. This has been a major factor in causing salinisation of many soils such as in the Sind Valley of Pakistan. Therefore great caution must be exercised when irrigating a soil with a deep saline ground water table.

SOLONETZES

[Plate IVd

Derivation of Name: From the Russian word *sol* = salt.

Approximate Equivalent Names: Natrustalfs, Natrixeralfs; Natrargids.

References: Kovda 1946; Stace *et al.* 1968.

General characteristics

The distinctive horizon of this class is the solon which occurs in the middle of the pedounit. This horizon is alkaline, saline and has a higher content of clay than the upper and lower horizons and usually has a well developed prismatic or columnar structure. Solonetzes are confined to the arid and semi-arid parts of the world.

Morphology

At the surface there may be a thin loose leafy litter resting on a black humifon about 2–3 cm thick. The humifon grades quickly into a brown granular tannon or mullon up to about 15 cm in thickness. In thin section the matrix in the upper mineral horizon is isotropic, forming coatings around sand grains or occurring as small granules. The upper mineral horizon changes sharply into the mottled greyish brown and brown solon with its prismatic or columnar structure and higher content of clay. In thin section the matrix of the solon is predominantly birefringent with abundant medium and large domains and thin zones of domains some with oblique orientation. Clay cutans occur on some of the surfaces of the peds as well as around pores within peds but their frequency seldom exceeds 3% of the soil. The solon grades with depth into the somewhat more mottled and massive saline horizon or there may be an intervening gypson with its characteristic clusters of gypsum crystals which are clearly seen in thin section (Fig. 113). In many places solonetzes are in a fairly advanced stage of development and a luvon is beginning to form; the example given in Plate IVd shows this type of situation.

Analytical data – Fig. 114

Perhaps the most conspicuous property of solonetzes is the abrupt and large increase in clay in passing from the upper horizon into the solon. The increases may be up to 3 fold with the greatest increase being in the fine clay ($< \cdot 2 \ \mu$). The amount of organic matter in the surface mineral horizon varies but is usually less than 10% with a C/N ratio less than 12, indicating a high degree of humification. The pH values are usually between 6·0–7·5 at the surface increasing to over 8·5 in the lowest horizons.

The cation exchange capacity varies with the texture and clay mineralogy but is usually between 15–35 me% and apart from the upper horizons the entire soil is saturated with basic cations. In the upper horizons calcium may be the principal exchangeable cation but in the solon, sodium and magnesium predominate, and together exceed the calcium content; alternatively sodium occupies at least 15% of the exchange complex. With depth calcium may again be the principal exchangeable cation.

Usually the upper horizons are non-saline but salinity increases with depth. The conductivity of the solon often attains 2·0 mmhos/cm but in the underlying horizons it may be as much as 15 mmhos/cm where there is $> \cdot 5\%$ salts and sometimes much carbonate. As with solonchaks the amount and type of salts vary from place to place, but ions of calcium, magnesium, sodium, carbonate, bicarbonate, chloride and sulphate predominate. When the parent material contains carbonates the surface horizon may be decalcified.

The type of clay minerals are mainly inherited from the parent material, but generally they are dominated by mica but kaolinite and montmorillonite can be present in high proportions. The

500μ

FIG. 113. A cluster of gypsum crystals in crossed polarised light.

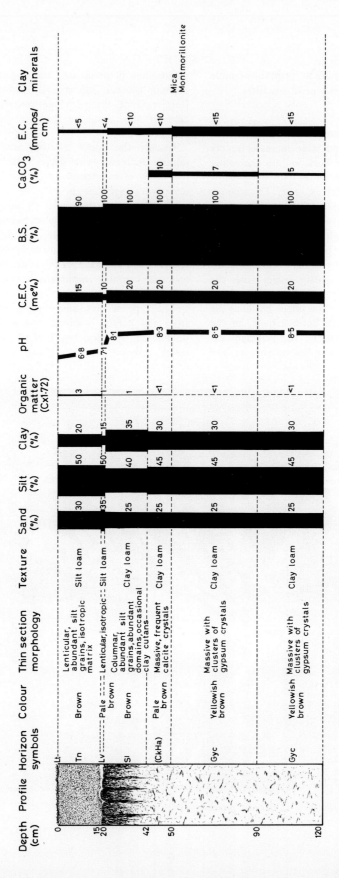

FIG. 114. Generalised data for solonetzes.

principal minerals produced by hydrolysis are mica, vermiculite and small amounts of montmorillonite. The upper horizons usually contain quartz in the clay fraction.

Genesis

It is generally accepted that solonetzes formed by the progressive leaching of solonchaks which are deficient in calcium but have a large amount of sodium ions. This leads to the migration of clay from the upper to the middle position of the soil to form the solon. When the data for these soils is examined carefully it is difficult for this theory to be sustained. The increase in clay in the solon over that in the parent material cannot be attributed to the small amount of clay cutans contained in the solon. Similarly the amount of clay cutans cannot account for the considerable amount of clay that has been removed from the upper horizon. Thus there appear to be three processes taking place in these soils, the most important is the destruction and removal of clay to form the luvon, evidence for which is afforded by the presence of quartz in the clay fraction. A small proportion of clay translocation is taking place hence the presence of clay cutans in the solon but often the greatest amount is in the lower part of the solon and does not coincide with the maximum amount of clay.

The increase in the clay content in the solon over that of the parent material seem to result mainly from weathering *in situ* in a somewhat similar manner to that formed in clamosols (see page 184). Thus the textural contrast is due to clay destruction in the upper horizon and clay formation in the middle horizon together with small amounts of clay translocation.

Although it is probable that some solonetzes may have formed by the progressive leaching of solonchaks it seems that many can form where sodium rich waters percolate through the soils.

Principal variations in the properties of the class

Parent material

Unconsolidated deposits including loess, alluvium and glacial deposits are the principal parent materials. In areas of old landforms solonetzes may form in the old soils or colluvial material derived from them, this is particularly common in parts of Africa and Australia. Solonetzes will not form in material containing calcium carbonate because the calcium displaces the sodium and magnesium but they will form following decalcification.

Climate

These soils are most widespread in the semi-arid parts of the world where there is sufficient precipitation to cause leaching of the upper horizons. On the other hand the amount of moisture passing through the soil is insufficient to reduce the alkalinity in the lower horizons.

Vegetation

Various succulents and xerophytically modified species form the normal plant communities on these soils. Generally the combination of low rainfall, high pH and poor structure restricts the development of a complete plant cover.

Topography

Solonetzes are confined to flat or gently sloping situations, they tend to be absent from depression where the water table comes near to the surface, then solonchaks form. The topographic relationships are given more fully on page 267.

Age

Although solonetzes may be found associated with old land surfaces their development seems to have taken place during the Holocene period and may have extended over most of that period since a considerable length of time is required for the destruction and removal of clay from the upper horizons.

Pedounit

A common variation is in the nature of the principal upper horizon which depends upon the character of the climate and the other soils with which solonetzes are associated. Therefore the upper horizons may have similarities with chernons, burons and other upper horizons. The individual properties of solonetzes can vary widely from one soil to the other. For example, there may be a 10 fold increase in clay between the luvon and solon and in some soils the solon is very dark in colour and contains up to 3% organic matter. At present there is still a measure of disagreement about what constitutes a solon, some workers use chemical criteria and others use morphological evidence while others require both sets of criteria. The situation is complicated further because the chemical criteria can vary from one worker to the other. Originally, it was suggested that solons should have $> 15\%$ of the exchange complex saturated with sodium but it was discovered that magnesium can be important. Then it was suggested that the exchangeable sodium plus magnesium should be greater than the exchangeable calcium plus hydrogen. At present both chemical and morphological criteria are required, therefore those horizons that have only one of these criteria are considered to be intergrades. When a horizon has only the chemical characteristics of a solon it usually occurs in the lower part of the soil and indicates some influence of saline underground waters and is fairly common in areas of chernozems, kastanozems, burozems and vertisols. In contrast there are a number of soils that have certain morphological features of solonetzes but do not have the chemical characteristics, these soils are considered to have been solonetzes which have lost their chemical characteristics by progressive leaching through irrigation or after a long period of leaching by rainfall. A somewhat unique set of conditions is found in cold semi-arid areas such as in Alberta in Canada, where intergrades between podzols and solonetzes develop.

In these soils the surface horizons are those of podzols with a solon in the middle or lower position.

There is a continuous gradational sequence between solonetzes and solods, morphologically expressed through a gradual development of a luvon and decreasing thickness of the solon.

Distribution – Fig. 115

These soils are of common occurrence in the arid and semi-arid regions of the world. They are widespread in Australia, west Pakistan, south western U.S.S.R., northern and southern Africa. In North America they occur spasmodically in a belt that starts in northern Alberta and continues to western Texas and Mexico. In South America they occur in the narrow coastal belt of northern Chile and also in certain parts of Argentina.

Solonetzes may form large contiguous areas but they are found more usually as part of a complex with other soils, especially burozems, clamosols and serozems.

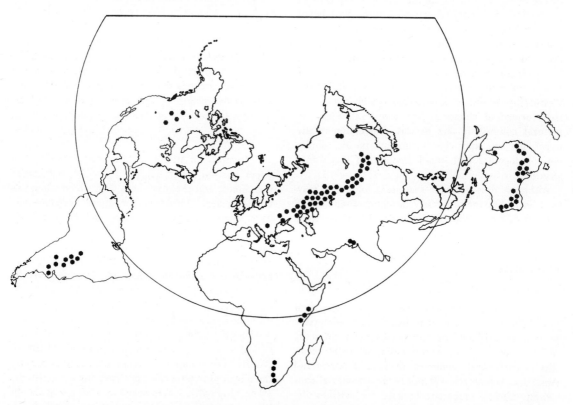

FIG. 115.   Distribution of solonetzes.

Utilisation

The value of these soils for crop production varies with the content of organic matter and climate. They are regarded as being fairly good for crop growth when they occur in a cool climatic area associated with chernozems or kastanozems. On the other hand when they occur in tropical and subtropical areas conditions are usually too dry for crop growth but they may produce forage for grazing.

Where irrigation is possible they are highly productive but precautions must be adopted to prevent the salts in the lower part of the soil from coming to the surface by capillary movement. Therefore adequate drainage has to be provided to remove the excess irrigation water and where necessary to lower the water-table.

The fine texture of the solon may cause impeded drainage but this can often be alleviated by deep ploughing.

In all cases these soils are deficient in plant nutrients which have to be supplied and often liming materials have to be applied.

SUBGLEYSOLS   [Plate Va

Derivation of Name: From the Latin *sub* = under, Russian *glei* = compact bluish-grey
material and French *sol* = soil.
Approximate Equivalent Names: Gley soils; Ground-water-gleys; Aquents; Aquepts.
Reference: Glentworth and Muir 1963

General characteristics

The principal horizon in this class of soils is the mottled, grey or olive, gleyson which forms as a result of prolonged periods of anaerobism. Therefore, these soils occur in moist situations particularly in cool climates where they are associated with plant communities containing wet habitat species.

Morphology

At the surface under natural conditions there is usually a spongy or matted litter about 1 cm thick often resting on a hydromoron up to about 20 cm thick. This is a very plastic, dark grey or black, organic horizon containing abundant fine roots and in thin section it is dense massive and isotropic. The hydromoron changes sharply into a very dark, grey, modon of variable texture but is commonly a loam about 15 cm in thickness. In thin section the modon is seen to be composed of small, amorphous, black areas of organic matter and sand grains, some of which may have a thin dark coating of material. The modon changes fairly sharply into the gleyson with its characteristic ochreous mottling which forms an apparently irregular pattern or is associated with pores and ped surfaces. When the pore spaces are old root passages they often develop iron pipes which are clearly seen by the concentration of brown or yellowish brown iron oxide. The matrix in gleysons is usually birefringent with abundant large random domains or random zones of domains. Cutans vary in amount, depositional cutans are often occasional to frequent on ped surfaces while diffusion cutans surround sand grains. Weathering sand grains are sometimes quite common in this horizon. The gleyson grades with depth into the completely grey, olive or blue cerulon.

Analytical data

The content of organic matter in the hydromoron commonly exceeds 50% but there is a rapid fall to about 10% in the modon followed by a further decrease to < 1% in the gleyson. The C/N ratio in the hydromoron is usually greater than 20, indicating a low level of humification which is a normal feature for most of the soils of this class. The pH value in the surface is about 4·5 increasing to about neutrality with depth. The low values in the surface are due in part to the acid litter produced by the vegetation and to its decomposition under cool conditions. The base saturation is commonly high and in many of these soils the ratio of exchangeable calcium:magnesium is < 4, and lower than the associated freely draining soils. This is more common when subgleysols develop in depressions receiving drainage, for magnesium is more readily released by hydrolysis and transported in solution. Usually the clay fraction is fairly uniformly distributed throughout the pedounit or there may be a tendency for a little more to be present in the surface mineral horizon.

Genesis

Gleysons are regarded as forming in a zone in the soil that is saturated with water for a part of the year but partially or completely aerated during the summer or drier period of the year. As a result, oxidation of iron takes place locally along those surfaces that freely receive a supply of oxygen. These are usually the larger fissures or cracks formed upon drying as well as down old root channels. In contrast, cerulons form in zones that are permanently saturated, hence the absence of mottling.

In addition to wet conditions it appears that organic matter is necessary for iron to occur in the ferrous state, for there are many subsurface geological strata that remain continuously saturated with water without developing grey or olive colours. When horizons at or near the surface are saturated and the products of organic matter decomposition are dissolved in the water, then the characteristic colour patterns form. It does not appear that any specific organic substance is responsible, for Bloomfield (1964) has demonstrated that a variety of organic substances including leaf leachates can accomplish the mobilisation of iron and its transformation to the ferrous state. The most dramatic reduction of iron is seen in parent materials originating from certain Triassic or Devonian sediments in which the original bright red colour of the material has been transformed into olive and grey colours.

In the presence of such large amounts of moisture, hydrolysis and solution proceeds at a faster rate than in adjacent, freely drained soils or parent material of a similar age. This has caused more rapid rock disintegration, for it is common to find in these soils, stones that are easily broken down to their constituent minerals with a slight blow from a hammer but similar stones are quite hard in the adjacent freely drained soils. Probably this is better illustrated by the distribution of total phosphorus which is lower in these soils than in their freely draining neighbours.

Principal variations in the properties of the class

Parent material

The greatest extent of subgleysols is found on glacial deposits; consequently they usually have a medium or coarse texture and contain large amounts of felspars and other primary minerals. Elsewhere they are developed on a wide range of sediment mainly of Pleistocene or Holocene age. Since the drainage status of the soil is the most important factor in their formation, their mineralogy varies from basic to acidic and in some instances there are high proportions of carbonates. However the mineralogy does influence some of the properties such as the type and amount of cations which will increase as the material becomes more basic or contains more carbonates.

Climate

The atmospheric climate has little effect upon the formation of these soils because their development is controlled largely by topography however they tend to be more common in areas of high rainfall or where evapotranspiration is much less than precipitation. These latter conditions are most common where the climate is marine, taiga or humid continental cool summer, but they are also widespread in certain situations in the wet and dry tropics, rainy tropics and monsoon tropics.

Vegetation

The plant communities on these soils usually differ strongly from those of the adjacent drier sites. In a marine climate, species of *Sphagnum, Eriophorum* and *Juncus* predominate.

Topography

These soils usually form where the water-table comes near to the surface, this is on flat sites, depressions or at the lower ends of slopes. However they can form over the surface of gentle undulations in some areas with high rainfall.

Age

The greatest areas of subgleysols have formed during the Holocene period. Even when they are found associated with strongly weathered soils in old landscapes it is usual to find that their parent materials are late Pleistocene or early Holocene superficial deposits. This is shown also by their degree of weathering which is usually small as compared with some of the surrounding soils which can be krasnozems or zheltozems in tropical regions.

Pedounit

Some of the most important variations are found in the upper horizons. Under natural conditions a modon or an anmooron are common but where cultivation has been practised for some time or where the parent material is basic or contains carbonate it is usual to find a mullon.

The pattern of mottling also shows important variations which seem to result from differences in the texture of the material. In sands and loams there are usually small irregular areas which as stated above are associated with large pores especially old root passages. In fine textured material with a well developed structure the surfaces of the peds are usually grey due to maximum reduction of iron. This results because moisture moves freely down the pore spaces. Part of the iron that is reduced diffuses into the inner part of the ped where it is oxidised during the succeeding dry season. Thus the pattern from the outside to the inside of the ped is grey, yellow and reticulate mottling on the inside often with black staining of manganese dioxide. Subgleysols intergrade into a number of other soils which may be freely draining such as podzols or altosols or they may grade into peat. Commonly they grade into solonchaks as the climate becomes more arid.

When the parent material is very basic or contains carbonates they are usually saturated with basic cations and the pH values are slightly above neutrality.

Distribution

These soils occur throughout a large part of the earth's surface. Their greatest extent is in the cool humid parts of the world particularly in Canada and northern Eurasia.

Utilisation

The major obstacle to the utilisation of subgleysols is the large amount of water which has to be removed by installing an adequate system of drainage. This may be in the form of open drains but tile drains are usually more effective and it is often advantageous to place the drains at some depth in order to intercept the ground-water thus preventing its rise to the surface. Where the content of organic matter at the surface is high and pH values are low, these properties can be altered by the addition of liming material and ploughing. The latter improves aeration while the former raises the pH values, providing a better habitat for the soil micro and meso organisms so that the decomposition of the organic matter will proceed. Where continuous agriculture is practised it is necessary to add fertilisers particularly phosphorus, potassium and nitrogen.

When adequate drainage is accomplished these soils can be adapted to the systems of agriculture of the adjacent sites. This may be a form of arable crop rotation, dairying or horticulture. In low lying situations drainage is extremely difficult so that cultivation is almost impossible except in the driest years. When an attempt is made to cultivate soils that are very wet, the implements may become stuck or large clods may be produced instead of a good seed bed, similar difficulties may be encountered at harvest time. Often wet soils are kept under permanent grass but the sward may be badly damaged by the hooves of the grazing animals, during the wettest period of the year.

SUPRAGLEYSOLS   [Plate Vb

Derivation of Name: From the Latin word *supra* = above, Russian word *glei* = compact bluish-grey material and French word *sol* = soil.

Approximate Equivalent Names: Surface water gleys; Pseudogleys; Stagnogleys; Aquents; Aqualfs; Aqults.

References: Mackney and Burnham 1966; Mückenhausen 1962; Racz 1964; Racz *et al.* 1967

General characteristics

These soils are not characterised by a specific horizon but by unique sequences of horizons in which the middle horizon or horizons show the influence of anaerobic conditions. This is due to the slow permeability of the lower horizons which causes moisture to accumulate in the middle part of the soil for varying periods during the year.

Morphology

Under natural conditions the surface may have a thin litter followed by a thin humifon which rests on a modon. In thin sections the humifon is very porous and composed of small irregular black particles. The modon is composed of small amorphous black areas and sand grains some of which have thin dark coatings of material. The modon grades into the pale coloured candon with similar or slightly finer texture and massive or coarse subangular blocky structure. In thin sections cutans are rare around pores while the matrix is predominantly birefringent with abundant random domains. Sesquioxide concretions are rare to occasional and randomly distributed. The candon passes fairly sharply into the brown glosson of similar texture, coarse prismatic structure, characteristic light coloured tongues and frequent manganiferous concretions. These tongues have a characteristic pattern and can be regarded as vertical continuations of the candon. In plan they form a subhexagonal pattern and have a thin ochreous border on either side. In thin section the matrix in the brown areas is predominantly birefringent because of abundant, small, random domains and occasional zones of domains. The manganiferous concretions are subspherical, opaque and have diffuse borders. The grey central part of the tongues have occasional clay cutans while the matrix in the yellowish brown border is strongly birefringent due to abundant, medium, random domains. The glosson often grades gradually into the relatively unaltered underlying material which is usually a sediment of Pleistocene age.

Analytical data

The clay content is usually uniformly distributed throughout the soil, but there is often a slightly lower content in the modon.

The content of organic matter in the modon is usually less than about 10% with a C/N ratio of 14 – 16. The organic matter decreases sharply to about 2–3% in the candon and there is also a fall in the C/N ratio indicating a higher degree of humification. The pH values increase steadily from about 5·0–5·5 at the surface to 6·0 in the lowest horizon. The percentage base saturation is usually < 20% in the surface increasing to over 50% in the lowest horizon and the SiO_2/R_2O_3 ratio of the clay fraction is fairly uniform throughout the soil. The minerals in the clay fraction are for the most part derived from the parent material and therefore are very variable. The usual tendency is for mica, kaolinite and vermiculite to predominate.

Genesis

Supragleysols are formed by the accumulation of moisture within the pedounit during winter, spring and the early part of summer. This causes iron to be reduced to the ferrous state aided by dissolved organic substances and to be reorganised within the soil. Some

of the products of hydrolysis may be lost in the drainage water but much of the iron and manganese that is mobilised during the wet period seems to be redistributed to form the small concretions in the candon and glosson. Some redistribution of clay also seems to take place as indicated by the presence of cutans in the candon and glosson, particularly on the faces of the large prismatic units of the latter.

The formation of the subhexagonal pattern of grey tongues may have resulted from shrinking and cracking upon drying or by the development of ice lenses during a cold phase of the Pleistocene. The latter explanation appears to be applicable to parts of Europe which have many well authenticated periglacial phenomena. When the ice disappeared the spaces that were left formed natural channels for water movement. Gradually the iron has been mobilised from their surfaces, some has been removed in solution but some iron has migrated into the soil forming the thin reddish brown zone on either side of the light coloured streaks. Presumably migration has taken place in the ferrous state during the wet period of the year followed by oxidation during the dry summer period.

Principal variations in the properties of the class

Parent material

This is usually material of medium or fine texture. In many cases it is loess or alluvium but these soils will also develop on loamy glacial drift. In a number of situations an old soil appears to form the parent material.

Climate

These soils are found mainly in cool moist continental areas where the upper part of the soil becomes wet during the spring and early summer but dries out during the summer.

Vegetation

The original plant communities on these soils were dominated by deciduous forests in which oak (*Quercus* sp.) was probably one of the principal species. Many sites have been cleared for agriculture which often failed. Now many supragleysols have permanent pasture or a scrub community; in Europe these latter are dominated by species such as heather (*Calluna vulgaris*) and bracken (*Pteridium aquilinum*).

Topography

Supragleysols are confined to flat or gently sloping

situations where moisture can accumulate in the upper part of the soil as a result of slow run-off and low permeability.

Age

The development of the candon has taken place mainly during the Holocene period but some supragleysols may have been developing continuously from the late Pleistocene period. In many cases they show polygenetic evolution from a Tertiary soil or sediments derived from Tertiary soils.

Pedounit

The principal variations in the morphology of the soils of this class occur in the middle part of the soil. Frequently a gleyson overlain by a minon develops when conditions are wet for long periods of the year. In such cases a marblon or glosson usually underlies the gleyson. Some soils have an ison in the lower position which is responsible for the excess moisture near to the surface. There are a number of complex pedounits in which the upper and middle horizons may be those of an altosol or podzol followed by a candon and then a glosson. A relationship between supragleysols and other soils is given on page 261.

Distribution

These soils occur mainly in a cool, humid, continental area; they are common in the central and eastern part of Europe, central U.S.A., various parts of Australia and New Zealand.

Utilisation

The annual saturation of the middle part of the soil severely restricts the utilisation of supragleysols but with drainage this problem can be overcome. It may be difficult to achieve adequate drainage because the zone of maximum moisture accumulation is near to the surface. When tile drains are used they are liable to be disrupted by cultivation, frost or plant roots. The low pH values at the surface and the massive structure of the horizons are also limiting factors requiring liming and thorough cultivation. When these ameliorative methods have been carried out and fertilisers applied these soils give good yields of a variety of crops.

THIOSOLS

Derivation of Name: From the Greek word *theion* = sulphur

Approximate Equivalent Names: Cat clays; Acid sulphate soils.

General characteristics

At the surface these soils usually have a dark brown or black plastic hydromoron or there may be peat underlain by the olive, grey or bluish grey thion containing pyrite or elemental sulphur, and varying amounts of partly decomposed organic matter. These soils are associated mainly with tidal mangrove swamps which usually have the distinctive pungent odour of hydrogen sulphide. When they are drained the pyrite is oxidised to form straw coloured crystals of basic ferric sulphate (jarosite) and sulphuric acid accompanied by the development of very acid conditions. The peat or hydromoron is also oxidised forming an anmooron. Such soils will have the following sequence of horizons: anmooron, jaron, thion.

Few micromorphological studies have been conducted on these soils however Eswaran (1967) has shown that the thion is massive with few pores. The matrix contains abundant, random, large domains and occasional small crystals of pyrite often associated with decaying roots. The jaron has no pyrite and the matrix has fewer domains.

Analytical data

These soils are usually formed from fine textured materials which upon drying develop a coarse angular blocky or prismatic structure, with large cracks. The most important chemical property is the high content of sulphur which is usually 2–3% but over 6% has been reported from Malaysia. Upon drying, oxidation takes place resulting in the formation of ferric sulphate and sulphuric acid. This leads to very acid conditions with pH values of less than 3. An interesting feature is the seasonal fluctuations of the pH values, when the soils are saturated with water and therefore anaerobic the pH may rise to 7 or over, but it drops rapidly to values of less than 3 upon drying. At the surface the organic matter has a C/N ratio of 30 or over indicating a low degree of humification. The cation exchange capacity is high in the organic surface but moderate in the mineral soil, however the base saturation is low.

Genesis

These soils are developed mainly in sediments which are usually bluish grey or grey marine muds, but they can be inland alluvium. The precise mechanism of formation is not understood but accepting the suggestions of van der Spek (1950) the following stages seem probable:

1. Reduction of sulphates to hydrogen sulphide by sulphate reducing bacteria under araerobic conditions. The sulphates are usually supplied by sea water but the reduction of elemental sulphur or organic sulphur can also be accomplished by microorganisms. In all cases an adequate supply of organic matter is necessary as a food supply for the bacteria.
2. Reaction of hydrogen sulphide with iron compounds present in the soil to form ferrous sulphide – FeS and pyrite – FeS_2. If the soil remains anaerobic the reaction stops here and there is a gradual build up of pyrite under the naturally wet conditions.
3. With drainage for cultivation, oxidation of the iron sulphide by oxygen of the atmosphere takes place resulting in the formation of ferric sulphate and sulphuric acid.
4. Hydrolysis of the ferric sulphate to form the straw coloured basic ferric sulphate (jarosite) and more sulphuric acid. Although these two final stages can be accomplished by simple chemical processes, microorganisms may also play a part.

Sulphuric acid has a number of very adverse effects on soils and plants; in addition to increasing the acidity the large number of dissociated hydrogen ions replace the base cations in the exchange complex. It attacks the primary minerals causing essential plant nutrients such as calcium and potassium to be released by hydrolysis, brought into solution and lost in drainage water. The clay minerals are also attacked with the result that large amounts of aluminium sulphate are formed followed by the precipitation of aluminium hydroxide. Since the acidity only develops upon oxidation, the pH in the soil may vary from less than 3 in the oxidised surface up to 7 in the thion.

Although the sequence of events outlined above usually takes place after drainage, Moormann (1963) has reported their occurrence on naturally drained old elevated river terraces.

Principal variations in the properties of the class

Since thiosols are associated mainly with coastal conditions, variations in the factors of soil formation have little influence upon their formation. However there are some differences in the nature of the pedounit. The upper horizon may vary widely in thickness and content of organic matter and so may the jaron which increases in thickness as drainage improves and oxygen penetrates deeper into the soil.

Distribution

These soils occur to a greater or lesser extent in most tropical areas adjacent to the sea. Extensive areas occur in south-eastern Asia particularly in Vietnam, Thailand, Indonesia, Sumatra and Borneo; they occur in East Pakistan and in East as well as West Africa. In the new world they have been reported from Surinam and doubtless they occur elsewhere. They also occur in temperate areas as in Holland where they were first described and intensively studied.

Utilisation

Thiosols are utilised only when there is great need for land, otherwise they are left in their natural state. The poor structure, low base saturation and wide C/N ratio are obvious limiting factors to plant growth but it is their high acidity following drainage that is the main factor preventing land use, for it is necessary to apply extremely large amounts of lime to raise the pH and this may not be economic. The extreme acidity causes iron, aluminium and possibly manganese to be present in toxic proportions also the high content of iron causes phosphate fixation which reduces their fertility still further. However some success has been achieved, for example, by keeping the soil saturated throughout the year, oxidation is prevented and the acidity is kept to a minimum, then some varieties of rice can be grown. When the soils are drained and become very acid some acid tolerant species such as pineapples, can sometimes be grown. Another difficulty encountered when drainage is carried out is the rapid silting or clogging of the drains by $Al(OH)_3$.

The most expedient method of amelioration seems to be a gradual improvement of drainage, coupled with the addition of small amounts of lime. In this way there is only a small annual production of sulphuric acid which is neutralised by the lime and the resulting salts particularly aluminium sulphate are lost in the drainage and not precipitated as aluminium hydroxide.

VERTISOLS

[Plate Vc

Derivation of Name: From the Latin word *vertere* = to turn
Approximate Equivalent Names: Black tropical soils; Tirs; Barros; Regurs.
Reference: Dudal 1965

General characteristics

These are dark coloured soils having fine or very fine texture and a low content of organic matter; but perhaps their most important property is the dominance in the clay fraction of montmorillonite which causes these soils to shrink and crack upon drying. Typically they occur in arid and semi-arid areas beneath tall grass.

Morphology

The surface may have a sparse litter but usually the bare, dark coloured, fine texture, soil is exposed. The uppermost mineral horizon can be a thin granular crumon (Fig. 57) or a thin dermon which is massive or platy. In a humid environment a luton may develop if there is repeated cultivation and the structure is allowed to deteriorate.

The dark coloured verton varies in thickness from 30–100 cm and comprises most of the pedounit. This horizon is also of fine texture and has a structure that varies from coarse prismatic or angular blocky at the surface to wedge at depth (Fig. 63). The coarse prismatic structure is due to shrinkage upon drying, forming cracks that may be up to 10 cm wide and penetrate to over 1 m in depth while the wedge structure units usually have shiny slickensided surfaces. The colour of the crumon, dermon and verton usually have hues of 2·5Y or 10YR with low values of 2 and 3, similarly the chromas seldom exceed 2. Concretions of calcium carbonate are often distributed through the soil and may form a thin layer on the surface. The consistence of these three horizons varies widely with moisture content, generally they are hard when dry, firm when moist, plastic and sticky when wet.

In thin section the verton has a dominantly birefringent matrix being composed of abundant random fine domains and frequently there are zones of domains some of which have an oblique orientation. Many of the peds have very thin birefringent surfaces which are interpreted as slickensides and pressure cutans (Fig. 71). A characteristic feature of the matrix is the abundance of opaque grains of fine sand and silt size. These are presumed to be composed of manganese dioxide which also forms coatings or dendritic inclusions in the carbonate concretions (Fig. 116).

The verton usually grades into the underlying material through a gradual change or there may be an irregular mixed pattern or commonly there may be interdigitations. The lower horizon may be weathered rock or relatively unaltered sediments but in most cases it is extremely difficult to determine where the unaltered material begins.

Analytical data

The content of clay in these soils is usually uniform throughout the pedounit being > 35% but in many cases it exceeds 80%. Although the mineralogy of the clay fraction is somewhat variable, it is dominantly montmorillonite or mixed layered minerals which have a high capacity for expansion and contraction following upon wetting and drying giving changes in volume of 25–50%. In a number of situations mica and kaolinite may be present but the amounts of these two minerals are fairly small.

These soils have density of 1·8–2·0 in the verton and therefore are more dense than most soils, probably resulting from repeated expansion and contraction which causes closer and closer packing.

The content of organic matter can be as much as 5% at the surface but usually it is not greater than 1–2% with a C/N ratio that is sometimes wide, but usually is 10–14.

As expected from the clay content and composition, the cation exchange capacity is high and varies from 25–80 me% with a high degree of base saturation which is seldom less than 50%, increasing with depth. A somewhat unique feature which is shared with some brunizems of the central U.S.A. is that for a given percentage of clay the cation exchange capacity decreases as the content of organic matter increases. Most vertisols contain free calcium carbonate in the form of powdery deposits or as concretions but many do not have this property; the content may be as high as 60%, but usually varies from about 5–10%. Exchangeable sodium is normally in the range of 5–10% and therefore higher than for soils of humid areas but is much less than in saline or alkaline soils.

Generally the salinity is low since salts seldom appear to accumulate in vertisols and when they do it is usually below about 30 cm. Vertisols seem to have a self-flushing mechanism, any salts which accumulate on the surfaces of the peds in one season are washed into the lower part of the soil by the rains of the next season. These chemical properties combine to give vertisols pH values that range from 6·0–8·5. However pH values increase as the exchange complex becomes more saturated with sodium and therefore the soils are intergrading to solonchaks or solonetzes.

Genesis

Some vertisols have formed by progressive hydrolysis of the underlying rock, others have formed in fine textured sediments which either contain large amounts of expanding lattice clay or else montmorillonite has formed in these sediments.

The principal process taking place in these soils is the constant churning of the upper horizons. When the soil dries and cracks some of the surface horizon falls into the crack; consequently when the soil becomes wet and expands, high pressures develop which are released by upward movement of the material.

500μ

Fig. 116.
Concretion of calcium carbonate in a vertisol. It is composed mainly of small crystals of calcite and has an outer coating of manganese dioxide which also occurs as dendritic growths within the concretion.

Annual repetition of this cycle results in churning of the soil down to the depth of cracking which is usually about 1 m, hence the relatively deep, uniform pedo-unit.

An additional result of the release of pressure is the differential displacement of the material, causing the formation of the wedge structure and slickensides, as a result of one part of the soil moving and slipping over another part.

It appears that two of the important requirements for the formation of these soils are a period of complete saturation with water, and secondly a marked dry season. The period of complete saturation causes an-aerobism and reducing conditions. On the other hand the marked dry season causes many of the basic cations to remain in the system thus producing suitable conditions for the formation of montmorillonite. In view of the small amount of organic matter it is difficult to determine the origin of the dark colour of these soils. Recently Singh (1956) has suggested that it might be due to a dark coloured complex of organic matter and montmorillonite which forms in the wet environment when these soils are flooded during the wet period of the year. Thin sections of some vertisols from several places show that the matrix contains a large amount of finely divided opaque material which may be a second-ary dark coloured mineral of iron or manganese which might form under anaerobic conditions. Therefore the dark colour may be due to a combination of properties.

Principal variations in the properties of the class

Parent material

Many vertisols have developed in superficial de-posits of fine or very fine texture; these are usually alluvium or lacustrine deposits, however in some cases it is not possible to be certain about their origin. Some deposits appear to be colluvium derived by the insidious erosion of soils and the accumulation of material in a depression on a flat site. Other vertisols have developed through progressive weathering of the underlying rock which may be basalt, shale, limestone, or volcanic ash.

Their development is encouraged by a high content of plagioclase felspars, ferromagnesian minerals and carbonates. Deficiencies of certain minerals parti-cularly magnesium in the parent material can some-times be made good by seepage.

In a number of instances the presence of vertisols is determined by the occurrence of a particular type of parent material. Usually they tend to become confined more and more to basic or carbonate parent material as the climate becomes more humid, the reason being that both of these materials produce large amounts of cations which maintain suitable conditions for the formation of montmorillonite. In some very humid environments they are confined to sedimentary rock containing montmorillonite and carbonates.

Climate

Vertisols develop within the climatic range that includes marine, humid continental warm summer, wet and dry tropics, and semi-arid climates. Their greatest extent seems to be in tropical and subtropical arid and semi-arid areas where leaching is at a mini-mum so that basic cations accumulate in the soil pro-viding conditions for the formation of montmorillonite. Under these conditions precipitation varies from 250 to 750 mm and there is a marked dry season of 4–8 months.

Vegetation

Since the greatest extent of these soils occurs in tropical semi-arid areas the predominant plant com-munities are dominated by tall grass, or acacia wood-land. In some cases the grass communities are considered to be secondary after forest. In a humid environment, as in parts of Java, teak forests occur on these soils.

Topography

Vertisols develop only on flat or gently sloping sites, usually on terraces, plains and valley floors; occa-sionally they occur on low smooth crests but they never occur on slopes > 8%. These soils are more common at elevations below 300 m but extensive areas in India are above this altitude. Commonly, vertisols have gilgai phenomena which form small but characteristic topo-graphic features whose frequency varies from country to country. They are common in Australia and the U.S.A. (Texas) but are less common in India and South Africa (Figs 45 and 46).

Age

Vertisols vary in age from Holocene to Pleistocene as indicated by the age of the material on which they are formed. However, where they are formed from the underlying rock they may date from the mid-Pleisto-cene or earlier but where they are formed in trans-ported old soil material or other sediments they are likely to be of mid-Pleistocene age or later.

Pedounit

Some of the principal variations in this class are induced by climatic differences. In a humid environment there is a tendency for large amounts of moisture to pass through the soil so that the content of soluble salts and exchangeable sodium is low. Also there may be a luton at the surface with its characteristic massive structure. In such situations the lower horizon may be a grey or olive clay rich celon showing the influence of large amounts of moisture. With progressive aridity vertisols show a gradual increase in the amount of exchangeable cations and more particularly calcium carbonate which may form a calcon or it may be fairly uniformly distributed throughout the soil in the form of concretions or as a fine powder. Gypsons also occur in the lower part of vertisols. These two horizons are often best developed in situations where the original material in the lower horizons was of a slightly coarser texture and therefore is not disrupted by churning. Some of the best examples are found where vertisols have developed from rock so that there is a lower

horizon of chemically weathered rock with little clay. It is in this layer that carbonate has been deposited to form the intergrade (CkBFw), thus indicating that some vertisols may be polygenetic. With an increase in the amount of sodium, the expansion and contraction of these soils upon wetting and drying increases and the development of slickensides often increases in size and may also cause an increased frequency of gilgai.

The stone content of vertisols is usually low but when the soils are shallow and formed from rock there may be an occasional stone in the pedounit. It is more usual for the stones to be carried to the surface by expansion and contraction to form a hamadon.

There are also a number of intergrading situations chiefly to pelosols, thiosols, rossosols and krasnozems.

Distribution – Fig. 117

Vertisols occur mainly in arid, semi-arid and wet and dry tropical areas but they also occur elsewhere making a total area of about 257 million hectares which is equivalent to the size of western Europe with

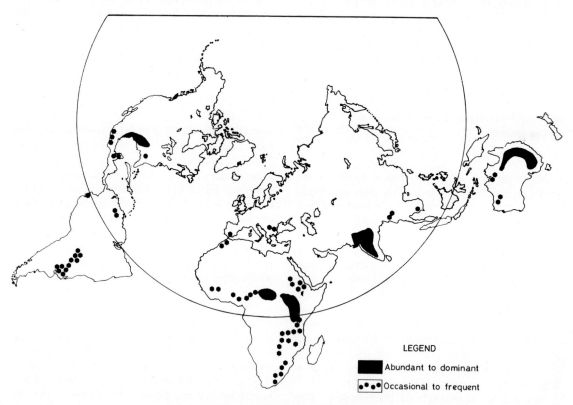

LEGEND

Abundant to dominant

Occasional to frequent

FIG. 117.   Distribution of vertisols.

the largest areas occurring in Australia, India and the Sudan.

Utilisation

The extent to which these soils are utilised depends very much upon the development of local technology. Generally in tropical areas the natural grassy vegetation is grazed but this is often a tenuous practice because of water shortages, so that a measure of shifting husbandry has to be practised particularly if there are periods of drought.

Where arable agriculture is carried out moisture conservation is essential through the improvement of infiltration and reduction of losses by excessive evaporation and transpiration. If there is an adequate supply of good water, irrigation is normally practised; perhaps the classical example of this is the Gezira in the Sudan. On the other hand, where moisture is a serious limiting factor, then dry farming is practised. The high content of clay in vertisols can impose severe limitations on their utilisation because the moisture range for cultivation is narrow. If cultivation is attempted when the soil is not at the optimum moisture level, either puddling will result if it is too wet or it will be very intractable if it is too dry.

Generally, however, the level of utilisation is fairly primitive being usually at subsistence level, employing hand tools without the addition of fertilisers or irrigation. Therefore it is not possible to be certain about the full potential of these soils but in some places such as Australia where a high level of technology is applied crop yields are high and a high standard of living results. Apart from grazing the main crop grown is cotton followed in importance by sugar cane and other grain such as sorghum, millet, rice and wheat.

Vertisols are usually deficient in many of the major and micro plant nutrients; the amounts of nitrogen, phosphorus and potassium are low and have to be supplemented by the addition of fertilisers if high yields are required but the response is sometimes disappointing. However the release of nutrients is usually sufficient to maintain a subsistence level of agriculture for crops like cassava and ground nuts.

Vertisols are highly susceptible to all forms of erosion, even slopes of $5°$ or less may develop deep gullies in a very short period. A characteristic form of erosion are landslides during which a large area will move as a unit down the slope as the upper horizons slide over the lower layers.

Finally it can be said that this class of soil has the greatest need for improved utilisation since a ten fold increase in production can result by the application of modern technology.

ZHELTOZEMS

[Plate Vd

Derivation of name: From the Russian words *zhelti* = yellow and *zem* = soil.
Approximate Equivalent Names: Red-yellow latosols; Orthox; Torrox; Ustox.
References: Gauld 1968; D'Hoore 1963; Maignien 1966.

General characteristics

The predominant horizon in the soils of this class is the thick, relatively uniform brown or yellowish brown zhelton which has been formed by profound hydrolysis and can be several metres thick. Zheltozems develop predominantly in humid tropical areas where the soils are very moist for long periods during the year and the vegetation is a forest community.

Morphology

Under natural conditions there is generally a sparse, loose, leafy, litter at the surface and exposed areas of mineral soil. The uppermost horizon is usually a brown or dark brown fine granular clay loam or clay tannon about 10 cm thick, grading into the brown or yellowish brown zhelton which commonly exceeds 1 m in thickness and may have a slightly finer texture than the tannon above. The structure of the zhelton is somewhat variable and seems to depend upon the nature of the faunal population. Where there are no mesofauna the soil has a massive structure with occasional linear pores. On the other hand where organisms such as termites are common the whole soil appears to have been churned by their activity, producing a marked vermicular structure composed largely of faecal material (Fig. 67).

The matrix in the tannon is isotropic or there may be occasional small randon domains. In the zhelton the matrix is dominantly birefringent being composed of medium or large random domains, also there may be clusters or narrow random zones of domains.

Where these soils are formed on sedimentary rocks such as shales and sandstones the zhelton usually grades sharply into the weathered rock with well preserved structure even although it may be very soft and easily dug with a spade.

Analytical data

One of the most important features of these soils is the high content of clay which has been formed by progressive hydrolysis of the primary silicates and the formation of clay minerals and oxides particularly kaolinite, gibbsite and goethite. Weathering has been so effective that there is very little material in the size range 0·5 to 50 μ, generally the greatest amount of the material is $< 0·5 \mu$. The small amount of material in the middle size range and therefore the small silt:clay ratio is a characteristic feature of many soils of the tropics, indicating that even the highly resistant primary silicates in this size range have been decomposed.

In recent years the percentage of water dispersable clay has been used as a criterion to distinguish between certain tropical soils. In these cases the percentage decreases with depth to zero indicating a high level of aggregation.

In spite of the luxuriance of the forest, the organic matter content in the upper horizon of zheltozems is only 2–3% with a C/N ratio of 8–12, indicating that humification is fairly complete and is doubtless due in large measure to termites. The cation exchange capacity is characteristically low being less than 15 me. per 100 gm of clay and it is even low in the surface horizon containing organic matter. The base saturation is also low because of small amounts of primary minerals and continuous leaching due to high rainfall. The pH values at the surface are about 4·5, remaining uniform or decreasing slightly in the zhelton and then increasing with depth to about 6·0. The partial ultimate analysis shows that silicon, aluminium and to a lesser extent iron are the main elements present. The small amount of iron is particularly significant, indicating that it does not accumulate in these soils as it does in other tropical soils. The sand fraction is dominated by quartz with small amounts of resistant minerals such as zircon, magnetite and rutile.

Genesis

These soils form by the progressive hydrolysis of the primary silicates with the formation of kaolinite, mica, goethite and gibbsite and the gradual loss of the structure of the weathered rock to form the homogeneous zhelton. It is not certain how this is accomplished; in some areas faunal activity may be responsible but it seems to take place in the absence of fauna, then expansion and contraction and differential removal of material may be responsible (see formation of krasnozems).

As hydrolysis proceeds under conditions of high rainfall most of the basic cations and iron are removed from the soil. This contrasts sharply with krasnozems in which there is an accumulation of iron and aluminium. It would appear that in soils that remain moist throughout the year removal of iron is more complete and the formation of kaolinite takes place more readily. The large number of well formed domains in the zhelton is evidence of the high content of well crystallised clay minerals which in this case is kaolinite. Further, the large domains and markedly birefringent matrix seem to be indicative of wet conditions since this type of fabric is only found in soils of areas with high rainfall or partially anaerobic horizons. The yellow colour is due to the presence of goethite in a finely divided state having a high degree of hydration. This is also a feature of soils formed in a moist environment (Soileau and McCracken 1967).

Principal variations in the properties of the class

Parent material

Zheltozems occur in a wide variety of parent materials, which include consolidated rocks such as granite or shales but they also occur on sediment. For some reason that is not fully understood they seldom form on very basic rocks such as basalt where krasnozems develop.

Climate

The greatest extent of these soils occurs in a rainy tropical or monsoon tropical climate. They may develop locally in moist situations at the lower ends of slopes in the wet and dry tropics.

Vegetation

Tropical rain forests are the principal plant communities but semi-deciduous forest is also found. When these soils occur outside their normal climatic environment there is a wider range of plant communities including thorn woodland and savanna.

Topography

Zheltozems are confined to the lowlands and do not appear to occur above 1000 m to 1400 m. Within their normal climatic range they occur on sites that range from flat to quite steeply sloping but as the climate becomes drier they occur in the moister topographic situations on flat sites or at the lower end of slopes.

Age

Probably the youngest members of this class occur on sediments and land surfaces of Pleistocene age; therefore they are all old soils. Where zheltozems are formed on rocks such as granite they are probably at least of late Tertiary age because of the considerable amount of hydrolysis required to decompose the primary silicates.

Pedounit

In spite of wide variation in the factors of soil formation many of the pedounits of this class display very similar horizonation in the upper and middle positions but there is usually some variation in the thickness of the zhelton which may vary from 1 m to over 4 m in thickness. It is the lowest horizons that display the greatest variability, there may be a flambon or there may be chemically weathered rock. In a number of flat situations a pesson or a veson may form. On sloping situations it is common to find cumulons within the upper metre of the pedounit.

Distribution

It has already been stated that these soils occur in the wetter parts of the tropics up to an elevation of about 1400 m. Therefore their greatest extent is in south eastern Asia, parts of West Africa and certain equatorial areas of eastern Brazil. Their distribution in the past must have been more extensive for it seems that many luvosols of the [LvZh] subclass which occur in the subtropics were zheltozems of the Tertiary period.

Utilization

The low nutrient status and low organic matter content impart an overall low fertility status for most crops. However since rice is one of the principal crops grown on these soils the high clay content is an advantage since it restricts percolation when puddled. When dryland crops are grown, large quantities of lime and fertiliser are required but even so there is a tendency to leave these soils under forest and many areas that were cultivated have been abandoned. However extensive areas are used for rubber, coconuts, oil palms and tea. Erosion is a constant hazard and protective measures must always be in operation.

7

SOIL CLASSIFICATION
A REVIEW

INTRODUCTION

Prior to the contributions of the Dokuchaev School, soil texture formed the basis of most soil classifications. These were only of local significance, but they served in some instances to differentiate between the good and bad agricultural soils. Except in a few cases texture is used today at a low level of categorisation but is important in certain situations, particularly with regard to land capability classification.

The biological sciences have exercised a strong influence on the construction of methods of soil classification. They employ a hierarchical system in which there are a number of levels of categorisation, each having fewer members than the category below, achieved by grouping the members of the lower category on the basis of mutually exclusive properties. This arrangement forms a triangle with a relatively small number of groups at the apex and when set out on paper, resembles a family-tree. This system works satisfactorily only in the biological sciences but even in these some compromises are necessary. Many pedologists have attempted to produce a hierarchical system of soil classification. Some of them are presented below together with the reasons for their lack of success. These repeated failures should be interpreted as clear evidence that soil classification is an unattainable goal while the designation of soils by formulae as suggested in Chapter 5 is regarded as a practical expedient which allows information to be communicated in a simple and straightforward manner.

THE PRINCIPAL SYSTEMS DEVELOPED
IN THE U.S.S.R.

The work of Dokuchaev (Afanasiev, 1947) and his students focused attention on the soil profile which quickly led them to realise that many soil characteristics are profoundly affected by differences in environmental factors. It was evident also that in the U.S.S.R. there is a fairly close relationship between soils and vegetation, but more particularly between soil and climate. These discoveries revolutionised soil science and inspired Dokuchaev to propose a classification of soils, the final form of which is presented in Table 18 below. Most workers in the U.S.S.R. accepted these suggestions and later Sibirtsev (1914) produced a refined system by introducing the terms zonal and intrazonal soils, to replace two of the original names of the classes put forward by Dokuchaev. In both systems the members of the first two classes are defined in terms of environmental factors, while those of the third class are defined mainly on the basis of their intrinsic soil characteristics. Zonal soils have well developed pedounits which reflect the influence of climate and vegetation. Intrazonal soils also have well developed pedo-units, but are formed as a result of the influence of some specific local factor other than climate; this may be parent material or topography. The azonal soils are poorly developed soils such as recent alluvium or stony mountain soils. Dokuchaev also emphasised that the definitions of soils should include a consideration of the properties of the soils themselves.

This system is well suited to the great gently undulating continental transect down through western U.S.S.R. but it does not apply satisfactorily to mountainous areas, maritime conditions, or to the old land surfaces such as Africa or Australia. In the latter areas climatic changes during the Tertiary and Pleistocene periods brought about a situation where many soils that developed under one climatic regime are present in an entirely different one today. In spite of these deficiencies this system is still to be used in the U.S.S.R with slight modifications to include the influence of man and also some indications of the agricultural potentialities of the soils. (Basinski 1959).

The most recent work on soil classification to emerge from the U.S.S.R. is by Kovda et al. (1967) who state that their scheme is 'a historical-genetic classification ... according to their properties and characteristics which reflect their evolution in time and not according to environmental conditions as was done in some earlier classifications'. This scheme relies heavily on the interpretation of facts rather than on the facts themselves, namely the soil properties. In this respect it is identical with classifications based on environmental factors. However it represents a substantial break with the past and many of the ideas and practices used in the U.S.S.R. at present.

Table 18

CLASSIFICATION OF SOILS BY DOKUCHAEV

CLASS A. Normal, otherwise dry land vegetative or zonal soils

Zones	I. Boreal	II. Taiga	III. Forest-steppe	IV. Steppe	V. Desert-steppe	VI. Aerial or desert zone	VII. Subtropical and zone of tropical forests
Soil types	Tundra (dark brown) soils	Light grey podzolised soils	Grey and dark grey soils	Chernozem	Chestnut and brown soils	Aerial soils, yellow soils, white soils	Laterite or red soils

CLASS B. Transitional soils

VIII. Dry land moor-soils or moor-meadow soils	IX. Carbonate containing soils (rendzina)	X. Secondary alkali-soils

CLASS C. Abnormal soils

X. Moor-soils	XI. Alluvial soils	XIV. Aeolian soils

EARLY SYSTEMS DEVELOPED IN THE U.S.A.

Subsequent to the initial contributions from the U.S.S.R. workers in various parts of the world have taken one of two paths: either they have accepted the morphogenetic approach or have insisted that the intrinsic properties of soils should form the basis of soil classifications. Among the first to attempt the latter approach was Coffey (1912) who produced a system for the U.S.A. purporting to be based on soil properties but his terms such as 'Dark coloured prairie soils' have genetic connotations.

Marbut can be singled out as one of the great pedologists of the first four decades of this century. Although his original ideas led him to state that soil should be classified on a morphogenetic basis, he was fully convinced at a later date that the properties of the soils themselves should be used, whereupon he presented the system given in Table 19.

In this system soils are divided at first into two main categories – Pedalfers and Pedocals. The Pedalfers are soils that accumulate sesquioxides while the Pedocals have an horizon of carbonate accumulation. Marbut placed great emphasis on mature freely drained soils and so little account is taken of peat, poorly drained soils, or the initial stages of soil development. A further criticism is that difficulties are encountered when trying to accommodate soils such as the Altosols, (Brown earths) and Zheltozems, most of which accumulate neither iron nor carbonate.

Marbut's (1928) system was superseded quickly by that of Baldwin *et al.* (1938) who reintroduced an embellished zonal system of classification by including more soil types and a few minor refinements to suit the particular conditions in the U.S.A. This cannot be regarded as a step forward because it does not continue the more realistic trend established by Coffey and Marbut whereby soils should be classified on their properties. Soil classifications founded on zonal concepts are very attractive to soil surveyors, geographers and ecologists and this possibly accounts for their continued use.

Table 19
CLASSIFICATION OF SOILS BY MARBUT

CATEGORY VI	PEDALFERS (VI–1) Soils from mechanically comminuted materals.	PEDOCALS (VI–2)
CATEGORY V	Soils from siallitic decomposition products. Soils from allitic decomposition products. Tundra. Podzols. Grey-Brown Podzolic soils.	Soils from mechanically comminuted materials. Chernozems. Dark-brown soils.
CATEGORY IV	Red soils. Yellow soils. Prairie soils. Lateritic soils. Laterite soils. Groups of mature but related soil series. Swamp soils. Gley soils. Rendzinas.	Brown soils. Grey soils. Pedocalic soils of arctic and tropical regions. Groups of mature but related soil series. Swamp soils. Gley soils. Rendzinas.
CATEGORY III	Alluvial soils. Immature soils on slopes. Salty soils. Alkali soils. Peat soils.	Alluvial soils. Immature soils on slopes. Salty soils. Alkali soils. Peat soils.
CATEGORY II	Soil series.	Soil series.
CATEGORY I	Soil units, or types.	Soil units, or types.

THE SYSTEMS DEVELOPED IN EUROPE

A major contribution to soil classification was made by Kubiëna (1953) in *The Soils of Europe*. He produced a hierarchical system using all of the data available at that time. This, he claims to be a 'natural' system and contrasts with systems based on a limited number of properties which are regarded as 'artificial'. Kubiëna established three major divisions using drainage characteristics as the principal criteria; these are,

1. Subaqueous or underwater soils
2. Semi-terrestrial or flooding and ground water soils
3. Terrestrial or land soils

The data used in this classification include details about the genesis of soils as well as about their intrinsic properties and, for the first time, information is included about soils in thin section. Therefore this system can be regarded as morphogenetic. Moreover, it goes further and contains a few features that deserve special mention. There are numerous references to intergrading situations which impart a dynamic aura, but no attempt is made to introduce a method to designate the intergrades. However they are recognised which is a positive step forward when compared with much of the preceding work. The second important feature of this work is the use of names for some horizons and the attempt to define them in terms of their macro and micromorphology as well as by their methods of formation. The third contribution is that Kubiëna defined many more classes of soil for Europe than his predecessors but he failed to appreciate some important features particularly the extent to which clay can be translocated to form an argillon. Perhaps the most serious criticism which can be made of this work is that he described, only what might be called modal, virgin soils. No attempt was made to define the range within the individual members of each class nor was any account taken of cultivated soils. Even with these deficiencies his work was an outstanding contribution. In Europe many workers were quick to adopt the ideas of Kubiëna particularly in Germany where Mückenhausen (1962) produced a modification of his scheme.

Aubert and Duchaufour (1956) have attempted to produce a system based largely on soil evolution but it tends to oversimplify the situation. Although it predates that of Kovda et al. (1967) the two systems are very similar in approach and have the same deficien-cies. A further criticism of both systems is it is doubtful whether any system based on evolution can be sustained since soils are not like plants and animals for which a common ancestry can be demonstrated.

Tavernier and Marechal (1962) have suggested the use of a coordinate system for classifying soils, this idea is supported by Webster (1968) and foreshadowed by Crowther (1935) who suggested a conceptual model with infinite coordinates in hyperspace. Although it seems that this approach is the one most likely to be useful for the quantitative characterisation of soils and ordering them by computer methods, ultimately it will have to be used to create classes which will most probably be classes of horizons. Pedologists already employ a coordinate system for classifying texture in the form of a ternary system in a triangular diagram. Even in this relatively simple system, arbitrary classes are created and as with horizon boundaries, any two points in different classes immediately adjacent to the class boundaries are more alike than each is to distant points within its own class. However with all its deficiencies it is considered essential to create these classes for ease of communicating information.

SOME OTHER EARLY SYSTEMS OF SOIL CLASSIFICATION

A number of other classifications based on a variety of features have been tried. Set out below is a list in order of date showing the basis on which each system is established and also the authors concerned.

1. Mode of weathering – von Richtofen (1886)
2. Climate – Hilgard (1914)
3. Climate and weathering – Ramann (1911)
4. Maturity of profile and intensity of leaching – Glinka (1914)
5. Climate – Lang (1915)
6. Climate – Meyer (1926)
7. Factors of formation – Vilensky (1927)
8. Climate and parent material – Neustreuev (1926)
9. Character of the absorbing complex – Gedroiz (1929)
10. Climate and parent material – Stebutt (1930)
11. Chemistry of the soil – De 'Sigmond (1933)
12. Environment and process – Zakharov (1946)

DEFINITIVE SYSTEMS OF SOIL CLASSIFICATION

In recent years there has been a tendency to produce definitive systems of soil classification based upon the intrinsic characteristics of the soil, but also using genesis where appropriate. Published systems of this type include those of Del Villar (1937), Leeper (1956), Northcote (1960), Soil Survey Staff of the United States Department of Agriculture (7th Approximation 1960 and the Supplement 1967) and FitzPatrick (1967).

The system of Del Villar

This is probably the first fully developed system in the form of a key based solely upon intrinsic characteristics. Although this work seems to have been overlooked, it is of some significance since it establishes a complete break with previous systems.

The system of Leeper

Leeper (1956) was the next person to suggest that properties should be used for classifying soils. He states that a fixed number of properties should be chosen for each horizon and ordered according to their importance and used on a presence or absence basis. He does not develop the system, so it is difficult to know exactly how it would operate but some of his ideas seem to be incorporated into the more recent work of Northcote (1960) which is discussed later.

There is some merit in Leeper's scheme because it is the first attempt to give a designation to soils based purely on their intrinsic properties. Further, it requires the quantitive characterisation of each property to be used. Also it focuses attention and concentrates on the soil itself and not upon attempts to produce a hierarchical system. However like its predecessors it fails to make any allowance for intergrades and by ordering the properties according to their importance it is assumed that a given property always has a certain level of importance.

When due allowance is made for the above deficiencies it is questionable whether the presence or absence system constitutes a classification since it is largely devoid of versatility. After the system is established, the discovery or introduction of a new property means that the entire scheme has to be reconstructed to accommodate this new property. Therefore this is not a classification but the formulation of a designation for each horizon but is worthy of a measure of praise as it is the first attempt to recognise and designate all horizons.

Leeper shows much dislike for Kubiëna's system and states that all systems are artificial since they are the creations of man. Whether classifications be natural or artificial it would seem correct to consider all of the known properties when a phenomenon such as soil is to be classified and not merely a relatively small fixed number.

The system of the Soil Survey Staff of the United States Department of Agriculture

The 7th Approximation as the scheme is called, has made a considerable impact on pedologists but they are divided about its value: there are the strong adherents and there are also those who vigorously oppose the scheme (Webster, 1968). A scrutiny of this method for classifying soils does not reveal many reasons for giving it support. The proposers tacitly assumed that soils can be classified in a hierarchical manner. Adopting this well trodden and fruitless path comes as a great surprise since it should be obvious that hierarchical systems have repeatedly failed to satisfy the demands of the soil continuum. In this scheme, soils are divided into ten *Orders* which are created in a subjective manner, i.e. there is no fixed set of principles involved. Thereafter the Orders are divided into *Sub-Orders* which in turn are divided into *Great Groups* and then into *Sub-Groups*. The subdivisions are created largely on a 'presence or absence' basis. The names of the orders are all coined words with the common ending *sol* such as *spodosol*. A formative element is abstracted from the name and used as an ending for the names of all Sub-orders, Great Groups and Sub-Groups of one order. In the case of the Spodosols the formative element is -*od*-. The names of the orders and their equivalents in the modified systems of Baldwin *et al.* (1938) and FitzPatrick (1967) are given in Table 20. Each Sub-order name consists of two syllables. The first indicates the property of the class and the second is the formative element of the Order. Thus the *Orthods* are the common Spodosols (Gk. *Orthos* = true). The names of the Great Groups are produced by adding one or more prefixes to the name of the sub-order. Thus the *Cryorthods* are the cold *Orthods* (Gk. *Kryos* = cold). The sub-group names consist of the name of the appropriate great group preceded by one or more adjectives as in *Typic Cryorthods* – the typical Cryorthods. An example of how the system operates as applied to the spodosols is shown in Fig. 118. Also included are the equivalent horizon designations for the sub-groups.

The most striking feature of this system is the overall appearance or orderliness. The subdivisions appear to follow logically one from the other and the names seem to be comprehensive and informative, but this is certainly not the case particularly with regard to the value of the names of the sub-groups. A careful inspection of the 7th Approximation soon reveals that even the names at the sub-group level are either not very informative or they are not specific. For example it is shown in Fig. 118 that a Typic Fragiorthod may have a friable or hard sesquon. Alfic Fragiorthods have a much wider range of possibilities; in fact the only horizon common to all the members is an ison which is not a diagnostic characteristic of spodosols. The number of possibilities is even greater since the fragi-

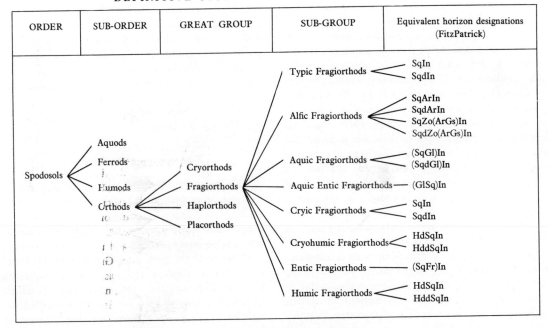

ORDER	SUB-ORDER	GREAT GROUP	SUB-GROUP	Equivalent horizon designations (FitzPatrick)

FIG. 118.   An illustration of the subdivisions
within the 7th Approximation.

Table 20

SOIL ORDERS OF THE 7TH APPROXIMATION AND THEIR APPROXIMATE EQUIVALENTS

PRESENT ORDER	APPROXIMATE EQUIVALENTS Baldwin et al. (1938) (modified)	FITZPATRICK (1967)
1. ENTISOLS	Azonal soils, and some Low Humic Gley soils.	Primosols, Rankers, some Subgleysols, Arenosols, Cryosols, Gelosols, Thiosols.
2. VERTISOLS	Grumusols.	Vertisols, some Pelosols.
3. INCEPTISOLS	Ando, Sol Brun Acide, some Brown Forest, Low-Humic Gley and Humic Gley soils.	Andosols, Altosols, some Subgleysols, Flavosols, some Pelosols, Rubosols, Rufosols, Suprageysols.
4. ARIDISOLS	Desert, Reddish Desert, Serozem, Solonchak, some Brown and Reddish Brown soils, and associated Solonetz.	Serozems, some Burozems, Solonchaks, Solonetzes, Solods.
5. MOLLISOLS	Chestnut, Chernozem, Brunizem (Prairie), Rendzinas, some Brown, Brown Forest, and associated Solonetz and Humic Gley soils.	Kastanozems, Chernozems, Brunizems, Rendzinas, some Altosols, Solonetzes and Subgleysols.
6. SPODOSOL	Podzols, Brown Podzolic soils, and Ground-Water Podzols.	Podzols, Placosols.
7. ALFISOLS	Grey-Brown Podzolic, Grey Wooded soils, Non-calcic Brown soils, Degraded Chernozem, and associated Planosols and some Half-Bog soils.	Some Argillosols, Burozems, Supragleysols, some Luvosols, Planosols.
8. ULTISOLS	Red-Yellow Podzolic soils, Reddish-Brown Lateritic soils of the U.S. and associated Planosols and Half-Bog soils.	Some Argillosols, Supragleysols and Luvosols.
9. OXISOLS	Laterite soils, Latosols.	Krasnozems, Zheltozems.
10. HISTOSOLS	Bog soils.	Peat soils.

pan of the 7th Approximation includes isons and fragons so that there may be twice as many horizon sequences if Fg could be substituted in every case for In. Thus, after descending through four levels of categorisation, very little positive information is available about the soil. This is inevitable when a presence or absence, or bifurcating system is used. An additional feature of this system is the great emphasis placed on so-called diagnostic horizons for which there are lengthy descriptions. Most of these are so broad that they are virtually meaningless. Perhaps an examination of the definition of the argillic horizon will demonstrate the weaknesses of the description of this particular horizon. It is stated that the argillic horizon is an illuvial horizon which contains more clay than the horizon above and usually has oriented clay which, when present is greater than one per cent. However the evidence presented does not demonstrate that an increase in clay in the middle position of the soil over that of the upper horizon is always due to translocation. In fact the descriptions of certain planosols do not even mention oriented clay. Furthermore it is well established that a marked textural change in some soils is often caused by differential weathering and does not result from the translocation of fine clay. Therefore it would seem that different argillic horizons not only have widely differing properties but also have different methods of formation. These ambiguities are not confined to the definition of

the argillic horizon, they are found in most others particularly in the definitions of the cambic horizon and the mollic epipedon. However the stress placed on recognising and defining a number of horizons is a *major contribution and emphasises the need for greater precision and more criteria for soil categorisation.*

A further severe criticism of this scheme is the absence of a simple and straightforward method for designating intergrades.

The system of Northcote

This system is entitled a 'Factual Key for the Recognition of Australian Soils' and judging from the evidence available, it accomplishes its objective in a most satisfactory manner. This is largely a bifurcating scheme using defined values of soil properties. Starting with the properties of whole 'profile form', thereafter it employs the properties of individual horizons. Figs. 119 and 120 illustrate how the scheme works. Now since this is a key and not a classification, soils such as Uc 2·3 and Um 2·3 that are morphologically and genetically similar will appear at widely different positions in the final column of the system but this is usually permissible in a key.

However, although this key appears to work in a highly efficient manner it can still be criticised on a number of very important points. As with all previous schemes little account is taken of intergrades. This

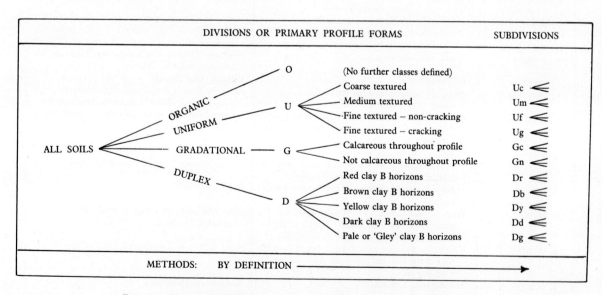

FIG. 119. Diagrammatic presentation of the division and subdivisions of the factual key.

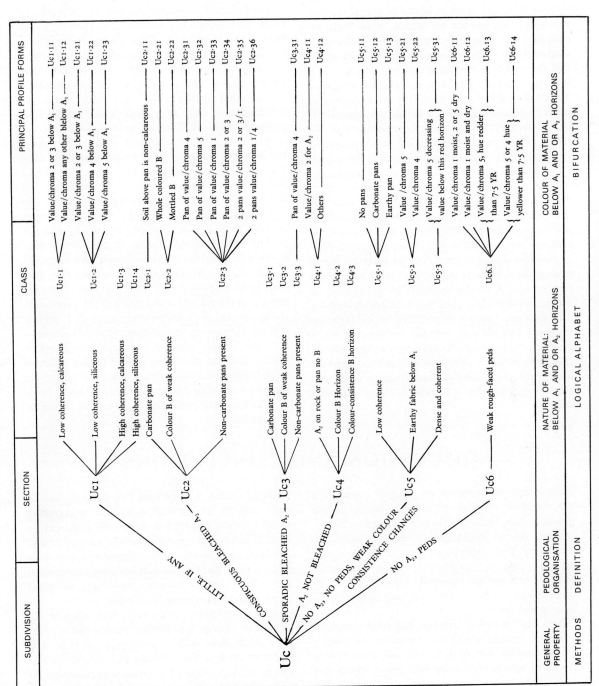

FIG. 120. Diagrammatic presentation of the sections, classes and principal profile forms of the subdivision Uc.

also causes many soils that are very similar to appear at widely different points in the last column of the scheme. A further point of weakness is that the code letters and numbers are not specific at all levels, for example, Uc 1·3 has high coherence and is calcareous, whereas Uc 3·3 has a non-carbonate pan. This shows that the final code number at the class stage has more than one meaning. The same is found at the stage of 'Principal Profile Form'. Thus it is not a system that could be memorised easily. Like the 7th Approximation, orderliness is apparent at first, but there are the same intrinsic deficiencies. It is not very informative. For example, the pedounit represented by the code Uc 2·31 can either have a zolon or candon overlying a hudepon since his scheme does not differentiate between these former two horizons. Thus after descending through five category levels an indication is made about the presence of only two horizons. At present there are about 300 codes, and any further subdivisions aimed at increasing its precision would enlarge the scheme to such a level as to make it too cumbersome. A further examination of the code Uc 2·31 shows that it is composed of five separate symbols. This does not compare favourably with soil formulae since any formula that contains five symbols states positively the presence of five horizons which is usually enough to designate fully most soils and too many for some. However in spite of the deficiencies mentioned above Northcote's key is superior to the 7th Approximation.

Numerical methods of soil classification

More recently numerical taxonomy has been applied to soil classification by a number of workers including Bidwell and Hole (1964). Their results are interesting but further refinements of their methods seem to be necessary. These authors were able to show that reddish prairie soils are very similar to chernozems thereby implying that colour is not a safe criterion for differentiating soils at a high level. This is now generally accepted by many workers hence the use of the term altosol in this book to replace the term brown earth. A further set of results shows that some Planosols appeared similar to brunizems. Bidwell and Hole recognise this as a serious defect and state that this 'suggests the need for reviewing the characteristics and re-evaluating them in order that the planosols might segregate more distinctly in keeping with our present concept of these soils'. In other words their desire is to express in numerical terms what is already established by subjective methods; further they state that some weighting might be necessary. This re-introduces the subjective element which does not seem to constitute an improvement upon our present systems. Perhaps one further criticism may be allowed – it is questionable whether such complicated mathematical analyses should be applied to data collected by subjective methods. Numerical methods would seem to have their greatest potential at lower levels of classification, particularly in helping to order intergrades.

DISCUSSION AND CONCLUSIONS

It is clear that attempts to produce hierarchical systems of soil classification based on morphogenetic principles have failed and the newer systems of Northcote and the 7th Approximation are little more than incomplete and cumbersome keys while numerical taxonomy has as yet not contributed any fundamentally new method but merely tries to establish by mathematical methods what has already been achieved by trial and error.

The most pertinent criticism that can be made of nearly all systems of soil classification is the paucity of information conveyed by the particular system employed whether it involves the use of names or a code. This is a serious deficiency in these newer schemes that employ long names and codes. There is also inadequate provision for intergrades in spite of

the large amount that has been written about them by many authors. In the final analysis an intergrading situation is not regarded as sufficiently important to be classified separately, consequently it is pushed into one pigeon hole or the other. This procedure amounts to depriving soils of one of their basic and fundamental properties – their continuum – Jones (1959) like many other workers has already stated this.

A new approach is necessary. Soils must be examined as separate and distinct phenomena in order to establish those features common to all soils, then to determine how these distinct and specific features can be used to construct a system or organisation for soils. The first feature possessed by all soils is their profile but as Jones (1959) rightly pointed out profiles have only two dimensions which make them unsuitable for any system of organisation. The second feature which is common to all soils are horizons which exist in three tangible dimensions, length, breadth and depth. The third feature is that horizons intergrade, laterally, vertically and in time with other horizons. Sometimes the lateral variations extend over a short distance; in other cases horizons intergrade gradually over long distances. These lateral intergrades are possibly the most important since there are many situations in which all the horizons do not intergrade laterally at the same rate. Further, detailed examination has shown that because of the continuum of horizons, soils have no properties other than horizons which can be used for their organisation. Any attempt to organise soils above the level of their constituent horizons can only be *ad hoc* and lacking in fundamental meaning since it leads to the sacrifice of intergrading situations, by doing this the spatial continuum is ignored and equally the concept of soil evolution is abandoned.

Because of this a fundamental system of classification is not possible. This inescapable reality and basic truth seems to have escaped the notice of pedologists who for the last seventy to eighty years have been striving to produce a hierarchical system of classification. These attempts must spring from the false assumption and incredible misconception that soils can be treated like discrete entities which have finite boundaries.

When the various branches of science are examined with classification in mind, one aspect is prominent and conspicuous. It is only in the biological sciences that one finds a hierarchical system and this is firmly based on Darwin's concept of evolution. There is no similar thread or theme linking soils together.

Repeated attempts have been made in previous chapters to demonstrate that soils form continua in space and time and that such a situation defies classification. However it is necessary to divide up the continuum into arbitrary units for ease of communicating information. The best method of achieving this has been demonstrated in Chapter 5 by trying to establish standard horizons and to use the horizon symbols to produce a formula for soils. This method focusses attention on the pedounit or the soil itself,

and achieves almost unlimited versatility while at the same time, the amount of nomenclature is reduced to a minimum. Intergrades are easily accommodated by using combined symbols, such as (SqAt). This is not a classification but a designation of each soil stating in a precise and concise manner its full character, including the nature of the relatively unaltered underlying material. It should be pointed out that even the most complicated names of the 7th Approximation do not give a full characterisation of the pedounit. Furthermore it is well known that any discussion about soils whether they be in writing or oral exchanges, quickly centres on the number and type of horizons. Therefore it seems logical to produce a system which, in the first instance, gives in a concise form, the number and type of horizons, and also some indication of their thickness. One of the principal aims of pedologists should be to produce a designation for soils and a simple but *ad hoc* system of classification. At the present state of knowledge a formula appears to be the best approach.

The work of Northcote has clearly demonstrated the need for a key to aid the identification of soils. Ideally this should be produced for all soils but the data available at present is insufficient to accomplish this aim. As a first step in this direction the Tables given on pages 109–153 have been prepared.

The system used in this book like any other which sets out to designate the soil, is suitable for punched cards and for the production of a dictionary of formulae. Information contained on punch cards as formulae is easily retrieved, it also allows relationships to be demonstrated with great ease as well as simplifying any *ad hoc* grouping or the construction of simple hierarchies.

In conclusion it should be pointed out that systems like those of Kubiëna, Northcote, and the 7th Approximation are all descending or bifurcating systems operating on a presence or absence basis. The system used in this book is an ascending system in which the soil formulae can be regarded as the starting point of a classification, which can be built up in a variety of ways to suit particular situations and allows almost unlimited versatility.

8

SOIL RELATIONSHIPS

The various relationships can conveniently be considered under the following headings:

    Spatial relationships.
    Relationships with parent material.
    Relationships with climate.

    Relationships with organisms.
    Relationships with topography.
    Relationships with time.
    Horizon relationships.
    Property relationships.

SPATIAL RELATIONSHIPS

The distribution of soils on the surface of the earth and their relationships one to the other is one of the chief interests of soil surveyors whose major concern is to present their distribution pattern in the form of maps. Since soils usually grade gradually from one to the other, the ideal method of investigation is to examine exposed transects. Usually this is not possible because of the immense amount of labour and time involved in such operations but some investigations of this type have been conducted when large sections are exposed by road works or similar operations. In a few cases individual investigators have produced large continuous sections but this is exceptional (Mattson and Lönnemark 1939). Normally, pedounits are studied at selected points in the landscape, chosen subjectively or at random. The subjective choice is the usual method but some investigators are now planning their operations on an objective basis choosing their sites either randomly or by using a fixed grid. Both of these methods can yield useful information which can be analysed by computer techniques. They suffer from serious disadvantages which are induced by the limitations of the time at the disposal of the operator. If one adheres strictly to statistical methods the number of samples required to give valid results about small areas of a particular soil are much greater than the time at the disposal of the surveyor. Therefore the error for

such areas may be very large and since some of these small areas may be important intergrading situations they may be missed during this type of investigation. On the other hand with a free survey the choice of sites is left to the discretion of the surveyor and the frequency of inspections is varied according to the variability of the soil.

In most cases the spatial relationships are extrapolations based on these points; therefore they are open to error. Thus sites must be sufficiently near to reduce the error to a minimum and at the same time they must not be so close that valuable time is wasted. Probably the most expedient method of investigation is by means of detailed surveys along random line transects.

For practical purposes the areas of relative uniformity are shown on maps where they are delimited by lines which are placed where the rate of change is most rapid. In reality however one area gradually merges into the other and it is common to find one area of soil merging into two or three others which may themselves merge into other soils or back into a similar soil at some other point forming a repeating pattern over the landscape. When this takes place the pattern is often related to the parent material or topography. Further it is common for the areas of individual soils to have similar shapes which prompted Hole (1953) to suggest a terminology for describing soils as three-dimensional bodies.

Finally it should be pointed out that the areas of apparent uniformity shown on maps may have a wide range of variability and in many cases may have small areas or inclusions of different soils which cannot be shown on the scale of the map in use. Thus it is very difficult to portray accurately the spatial distribution of soils but in the discussions to follow a number of different types of spatial relationships are illustrated.

RELATIONSHIPS WITH PARENT MATERIAL

The influence of parent material is expressed through its composition, permeability and surface area of the particles and is usually most strongly manifest in the initial stages of soil formation. Ultimately soil forming processes tend to minimise and may even nullify the influence of parent material. For example, in areas with intense and prolonged hydrolysis and solution, most rocks are transformed into clay minerals, oxides and resistant residues. Thus the chemical composition of the soil in these cases bears little resemblance to that of the original rock. Often this is important in tropical areas where nutrient rich rocks such as basalts and limestone are transformed to almost sterile soils. However there can be important differences among such soils particularly those that develop from rocks which vary widely in their original composition. This is illustrated in the first example given below. Probably it is in areas of Pleistocene sediments that the chemical composition of the parent material has its strongest influence on soils. In a present-day marine or humid continental cool summer environment the occurrence of altosols, podzols or placosols is determined in large measure by the composition of the sediments which are mainly till, solifluction deposits and loess. Such a relationship is described.

In a marine or humid continental cool summer environment, the permeability of the parent material can be an important factor; on fine textured glacial drift or lacustrine deposits the vertical movement of moisture is restricted so that supragleysols develop. This relationship is the third example described below. One other relationship demonstrating an influence of limestone is given and finally the relationship between geological structure, topography, and soil is illustrated in the fifth example.

1. Krasnozem – zheltozem – podzol relationship

In a rainy tropical environment as in certain parts of south-eastern Asia, fine grained basic rocks including basalt and andesite normally give rise to krasno-zems; intermediate and acid rocks result in zheltozems while podzols develop on Pleistocene deposits which are mainly deep quartzose sands of raised beaches. This is shown diagrammatically in Fig. 121.

2. Altosols – podzols – placosols – peat relationship

These four soils are common in many areas with a marine climate, where their distribution is often determined by the nature of the parent material. Fig. 122 shows the hypothetical relationship in which the character of the drift (parent material) changes gradually from basic on the left to acid on the right and where there is an accompanying change in the soils. Altosols occur on basic drift while podzols and placosols occur on the acid drift. As altosols grade towards podzols the surface organic layers increase in thickness, and gradually become pronounced. The mullon also changes into a modon, passing through intergrade stages. In the intergrading stages the zolon is either poorly developed or exists as small lenses beneath the modon but it is quite prominent in the podzol. The most significant change takes place in the middle position where the alton gradually changes into a sesquon but in some respects the further change to a placosol is more dramatic. The placon develops at some position in the soil and is quickly followed by the formation of an overlying candon because of increased wetness caused by the low permeability of the placon. Where rainfall is sufficiently high, usually over 750 mm the organic layers develop into peat.

These changes in morphology are accompanied by a number of changes in the physical and chemical properties. The natural vegetation associated with these soils varies from deciduous forest on the altosols through coniferous forest on the podzols to heath on the placosols and on the peat.

There is a certain amount of variation in this relationship particularly due to the wide differences in precipitation found in a marine climate. Where the precipitation is fairly low the relationship can shift

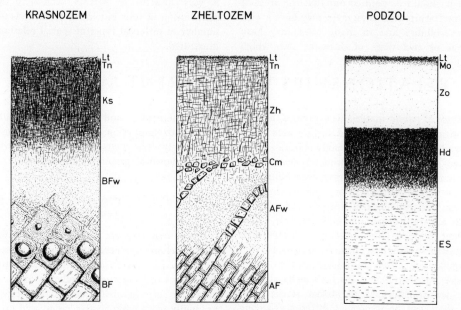

FIG. 121.   Krasnozem – zheltozem – podzol relationship.

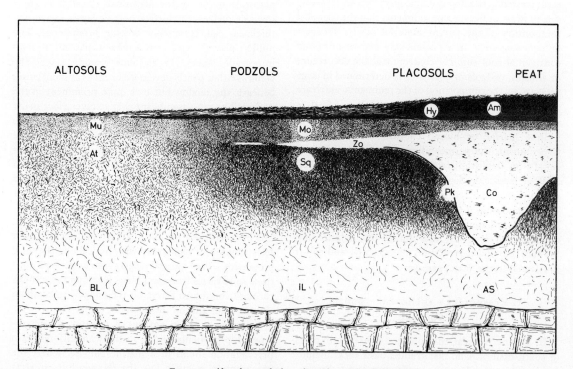

FIG. 122.   Altosols – podzols – placosols – peat relationship.

towards the right with altosols forming on acid parent material. Where the rainfall is high there is a shift towards the left, so that a podzol with a well developed modon and sesquon may develop on intermediate or basic parent material.

A similar relationship can develop through time for there are many situations where altosols have changed into podzols through progressive leaching. Sometimes this process is accelerated by a change of vegetation from deciduous to coniferous species or by an increase in precipitation.

3. Pelosols – supragleysols – peat relationship

Pelosols usually develop in a fairly dry environment as they can be changed quickly into supragleysols either because of an increase in rainfall or due to a slight change in topography. In Fig. 123 is shown the sequence of soils that develops when there is an increase in the surface-moisture status of pelosols due to a change in topography. At the left of the diagram is a pelosol in which the first change is marked by the for-

mation of a marblon or candon immediately beneath the upper mineral horizon which may be a mullon or a luton. This is often accompanied by the formation of manganiferous concretions in the upper part of the pelon as well as in the candon itself.

This stage is a supragleysol which may have the additional feature of tongues extending into the underlying horizons to form the glosson as shown in the diagram. With an increase in wetness the candon increases in thickness and gradually changes into a gleyson. This, in turn changes into a cerulon which is usually underlain by a gleyson or the relatively unaltered material. In cases of extreme wetness, the main upper horizon is an anmooron, hydromoron or an amorphon.

In addition to being found in the same locality due to changes in topography, the various stages of this sequence can be found at points distant from each other owing to differences in precipitation. Supragleysols gradually develop into planosols as hydrolysis increases the clay content of the gleysons which are transformed into planons. Restricted drainage and the formation of supragleysols can also follow the forma-

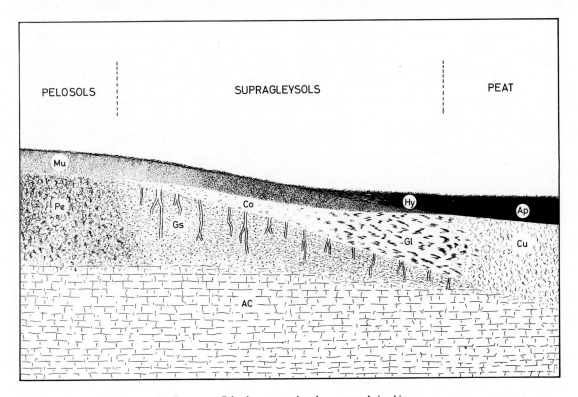

FIG. 123. Pelosols – supragleysols – peat relationship.

tion of a prominent argillon or clamon which reduces permeability.

4. Argillosols – altosols – rendzinas relationship

Avery *et al.* (1959) have described a sequence of soils developed on chalk in the Chiltern Hills of southern England. The chalk was strongly weathered during the Tertiary period to produce a landscape with flat plateaux, steep sided valleys and soils. These soils were possibly flavosols and are now referred to as clay with flints. During the Pleistocene this area lay outside the limits of glaciation, but within the area subjected to repeated periglacial processes, which caused solifluction of the Tertiary soil down the slopes and in many cases completely removed it to expose the underlying chalk. The resulting solifluction deposits which accumulated in the valleys contain a mixture of the old weathered material and chalk. Superimposed upon the whole landscape is a varying thickness of loess which must have been deposited during or prior to the solifluction since much of it is incorporated into the solifluction deposits. Thus, at the end of the Pleistocene period, the flat or gently sloping plateaux and valleys were covered with the truncated remnants of the Tertiary soils and loess. The plateau edges and upper slopes had exposed chalk and in the valleys there was a solifluction mixture of

the old weathered material, loess and chalk. During the Holocene period, soil formation under a marine climate has produced argillosols in the loess overlying the weathered materials, rendzinas on the chalk and altosols on the calcareous solifluction deposits. Locally among the argillosols where the material is more sandy, podzols have developed. The argillosols are possibly the most interesting class of soils in this area since their horizon sequence is influenced by the nature of the material. Where the loess is thick the horizon sequence is fairly normal, but where it is thin, part of the sequence is formed in the loess and part in the underlying clay with flints.

5. Relationships between geological structure and soil pattern

An interesting sequence of soils has been described by Mackney and Burnham (1966) for a part of central England where moderately dipping interbedded strata of limestones, shales and siltstones have produced a scarp landscape with a succession of soils that is related to topography as well as to rock type. This relationship is shown in Fig. 124. The softer shales were weathered and differentially eroded during the Tertiary period leaving the harder limestones and siltstones which due to their dip have produced two escarpments and long, continuous dip slopes. During the Pleistocene period the area was not strongly moulded by erosion or

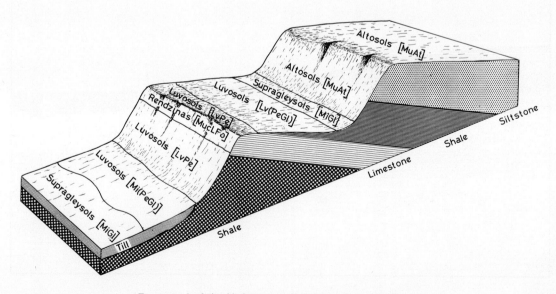

FIG. 124. A relationship between geological structure and soil pattern.

deposition. The valley floors have a covering of till and solifluction deposits but the latter are more widespread on the scarp and dip slopes where they are derived largely from the underlying material. Therefore the parent materials of the soils are strongly influenced by the underlying rock in the upland situation but, in the valleys mainly glacial drift occurs. Altosols are developed on the dip slopes and colluvial slopes of the relatively coarse siltstones. Luvosols are found mainly on the dip slope materials and in colluvium over limestones and shales. Towards the bottom of the slope and on flatter sites the luvosols intergrade to supragleysols which occupy the lower slope position and the floors of the valleys. Thus the resulting soil pattern is determined by the parent material as well as by the slope. A further characteristic of this relationship is the spasmodic occurrence of rendzinas on the limestone escarpment.

RELATIONSHIPS WITH CLIMATE

Perhaps the most obvious and striking relationships are those that exist between climate and soils. Frequently the influence of climate is so strong that many authors consider it as the dominant factor of soil formation, but in any one climate area there are many soils that owe their principal characteristics to other factors therefore this principle can be applied only in a very general way on the macro or continental scale. Nevertheless, climate can be used as a suitable basis for showing geographical distribution patterns and certain relationships between soils, bearing in mind the wide climatic changes that have occurred during the last two million years. Set out in Fig. 125 are the climatic regions according to Critchfield (1966) and a frequency estimation of the principal soils that occur in each of these areas. These estimates should not be treated as absolute because of the general dearth of survey information.

The second soil-climate relationship that is given is an example of the latitudinal changes in soils and climate through continental land masses.

Latitudinal zonation of soils

Possibly the best example of latitudinal change of soil with climate is seen in the great continental transects that stretch from Novaya Zemlya in the north of the U.S.S.R. to the Caspian Sea in the south and from the centre of the Sahara desert to the Congo. These sequences of soils together with their characteristic types of vegetation and climate are shown in Fig. 126.

At the left of the diagram there are cryosols of the northern latitudes with tundra vegetation, tundra climate, low rainfall and prolonged periods of low temperatures. This zone grades into a moister belt with subgleysols and peat, which correspond to the taiga climate where spruces predominate and there is an increase in rainfall and mean annual temperature. The latter is sufficient to increase the annual growth of vegetation but does not augment evapotranspiration sufficiently to cope with the extra precipitation; hence the wetter conditions. Next comes the area dominated by podzols carrying a cover of coniferous vegetation where there is a further increase in precipitation and mean annual temperature. Here evapotranspiration is relatively high so that accumulations of moisture take place only in depressions where subgleysols develop as part of a hydrologic sequence (see page 266 et seq.).

Continuing the transect, precipitation is maintained but there is an increase in temperature so that podzols give way to a complex of argillosols and supragleysols under deciduous or mixed forest. Then comes a narrow belt of brunizems in the tension zone between deciduous forest and the steppes. Further to the south there are chernozems and with increasing aridity are found kastanozems, burozems, serozems and finally the desert. Similarly the vegetation changes from tall grass to short grass then to bunchy grass and finally to xerophytically modified succulents on the serozems. The vegetation changes result from a steady decrease in precipitation accompanied by an equally steady increase in mean annual temperature and increased transmission of radiant energy through the atmosphere, causing an increase in evapotranspiration. The characteristic feature of this part of the transect is the steady decline in the content of organic matter in the soils and the equally steady increase in the amounts of soluble salts and carbonates. These reach very high proportions in the areas of burozems where solonchaks, solonetzes, and solods are found as parts of soil complexes. In a number of depressions large amounts of salts may accumulate and the soil surface develops small characteristic puffs or takyrs.

To the south of the desert both the precipitation

and the mean annual temperature increase rapidly but because of interception the amount of radiant energy reaching the soil surface is much less than in the desert. Here the beginning of the sequence starts with serozems, in complexes with solonchaks, solonetzs, and solods. Thereafter there are marked differences

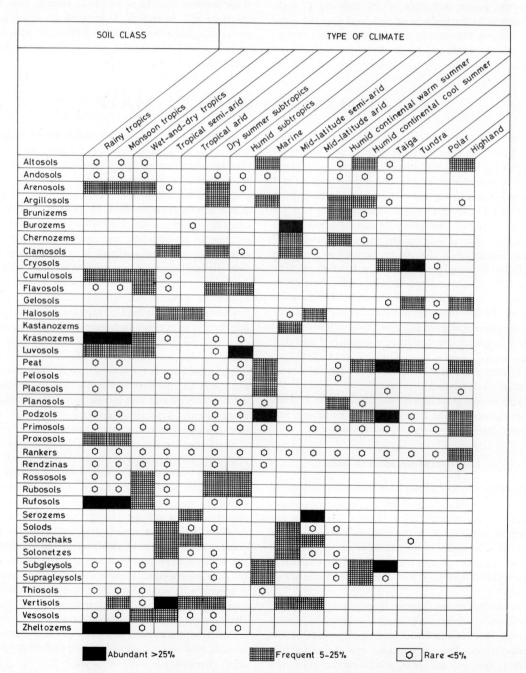

FIG. 125. The frequency distribution of the major soil classes in the main climatic areas.

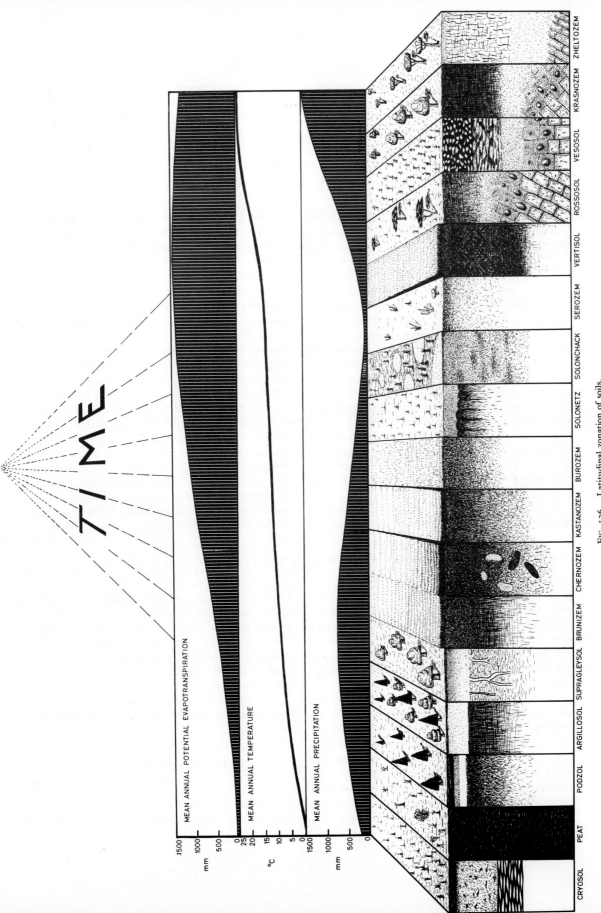

FIG. 126. Latitudinal zonation of soils.

induced by climate and age, for the soils are much older and more highly weathered than those to the north. Rossosols and vertisols form the next part of the sequence which then continues with krasnozems and eroded krasnozems (see page 276) so that hardened vesons occur at the surface over large stretches of landscape (see page 277). Finally zheltozems dominated by tropical rain forest occupy the areas with the highest rainfall.

RELATIONSHIPS WITH ORGANISMS

The influence of organisms is usually linked with that of climate which strongly influences the character of the world's flora and fauna. It is in transition zones that the nature of the plant community has a strong influence on the character of the soil. Perhaps one of the best illustrations of this is seen in the marine environment of north-western Europe where an altosol may be maintained under oak forest but if plant communities dominated by heather (*Calluna vulgaris*) or Scots pine (*Pinus sylvestris*) become established the soils may be changed to podzols by the increased hydrolysis and downward translocation of material caused by the acid litter from these species. Another interesting transition zone lies between the argillosols and chernozems in central Russia. In this zone brunizems develop beneath deciduous forest but there is a delicate balance and intricate interfingering of these three soils, for if trees invade the grassland brunizems develop. The balance can be pushed in the other direction if grasses replace trees.

A conspicuous effect of a change in soil fauna is found in Canada where earthworms were introduced into an area of argillosols with well developed luvons. The earthworms found favourable conditions for their growth in these soils and multiplied rapidly so that their normal churning activity soon mixed the upper and middle horizons.

The redistribution of material by termites to form krasnons is another example which has already been discussed on page 197 *et seq*. Thus it would appear that quite different soils will develop in the presence or absence of mixing organisms. This is a very important aspect of soil development and may help to account for the presence of well differentiated horizons in soils that have no worms or other churning organisms.

A dramatic example of the influence of vegetation is seen in New Zealand where the Kauri pine (*Agathis australis*) produces a very acid litter which causes podzol formation to proceed more vigorously near the tree so that the zolon becomes very thick beneath the tree and forms the so-called 'egg cup' podzol. A similar phenomenon develops in some luvosols of south-eastern Asia where a thick luvon is associated with a species of casuarina (*Gymnostoma nobile*).

RELATIONSHIPS WITH TOPOGRAPHY

Differences in topography influence the soil in many ways. Firstly the amount of moisture entering the soil system usually increases from the top to the bottom of slopes so that valleys and depressions are often wet for long periods of the year. This is illustrated below in the podzols-subgleysols-peat relationship and in the solonetzes-solods relationship.

When there are great differences in elevation between the valley floor and the highest points in the landscape there are differences in climate which can produce altitudinal sequences in the soils and vegetation such as that given below in the third example which also illustrates the influence of aspect.

In polar areas there is often a close relationship between the angle of slope and the nature of the surface phenomena; this is given in the fourth example.

Nearly all landscapes show evidence of erosion and deposition as well as areas of relative stability, thus within any one area there may be soils of widely differing ages and degree of development. This situation is very evident in tropical and subtropical areas such as Nigeria and Australia which supply the fifth and sixth examples that are described.

Topography is being used to an increasing extent in land capability classification which is discussed in the seventh example.

1. Podzols – subgleysols – peat relationship

In areas of fairly strongly undulating topography and uniform parent material it is common to find a

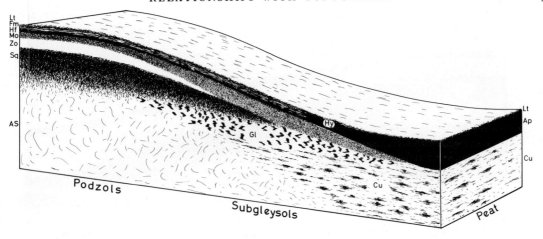

FIG. 127. Podzols – subgleysols – peat relationship.

sequence of soils that becomes progressively wetter on descending the slopes. This is known as a hydrologic sequence and is particularly common in areas of hummocky glacial deposits and in certain areas of stabilised sand dunes.

The sequence shown in Fig. 127 illustrates the morphological changes from a podzol at the top of a moraine through a subgleysol on the slopes to peat in the depression. The increase in wetness is partly caused by surface run-off down the slope but more especially by the water-table which gradually comes closer to the surface on the lower part of the slope. Finally water comes to the surface in the depression between the moraines where peat forms as a consequence (Mattson and Lönnemark 1939).

An interesting hydrologic sequence also exists between chernozems and solonchaks.

2. Solonetzes – solods relationship

Usually there is a simple topographic relationship in which the solonetzes occupy the upper slope positions and the solods the adjoining depression where moisture accumulates and leaching is greatest. In some situations the reverse relationship is found with the solods in the drier knolls and the solonetzes in the depression. This occurs where there is seepage, a high water-table and reduced leaching in the depressions. The normal relationship is shown in Fig. 128.

The most conspicuous feature in the change from a solonetz to a solod is the formation of the luvon which starts with the development of bleached sand

grains on the tops of the columnar ped in the solonetz. Then gradually the luvon gets thicker and the solon occurs deeper in the pedounit. This is followed by a stage during which the upper part of the solon is destroyed to form the luvon which retains vague outlines of the original columnar structure. This indicates an evolution from solonetz to solod. Finally in the depression there is a well developed solod which may have evolved from a solonetz but the evidence for this varies. Alternatively there is the direct development of the solod and only in the transition stage does the luvon expand at the expense of the solon.

3. Altitudinal zonation of soils

Significant changes in soil development take place with increasing elevation but since mountains occur in widely differing environments it is not possible to make many generalisations about the vertical zonation of soils. The example given below is described by Zakharov (1931) who demonstrated for the Caucasus Mountains (Fig. 129) a marked zonation of soils and vegetation which parallel the vertical zonation of climate. To some extent the vertical change in soils is similar to the horizontal latitudinal changes described on page 263 *et seq.* but there are variations induced by aspect and differences in day length. Also, mountain soils are shallower and have a coarser texture. On the warmer south side of the mountains luvosols occupy the lowest positions above which are altosols and brunizems. Serozems occur at the base of the mountains on the north facing slopes and are con-

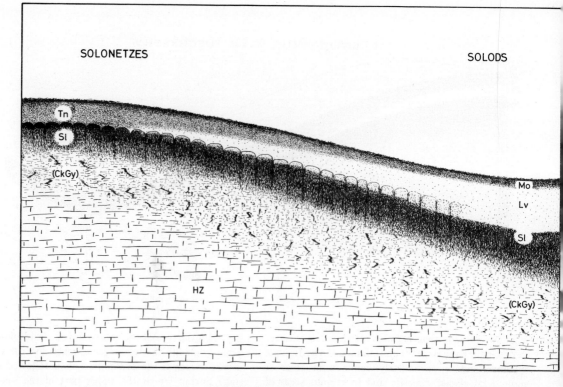

SOLONETZES

SOLODS

Fig. 128. Solonetzes – solods relationship.

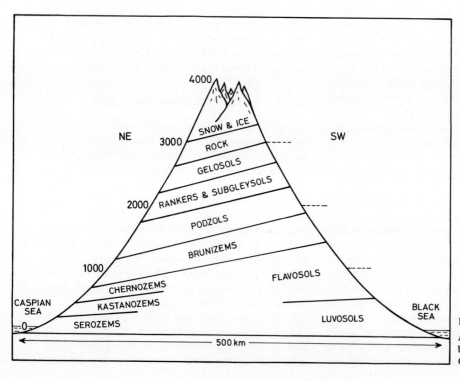

Fig. 129.

Altitudinal zonation of soil between the Black Sea and the Caspian Sea in southern U.S.S.R.

tinuous with those of the plains to the north. With an increase in elevation on the north slopes there is a change to kastanozems, chernozems and brunizems which form the first continuous belt from north to south. With increasing altitude one finds podzols, rankers, gelosols and finally bare rock outcrop. It should be noted that the zonation runs diagonally across the mountain range because the mean annual temperature increases towards the south. Further, the soils on the upper part of the mountains form complete belts whereas the lower ones are found only on one side of the mountain range and in large measure reflect the present differences in climate between the north and south facing slopes – that to the north being generally drier and cooler. The differences in the soils at the lower elevations are due also to the contrasting Tertiary and Pleistocene development that took place on opposite sides of the mountains. The northern side and upper southern slopes were strongly affected by Pleistocene erosion which removed most of the Tertiary soils whereas in the area of the Black Sea there are extensive remnants of the Tertiary soils so that many luvosols have evolved from the krasnozems and zheltozems. Another altitudinal relationship has been described by Thorp (1931).

4. A tundra sequence

Fig. 130 shows a sequence from Svalbard illustrating the relationships that exist between the various surface phenomena, slope and altitude. At the highest elevations the surface is covered with bare angular stones produced by frost action. On the steeper slopes the angular fragments accumulate as scree particularly at the bottoms of the slopes, while differential frost weathering leaves promontories or tors jutting out from the sides of the hills or mountains. This is particularly marked in areas of horizontally bedded sedimentary rocks. Alternating stripes of vegetation and bare soil occur on moderate slopes while at lower elevations on flat or gently sloping situations, there is usually abundant vegetation dominated by mosses, grasses and lichens. Such sites tend to be more stable and encourage cryosols to form. These soils occur as part of a complex which includes mud polygons (Fig. 38), tundra polygons (Fig. 43) and ice wedges (Fig. 44). Occasionally pingos (Fig. 40) are found in this situation and at the lowest position peat may develop if conditions are sufficiently wet. Stone polygons Fig 39 also form part of the tundra catena, developing at all elevations on many flat stony sites.

5. Primosols-rankers-cumulosols-arenosols-primosols relationship

This relationship is of frequent occurrence in certain parts of West Africa and is illustrated by the Iwo association from western Nigeria (Smyth and Montgomery 1962). In this area there was profound

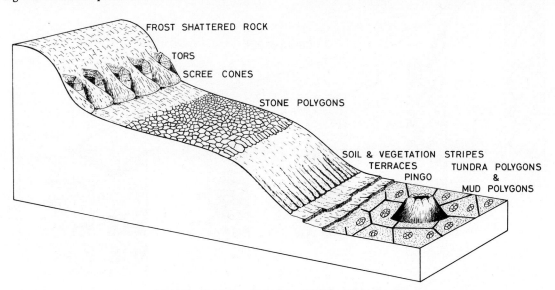

FIG. 130. A tundra sequence showing the relationship between slope and surface phenomena.

deep chemical weathering of the basement granitic gneiss in the Tertiary period followed by a considerable amount of erosion leading to the formation of numerous inselberge. It is possible that weathering and erosion proceeded simultaneously but the latter seems to have been fairly vigorous during the Pleistocene period causing much disturbance and redistribution of the surface soils. In fact most of the middle and upper horizons in the soils of this and other associations in this area have derived many of their characteristic features from the Pleistocene period. This relationship includes the soils developed on inselberge and those of the associated pediment slopes. Also included are the soils of the valley terraces Fig. 131. The highest elevations are inselberge which have primosols and many rock outcrops on their upper surfaces. On their steep sides rock outcrops are dominant but in places there are rankers. On the gently sloping pediments surrounding the inselberge there are cumulosols formed in part by the differential erosion and the concentration of gravel and concretions in the upper part of the soil. Usually at the lower end of the pediment slope there has been the greatest amount of

erosion which may have extended down to the bed rock followed by a deposition of alluvium in the form of terraces in which arenosols have formed. Commonly in this situation lateral movement of water has produced a seepage iron pan or gluton which may occur in each of the terraces forming a conspicuous break in slope. On the lowest terrace subgleysols and primosols are dominant.

6. K-cycles in Australia

In south-eastern Australia there is a succession of buried soils that is attributed to periods of erosion and deposition followed by periods of surface stability and accompanying soil formation. This periodicity is ascribed to changes in climate from humid – when soils formed – to arid – when erosion and deposition was active. Each buried surface has a specific soil which differs from those of all the other surfaces, but in some cases the differences are not at class level; nevertheless they are sufficiently marked to be used for purposes of differentiation and correlation.

Around Canberra, each cycle commenced with

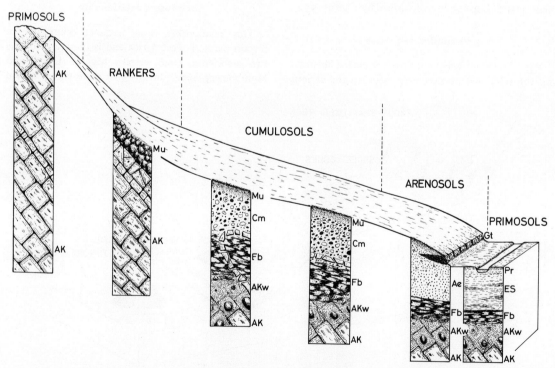

FIG. 131.   Primosols – rankers – cumulosols – arenosols – primosols relationship (adapted from Smyth and Montgomery 1962).

fairly vigorous erosion as indicated by channels containing current bedded material. Later, material sloughed off the upper slopes, filled the channels and buried considerable portions of the land surface, but some surfaces remained relatively unaffected. Thus within any one period of erosion and deposition some of the land surfaces were affected while the rest remained relatively unchanged. However an area that was unaffected during one cycle might be affected during the next. The material of which these deposits are composed is poorly sorted, unbedded and unorientated and since erosion and deposition is never uniform, the thickness of the mantle varies from slope to slope but on any one hillside it is relatively uniform. Together, these processes have produced three zones on the hillsides; there is the upper part from which material moved *en masse* and which therefore carries no buried soils. The second zone or middle slope position gained material during deposition but also tended to lose it during erosion, therefore there are not many buried soils in this position. The lower zone is the region of accumulation where many full sequences of buried soil occur. In some cases the whole soil was removed while in others the existing soil was buried by new material. Surfaces that were not affected by either erosion or deposition experienced changes in climate and processes of soil formation. These latter areas show the effects of several superimposed sets of soil forming processes. The result of all of these widely varying processes is a high degree of pedological and geomorphological complexity.

Generally there are five soils including the present phase, but they are not clearly displayed in all cases. Where remnants of previous soils or superimposed soil formation occur, the true character can be discovered only by means of a lateral traverse to sites where the full soil is preserved or where the two soils become separate due to the deposition of material. These changes from stable periods to erosional phases are termed K-cycles by Butler (1959), cycle K_1 being the youngest. Set out below are the freely drained soils associated with the various cycles in the Canberra area.

K_1 – Brunizems and kastanozems.
K_2 – Clamosols, rubosols and vertisols.
K_3 – Luvosols, rubosols and solods.
K_4 – Planosols and solods.
K_5 – Old weathered remnants.

Cycle K_1 occupies the upper parts of the slopes and is of relatively restricted distribution, while K_2 and K_3 each occupy about one third of the surface and are more extensive as buried soils. K_4 is also present at the surface but is more often buried. K_5 is a flambon and most probably represents highly weathered Tertiary material while the four upper soils are probably Pleistocene in age. Fig. 132 is an idealised cross section showing the surfaces in relation to each other and to slope.

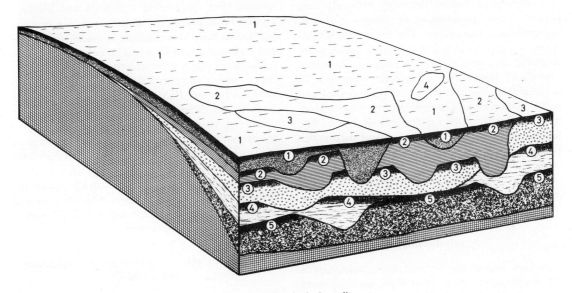

FIG. 132. K-cycles in Australia.

To the west of Canberra, loess (or parna as it is known in Australia) and sand dunes are important elements in the landscape, therefore the buried soils are not confined to lower valley slopes, but also occur on the hill tops.

These landscapes are comprised of many different soils in an intricate mosaic with soils of different ages lying side by side. This contrasts sharply with other land surfaces where differences in the soil pattern may be due only to moisture as in the case of the hydrologic sequences or to parent material. Landscapes with soils of widely differing ages are relatively common; other examples are found in the south-western U.S.A. where fossil luvosols in the desert are associated with serozems and also in certain parts of the central U.S.A. where gumbotil is associated with luvosols and other soils (see page 274 *et seq.*). Other instances are found in areas of repeated mass movement as in certain parts of southern and western Scandinavia where thixotropic marine clays frequently cause catastrophic landslides.

7. Land systems

In recent years great emphasis has been given to making quantitative assessments of topography leading to the concept of 'Land Systems' introduced by Christian and Stewart (1953) and Gibbons and Downes (1964). In each land system there is a recurring pattern of topography, soils and vegetation and is somewhat similar to the concept of the 'Catena' introduced by Milne (1935). This approach is being applied by a number of large organisations concerned with producing maps upon which are based interpretations about land utilisation in its broadest sense. Very briefly the system attempts to divide the land surface into a number of topographic facets and to describe each using parameters that include, soil, elevation, angle of slope, length of slope and curvature of the surface.

With this technique it is possible to make a quantitative statement about these areas without invoking the speculations that arise when interpretations are based on process. Many of the facets that are recognised are similar to those geomorphological features recognised by the more conventional methods with the addition of some quantitative assessment.

RELATIONSHIPS WITH TIME

Undoubtedly, the most fascinating relationships are those that are related to the age of soils. These vary from relatively simple developmental sequences to the very complex evolution involving extremely long periods of time accompanied by changes in climate, vegetation and topography. The first example illustrates a fairly simple development sequence from relatively unaltered material to podzols. The second example illustrates an evolutionary sequence in Scotland from the Tertiary period through the Pleistocene to the present. The third example shows a somewhat similar sequence for the central U.S.A. The fourth example attempts to show some of the complex evolutionary developments encountered in tropical areas. Finally a hypothetical example is given of the diversification and changes in the nature of the soil in an area of landscape in a marine environment.

1. Primosol – ranker – podzols relationship

Perhaps one of the simplest age relationships is the development of podzols from unconsolidated siliceous sands as shown diagramatically in Fig. 133.

Initially there is the primosol stage which is quickly followed by an accumulation of litter at the surface and the formation of a fermenton, humifon and modon. With progressive hydrolysis and leaching, a zolon forms and at the same time humus, iron and aluminium begin to accumulate in the middle position and initiate the formation of a husesquon. This is the ranker stage which starts with the first fully developed upper horizon and terminates when a part of the acid sand reaches the (HsAS) stage. Further development leads to the differentiation of a husesquon and an underlying sesquon and is sometimes accompanied by an increase in the thickness of the zolon at the expense of the upper part of the husesquon which is destroyed. The soil may remain in this stage, which is of frequent occurrence at present, or there may be a gradual hardening of the husesquon as more material is washed in. This latter stage is common in older soils particularly those of cool, moist, oceanic areas.

2. Tertiary soil – cryosol – podzol – placosol relationship

It is now generally recognised that within many

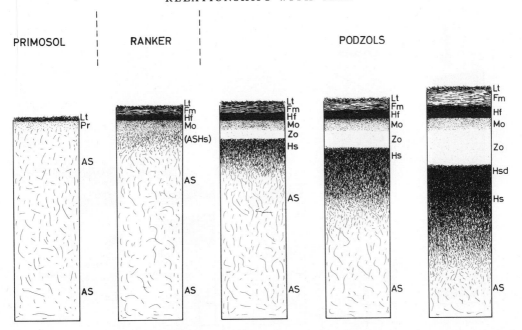

FIG. 133.   Primosol – ranker – podzols relationship.

areas that were glaciated during the Pleistocene period, there are many places that were not intensively glaciated due either to a thin cover of ice or probably to a single glaciation. In these situations it is possible to trace the evolution of the soil from the Tertiary period through the Pleistocene period into the Holocene period and up to the present. The evolution sequence of Placosol Charr illustrated in Fig. 134 probably exemplifies this sequence better than any other soil. Stage 1 shows the reconstructed Tertiary soil at the surface, passing down into chemically weathered rock which has core stones at its base and finally into granite. Stage 2 shows the Tertiary soil truncated by glacial erosion and a deposit of till rests on the remnants of weathered rock. In addition a cryosol has developed, transforming the lower part of the till and the underlying material into a cryon. It should be noted that the new total thickness is less than at Stage 1 indicating a net reduction in the level of the surface. During a single or successive tundra period solifluction and plastic flow removed the glacial deposits causing the weathered rock to come close to the surface so that the cryosol was developed entirely in the old Tertiary material as shown at Stage 3 which also shows oriented stones and solifluction arcs. With the amelioration of climate a podzol with an ison

developed, and below it there are the inherited arcs and oriented stones. With a change to cooler and moister conditions of the Atlantic period a placon formed within the podzol as shown at Stage 4. This new horizon caused the accumulation of moisture at the surface leading to the formation of a candon and peat. Somewhat similar sequences have taken place on a variety of other rock types; a common type occurs on basic or intermediate rocks but this usually terminates at Stage 3 as an altosol.

3. Pleistocene succession in the central U.S.A.

Repeated oscillations from cold glacial periods to warm interglacial conditions during the Pleistocene have produced a most fascinating and complex sequence of soil development in the central part of the United States of America, where there is evidence of four major glaciations. During each glaciation, erosion and deposition were the dominant processes, whereas during the warm interglacial periods intense weathering prevailed and deep soils developed. In the poorly drained positions, weathering produced thick clay soils which have been called gumbotil by Kay (1916); a term derived from the African word 'gumbo' which refers to the mucilaginous content of okra

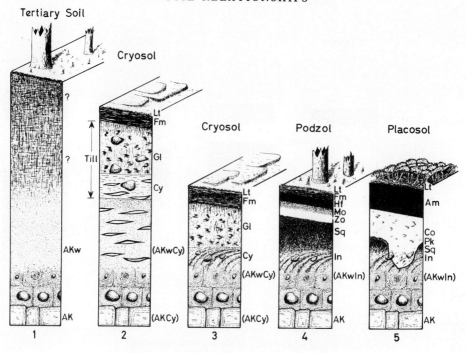

FIG. 134.   Tertiary soil — cryosol podzol — placosol relationship.

pods. Also there was profound sheet erosion during the interglacial periods to produce pediment surfaces (Ruhe *et al.* 1967) and for the accumulation in depressions of thick deposits of clay sediments which were at one time thought to be gumbotil but their sedimentary characteristics have been demonstrated by Frye *et al.* (1960a) Frye *et al.* (1960b) and Frye *et al.* (1965) who have called them accretion gleys. The freely drained soils of the interglacial periods are not dissimilar to those at the present time and in Illinois they are argillosols but they are relatively infrequent as compared with gumbotil and accretion gleys.

The simplified sequence with approximate dates shown in Fig. 135 starts at the bottom with the Nebraskan glaciation which left a deposit of till, outwash deposits and silts some of which are presumed to be loess. This was followed by an interglacial period and the development of Afton soils which are mainly gumbotil and peat. This initial part of the succession is fairly widespread but it is usually buried deeply beneath deposits of the succeeding glaciations.

The Kansan glaciation was the next, which also left till, outwash deposits and loess followed by the Yarmouth interglacial period and the development of

the Yarmouth soils including argillosols, gumbotil and accretion gleys. Interesting features of this period are the absence of loess overlying the till and the presence of a layer of volcanic ash which has become very important as a stratigraphic marker for the Kansan period. The third glaciation is known as the Illinoian because of its great extent into the state of Illinois where it covered surfaces that were not glaciated previously. Great thicknesses of loess, known as Loveland silts were formed during this period. During the succeeding Sangamonian interglacial period similar soils to those of previous interglacial periods developed but there seems to be a greater extent of freely drained soils from this period.

The final or Wisconsinan glaciation produced great thicknesses of till and loess particularly the latter which blankets almost the complete surface of the central states and buries the Wisconsinan till and Sangamonian soil. However Wisconsinan till is exposed at the surface in the states to the north.

The Wisconsinan period was also characterised by a number of interstadial periods some of which were relatively warm and stable so that soils developed; Chernozem Brady is probably the best known and most fully developed example. This period terminated about

10,000 years BP and during the post glacial time, argillosols, brunizems, supragleysols and planosols have developed on the present land surface.

The extent and distribution of the ice was not always the same so that many areas do not have the full sequence and some have only one layer of till. A significant feature of the glaciations in the central U.S.A. is the somewhat ineffective erosion, which has caused relatively little removal of the interglacial soils. This is due to the weak erosive power on the flat land surface as well as to the fact that these areas were at the margins of the ice sheet. However lumps of till and soil of one glaciation are often incorporated in the deposits of succeeding glaciations. To the north, the erosive power of the glaciers was greater so that pre-Wisconsinan tills and soils become less frequent and are completely absent from the northern parts of Michigan and Wisconsin. Although the tills and their interglacial soils may have been frozen during the maximum of each succeeding

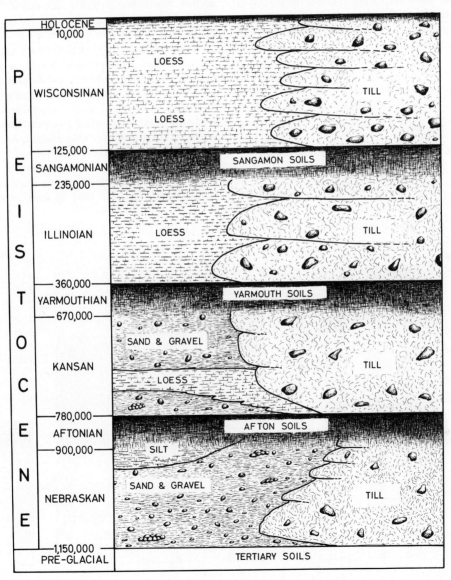

FIG. 135.   Simplified sequence of Pleistocene events in north-central U.S.A.

glaciation, periglacial phenomena and signs of solifluction of these materials are rare. Only during the final
phase of the Wisconsinan is there evidence for
periglacial processes and this is extensive only in
the northern states such as Wisconsin (Black 1964).

It should be pointed out that the interglacial soils
were not confined to the deposits of the preceding
glacial period but formed on any exposed surface.
In some places strongly developed luvosols that occur
in the southern states can be traced northward beneath
Wisconsinan loess indicating that soil formation in
the south has proceeded almost without interruption
for a long period and also that the interglacial periods
were relatively warmer.

In addition, there are places where soil formation
of one interglacial period was superimposed upon the
soils of the previous interglacial, the most common
is the development of a Sangamonian soil in a Yarmouth soil.

4. Rock – ranker – altosol – krasnozems – vesosol relationship

It is not possible to state with certainty the various
stages through which krasnozems pass because the land
surfaces on which they are found are very old and there
may have been many contrasting phases of pedogenesis, accompanied by changes in topography as
weathering and erosion proceeded. Therefore the
various stages given below in Fig. 136 should be
regarded as schematic. At the outset there is the fresh
rock surface produced by erosion, uplift or by volcanic
activity as shown at Stage 1. Rankers develop quickly
(Stage 2) in small depressions and pockets in the rock
surface where fine particles accumulate. Thereafter
development involving the progressive hydrolysis
of the rock is very slow but there is a gradual change
to an altosol (Stage 3) in which there are many primary
minerals. With further development, a krasnon is
formed followed by a thickening of the krasnon and the
layer of weathered rock (Stage 4). Accompanying the
development of the soil there is natural erosion which
gradually lowers the surface but weathering and soil
formation proceed faster than erosion. At each stage
the original land surface and volume of material
removed is shown by dotted lines. Stage 4 is the end
point on many moderately sloping situations but on
gentle slopes and flat sites at least three further

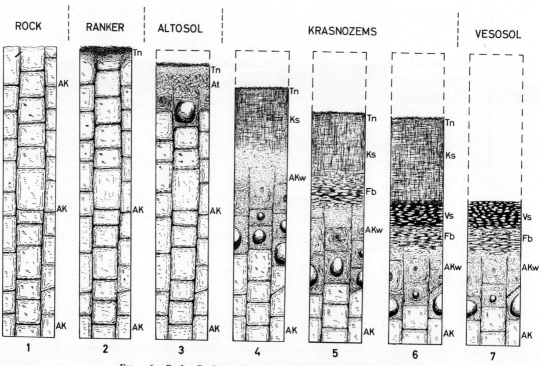

FIG. 136.   Rock – Ranker – altosol – krasnozems – vesosol relationship.

stages can develop. In these situations the lower parts of the soil tend to remain wet for long periods of the year and result in the formation of the characteristic mottled clay or flambon (Stage 5). Alternatively a light coloured kaolinitic pallon may form as is found in Australia. The next stage is the formation of a veson just above or within the upper part of the flambon or pallon. At first the veson forms as separate patches but gradually they increase in size and finally coalesce to form a continuous sheet as shown at Stage 6.

The stages that follow are extremely varied and display one of the most complex and fascinating evolutions to be found in soils. The simplest development that can follow is for erosion to remove the upper horizons and to lay bare the veson which hardens upon exposure to the atmosphere and forms a protective capping which prevents or reduces the rate of further erosion. This is shown at Stage 7 and Fig. 137 is a map showing the distribution of the cappings in various parts of the world. These vesosols are particularly common in arid and semi-arid areas where their

presence is due to climatic change. This is responsible for the erosion through reduced vegetative cover and surface protection. Where erosion cuts through the veson into the underlying flambon or pallon it usually proceeds at an extremely rapid rate often cutting down to the underlying bed rock. Erosion is enhanced if there has been an uplift of the surface and an increase in the gradient. Such a situation is described for western Australia by Mulcahy and Hingston (1961).

In many instances the climatic change has not been very marked so that the type of soil formation has remained unchanged; nevertheless erosion has proceeded. This is shown in Fig. 138 which is taken from the work of Brammer (1962). This sequence starts on the gently sloping upland area with a krasnozem containing a veson. In this case it appears that even this stage may be polygenetic with the upper horizons formed in a sediment and the veson and lower horizons developed from the underlying rock. In such cases it is possible that the development of the veson may have formed subsequent to the main dissection.

The change from the upper surface to the valley

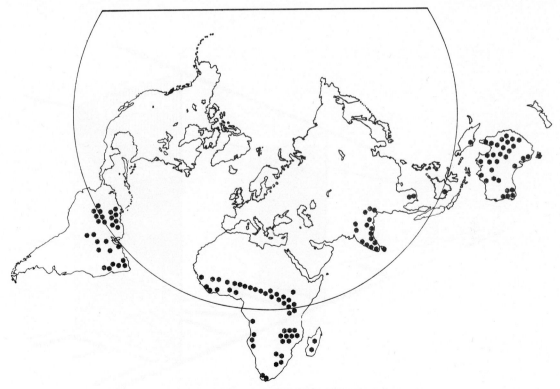

FIG. 137.   Distribution of vesosols occurring at the surface.

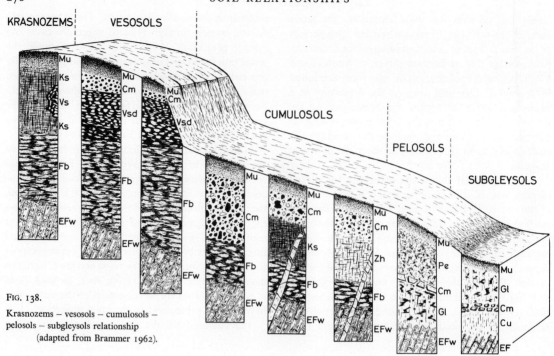

Fig. 138.

Krasnozems – vesosols – cumulosols –
pelosols – subgleysols relationship
         (adapted from Brammer 1962).

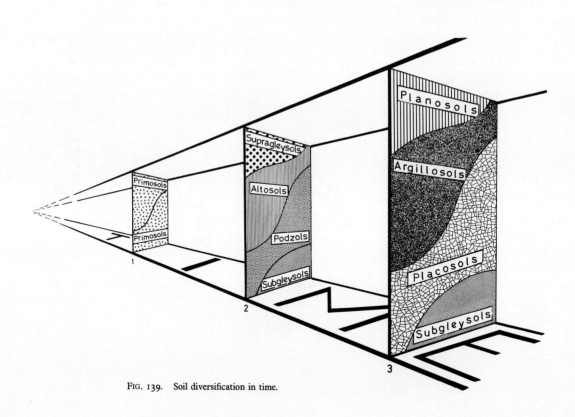

Fig. 139.   Soil diversification in time.

slopes is delimited by a small escarpment formed by the hardened and more resistant veson which comes near to the surface. Below this series there is a well developed cumulon formed by differential erosion of the fine material, resulting in an accumulation of iron concretions and fragments of the veson in the upper part of the soil. The next two series are separated on the basis of colour and are characterised also by a cumulon overlying a krasnon or zhelton which in turn overlies a flambon above weathered phyllite. The lowest sites are occupied by soils that have developed in material removed from above.

These evolutionary sequences help to explain some of the complexity of the landscapes in tropical and subtropical areas.

5. Soil diversification in time

The various soils which comprise any particular area of landscape usually evolve in different directions as determined by differences in the factors of soil formation and variations in the intensity of pedogenic processes. Fig. 139 illustrates three stages in the evolution of a small area of undulating landscape. Stage 1 shows the area as having two different types of parent material, one basic silt and the other acid sand. Also, soil formation has just begun so that primosols occur over the whole area. As soil development proceeds Stage 2 is reached when an altosol has formed on the higher and drier area of basic silt and a supragleysol has formed in the wet depression. Correspondingly a podzol and subgleysol have formed on the acid sand. At Stage 3 which can be regarded as the present time the soils show a further development. The altosol has been changed into an argillosol because the progressive leaching has caused acidification, with clay destruction or translocation downwards. The supragleysol has become planosol because progressive hydrolysis has led to increased clay formation in the gleyson. On the acid sand the podzol has changed into a placosol while peat has accumulated at the surface of the subgleysol.

HORIZON RELATIONSHIPS

It is estimated that within any one area of about 10,000 sq km there may occur about 35–40 different horizons together with numerous intergrades. Set out in Fig. 140 are the principal integrading pathways among the horizons in the soils of Scotland. Since nearly every horizon can intergrade to parent material this aspect is not shown.

In tropical and subtropical environments where the time factor has been greater than elsewhere, horizons intergrade in time through weathering stages as well as in space through different climates so that the end products of weathering vary from place to place. Some of these relationships are presented in Figs. 141, 142, 143. Fig. 141 shows the transformation of an acid rock such as granite under conditions that range from hot humid to hot sub-humid conditions. During its transformation under hot, humid, freely draining conditions the felspars and any ferromagnesian minerals are decomposed, their places being taken by kaolinite, goethite and gibbsite. The amount of quartz and accessories is reduced thus the end product is a typical krasnon. As climatic conditions become drier the gibbsite content diminishes and mica becomes increasingly important. Thus the diagram shows a variety of end products of weathering and a number of developmental stages which contain varying proportions of weatherable minerals. By varying the initial composition of the rock the content of quartz and weatherable minerals in the developmental stages is also varied with consequent changes in the end products.

While Fig. 141 shows only one set of end products for the areas of higher precipitation, Fig. 142 attempts to show the variability that is possible on account of variations in local conditions. The amounts of kaolinite, gibbsite and goethite are to be regarded as ratios and not absolute amounts since there are usually variable amounts of quartz and other resistant minerals. This is a theoretical diagram therefore it is likely that the end members do not exist but most of the other combinations appear to be possible with area A, including the most common combinations. The frequency of horizons with small amounts of kaolinite is low but they do exist therefore it may be necessary to create horizons to accommodate these situations. Alternatively one can construct a similar diagram to show the continuous variability that exists between quartz, weatherable minerals and weathering products.

Fig. 143 is a coordinate diagram in which an attempt is made to rationalise the situation and to show the relationships between some strongly weathered

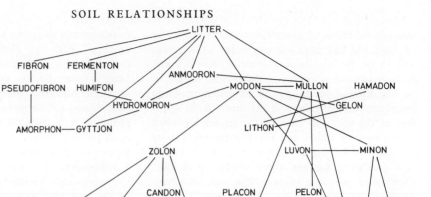

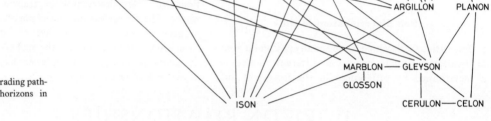

FIG. 140.

Principal intergrading pathways among horizons in Scottish soils.

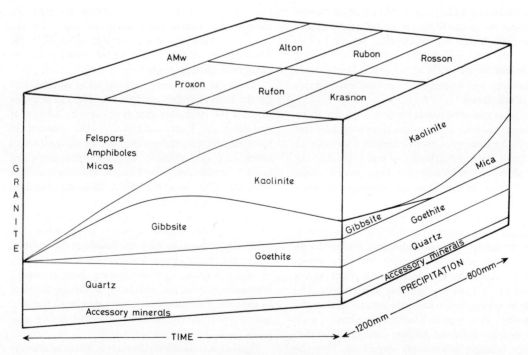

FIG. 141. The relationship between certain horizons in tropical areas as influenced by variations in age and precipitation.

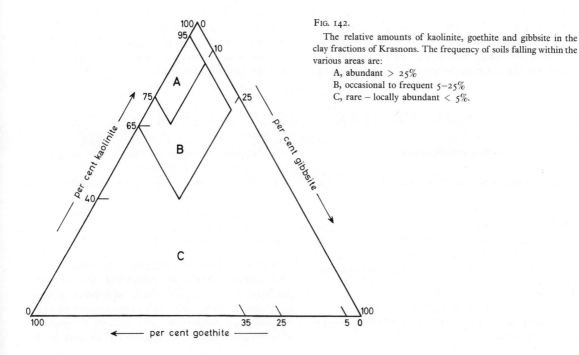

FIG. 142.

The relative amounts of kaolinite, goethite and gibbsite in the clay fractions of Krasnons. The frequency of soils falling within the various areas are:

A, abundant > 25%

B, occasional to frequent 5–25%

C, rare – locally abundant < 5%.

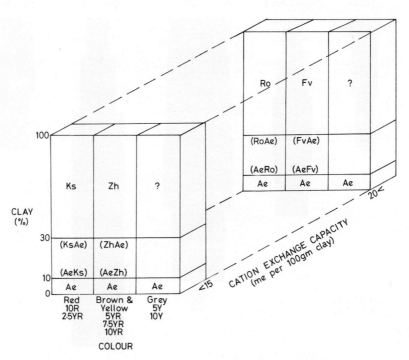

FIG. 143. Horizon relationships in humid tropical areas.

horizons of the tropical and subtropical areas. In this diagram it is assumed that colour, clay content and cation exchange capacity are the three most important coordinates – colour indicating the state of the iron oxides and the cation exchange capacity indicating the type of clay minerals present. It is apparent that the type of clay is not regarded as being very important when the content of sand is high. Being a theoretical diagram it is doubtful if some of these possibilities exist hence the question marks.

SOIL PROPERTY RELATIONSHIPS

Clay relationships

The amount, type, and distribution patterns of clay are probably three of the most important properties of soils for they are used more generally than any of the others as distinguishing and differentiating criteria. In Fig. 144 the quantitative distribution patterns for the clay fraction ($< 2\mu$) are given for a selected number of soils. These clay patterns fall into the following six groups.

1. Gradual decrease with depth – Altosol.
2. Sharp increase followed by a gradual decrease – Argillosol.
3. Gradual increase followed by a gradual decrease – Krasnozem.
4. Sharp increase, then uniform, and then a decrease – Luvosol.
5. Decrease followed by a sharp increase, then a gradual decrease – Solonetz.
6. Uniformly distributed – Vertisol.

Northcote (1960) has suggested that three types of clay pattern should be recognised viz. uniform, gradational, and duplex which correspond to types 6, 3 and 2 above. But this suggestion is an over-simplification since there are at least six patterns of clay distribution. Of particular interest is the

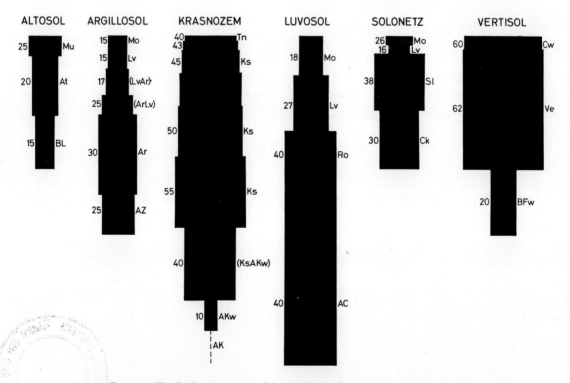

FIG. 144.   The distribution patterns of the clay fraction for a number of selected soils.

fact that as certain soils develop the clay content of the upper horizons becomes less than that of the middle horizons. This is found even in chernozems and kastanozems in spite of vigorous churning, particularly in the latter. This increase in clay with depth has been interpreted by most workers as resulting from translocation but this is not supported by detailed laboratory analyses which indicate that at least two other processes are involved. They are the destruction of clay in the upper horizon and *in situ* weathering in the middle horizon. These three processes may proceed singly, simultaneously or successively; therefore the significance attached to an increase in clay with depth will vary from soil to soil.

There is a need for more precise criteria for the recognition of translocated clay. Some soils may show a greater amount of clay in the middle horizon than above and below, yet show little or no evidence for illuvial clay in the form of cutans. Therefore those soils that contain less than about 5% of clay cutans in the middle horizon are not regarded as having an argillon.

GLOSSARY

ACID SOIL: A soil < pH 7·0.

ACTINOMYCETES: A group of organisms intermediate between the bacteria and the true fungi, mainly resembling the latter because they usually produce branched mycelium.

AEOLIAN: Pertaining to or formed by wind action.

AEOLIAN DEPOSITS: Fine sediments transported and deposited by wind, they include loess, dunes, desert sand and some volcanic ash.

AEROBIC: Conditions having a continuous supply of molecular oxygen.

AEROBIC ORGANISM: Organisms living or becoming active in the presence of molecular oxygen.

ALKALINE SOIL: A soil > pH 7·0.

ALLUVIUM: A sediment deposited by streams and varying widely in particle size. The stones and boulders when present are usually rounded or sub-rounded. Some of the most fertile soils are derived from alluvium of medium or fine texture.

ALLUVIAL FAN OR ALLUVIAL CONE: Sediments deposited in a characteristic fan or cone shape by a mountain stream as it flows onto a plain or flat open valley.

ALLUVIAL PLAIN: A flat area built up of alluvium.

AMMONIUM FIXATION: Adsorption of ammonium ions by clay minerals, rendering them insoluble and non-exchangeable.

ARÊTE: A sharp, jagged mountain ridge.

AUTOTROPHIC ORGANISM: Organisms that utilise carbon dioxide as a source of carbon and obtain their energy from the sun or by oxidising inorganic substances such as sulphur, hydrogen, ammonium, and nitrite salts. The former include the higher plants and algae and the latter various bacteria (cf heterotrophic).

BAD LANDS: An arid region with innumerable deep gullies caused by occasional torrential rain. The distribution and total precipitation are insufficient to support a protective vegetative cover.

BASE SATURATION: The extent to which the exchange sites of a material are occupied by exchangeable basic cations; expressed as a percentage of the cation exchange capacity – me%.

BIREFRINGENT: Alternately bright and dark in cross polarised light.

BOG IRON ORE: A loose nodular or massive ferruginous deposit formed under wet conditions by the oxidation of solutions containing salts of iron by algae, bacteria or the atmosphere. When decomposing organic matter is present, reducing conditions prevail and siderite is formed.

BOULDER CLAY: See Till.

BULK DENSITY: Mass per unit volume of undisturbed soil, dried to constant weight at 105°C. Usually expressed as g/cc.

CALCITE: Crystalline calcium carbonate, $CaCO_3$. Crystallises in the hexagonal system, the main types of crystals in soils being dog-tooth, prismatic, fibrous, nodular, granular and compact.

CATENA: A sequence of soils developed from similar parent material under similar climatic conditions but whose characteristics differ because of variations in relief and drainage.

CATION: An ion having a positive electrical charge.

CATION EXCHANGE: The exchange between cations in solution and another cation held on the exchange sites of minerals and organic matter.

CATION EXCHANGE CAPACITY: The total potential of soils for adsorbing cations, expressed in milligram equivalents per 100g. of soil. Determined values depend somewhat upon the method employed.

CEMENTED: Massive and either hard or brittle depending upon the degree of cementation by substances such as calcium carbonate, silica, oxides of iron and aluminium, or humus.

CHALK: The term refers to either (a) soft white limestone which consists of very pure calcium carbonate and leaves little residue when treated with hydrochloric acid, sometimes consists largely of the remains of foraminifera, echinoderms, molluscs and other marine organisms, or (b) the upper or final member of the Cretaceous System.

CHRONOSEQUENCE: A sequence of soils that change gradually from one to the other with time.

CIRQUE: A large semi-amphitheatre or arm-chair excavation in mountains formed by ice erosion.

CLAY: Either 1) Mineral material < 2μ
2) A class of texture
3) Silicate clay minerals.

CONIFEROUS FOREST: A forest consisting predominantly of cone-bearing trees with needle-shaped leaves: usually evergreen but some are deciduous, for example the larch forests (*Larix dehurica*) of central Siberia. Their greatest extent is in the wide belt across northern Canada and northern Eurasia. Coniferous forests produce soft wood which has a large number of industrial applications including paper making.

COLLUVIUM: Soil materials with or without rock fragments that accumulate at the base of steep slopes by gravitational action.

CORRIE: See cirque.

CUESTA: A ridge, or belt of hilly land which has a gentle dip slope on one side, and a relatively steep escarpment slope on the other.

DECIDUOUS FOREST: A forest composed of trees that shed their leaves at some season of the year. In tropical areas the trees lose their leaves during the hot season in order to conserve moisture. Deciduous forests of the cool areas shed their leaves during the autumn to protect themselves against the cold and frost of winter. Deciduous forests produce valuable hardwood timber such as teak and mahogany from the tropics, oak and beech come from the cooler areas.

DEFLATION: Preferential removal of fine soil particles from the surface soil by wind.

DEFLOCCULATE: To separate or disperse particles of clay dimensions from a flocculated condition.

DELTA: A roughly triangular area at the mouth of a river composed of river transported sediment.

DENITRIFICATION: The biological reduction of nitrate to ammonia, molecular nitrogen or the oxides of nitrogen, resulting in the loss of nitrogen into the atmosphere and therefore is undesirable in agriculture.

DENUDATION: Sculpturing of the surface of the land by weathering and erosion; levelling mountains and hills to flat or gently undulating plains.

DESERT CRUST: A hard surface layer in desert regions containing calcium carbonate, gypsum, or other cementing materials.

DEVONIAN: A period of geological time extending from 280 million – 320 million years B.P.

DOLINE OR DOLINA: A closed depression in a karst region often rounded or elliptical in shape, formed by the solution and subsidence of the limestone near the surface. Sometimes at the bottom there is a sink hole into which surface water flows and disappears underground.

DUNES, SAND DUNES: Ridges or small hills of sand which have been piled up by wind action on sea coasts, in deserts and elsewhere. Barkhans are isolated dunes with characteristic crescentic forms.

DRIFT: A generic term for superficial deposits including till (boulder-clay), outwash gravel and sand, alluvium, solifluction deposits and loess.

DRUMLIN: A small hill, composed of glacial drift with hog-back outline, oval plan, and long axis oriented in the direction of ice moved. Drumlins usually occur in groups, forming what is known as basket of eggs topography.

DRY FARMING: A method of farming in arid and semi-arid areas without using irrigation, the land being treated so as to conserve moisture. The technique consists of cultivating a given area in alternate years allowing moisture to be stored in the fallow year. Moisture losses are reduced by producing a mulch and removing weeds. In Siberia, where melting snow provides much of the moisture for spring crops, the soil is ploughed in the autumn providing furrows in which snow can collect, preventing it from being blown away and evaporated by strong winds. Usually alternate narrow strips are cultivated in an attempt to reduce erosion in the fallow year. Dry farming methods are employed in the drier regions of India, U.S.S.R., Canada and Australia.

ELUVIAL HORIZON: An horizon from which material has been removed either in solution or suspension.

ELUVIATION: Removal of material from the upper horizon in solution or suspension.

EQUATORIAL FOREST OR TROPICAL RAIN FOREST: A dense, luxuriant, evergreen forest of hot, wet, equatorial regions containing many trees of tremendous heights, largely covered with lianas and epiphytes. Individual species of trees are infrequent but they include such valuable tropical hardwoods as mahogany, ebony and rubber. Typical equatorial forests occur in the Congo and Amazon basins and south eastern Asia.

EROSION: The removal of material from the surface of the land by weathering, running water, moving ice, wind and mass movement.

ESKER: A long narrow ridge, chiefly of gravel and sand, formed by a melting glacier or ice sheet.

EVAPOTRANSPIRATION: The total amount of water lost from the soil by the combined processes of evaporation and transpiration.

EXFOLIATION: A weathering process during which thin layers of rock peel off from the surface. This is caused by heating of the rock surface during the day and cooling at night leading to alternate expansion and contraction. This process is sometimes termed 'onion skin weathering'.

FIELD CAPACITY OR FIELD MOISTURE CAPACITY: The total amount of water remaining in a freely drained soil after the excess has flowed into the underlying unsaturated soil. It is expressed as a percentage of the oven-dry soil.

FINE TEXTURE: Containing $> 35\%$ clay.

FIORD: A long, narrow coastal inlet, usually having steep sides. They have been formed by glaciers over-deepening valleys which were previously cut by streams.

FLOOD PLAIN: The land adjacent to a stream built of alluvium and subject to repeated flooding.

FLUVIO-GLACIAL: (see Glacio-Fluvial Deposits).

GEOMORPHOLOGY: The study of the origin of physical features of the earth, as they are related to geological structure and denudation.

GLACIAL DRIFT: Material transported by glaciers and deposited, directly from the ice or from the meltwater.

GLACIO-FLUVIAL DEPOSITS: Material deposited by meltwaters coming from a glacier. These deposits are variously stratified and may form outwash plains, deltas, kames, eskers, and kame terraces (see Glacial Drift; Till).

HALOPHYTE: A plant capable of growing in salty soil.

HARDPAN: An horizon cemented with organic matter, silica, sesquioxides, or calcium carbonate. Hardness or rigidity is maintained when wet or dry and samples do not slake in water.

HETEROTROPHIC ORGANISMS: Those that derive their energy by decomposing organic compounds (cf. autotrophic).

HOLOCENE PERIOD: The period of time extending from 0 – 10,000 B.P.

HORIZON: Relatively uniform material that extends laterally, continuously or discontinuously throughout the pedounit; runs approximately parallel to the surface of the ground and differs from the related horizons in many chemical physical and biological properties.

HUMIFICATION: The decomposition of organic matter leading to the formation of humus.

HUMUS: The well-decomposed, relatively stable part of the organic matter found in aerobic soils.

HYDROLOGIC CYCLE: Disposal of precipitation from the time it reaches the soil surface until it re-enters the atmosphere by evapotranspiration to serve again as a source of pre-cipitation.

HYDROMORPHIC SOILS: Soils developed in the presence of excess water.

ILLUVIAL HORIZON: An horizon that receives material in solution or suspension from some other part of the soil.

ILLUVIATION: The process of movement of material from one horizon and its deposition in another horizon of the same soil; usually from an upper horizon to a middle or lower horizon in the pedounit. Movement can take place also laterally.

IMMATURE SOIL: Lacking a well-developed pedounit.

IMPEDED DRAINAGE: Restriction of the downward movement of water by gravity.

INFILTRATION RATE: The volume of water which will pass into the soil per unit area per unit time.

INSELBERG: (pl. INSELBERGE) A steep-sided hill composed predominantly of hard rock and rising abruptly above a plain; found mainly in tropical and subtropical areas.

INTERGLACIAL PERIOD: A relatively mild period occurring between two glacial periods.

INTERSTADIAL PERIOD: A slightly warmer phase during a glacial period.

ISOTROPIC: Not visible in crossed polarised light.

KAME: A small hill of stratified gravel or sand formed by a melting glacier or ice sheet.

KARST TOPOGRAPHY: An irregular land surface in a limestone region. The principal features are depressions (e.g. dolines) containing thick soils which have been washed off the rest of the surfaces leaving them bare and rocky. Drainage is usually by underground streams.

KETTLE HOLE OR KETTLE: A hollow or depression in a previously glaciated area, probably formed by an ice block which was covered by gravel and subsequently melted, allowing the debris to settle.

LACUSTRINE DEPOSITS: Materials deposited by lake waters.

LANDSLIDE OR LANDSLIP: The movement down the slope of a large mass of soil or rocks from a mountain or cliff. Often occurs after torrential rain which soaks into the soil making it heavier and more mobile. Earthquakes and the undermining action of the sea are also causative agents.

LOESS: An aeolian deposit composed mainly of silt which originated in arid regions, from glacial outwash or from alluvium. It is usually of yellowish brown colour and has a widely varying calcium carbonate content. In the U.S.S.R., loess is regarded as having been deposited by water.

MANGROVE SWAMP: A dense jungle of mangrove trees which have the special adaption of extending from their branches long arching roots which act as anchors and form an almost impenetrable tangle. They occur in tropical and subtropical areas particularly near to the mouths of rivers.

MICRORELIEF: Small differences in relief, including mounds, or pits that are a few metres across and have differences in elevation not greater than about two metres.

MILLIEQUIVALENT: A thousandth of an equivalent weight.

MINERAL SOIL: A soil containing less than 20% organic matter or having a surface organic layer less than 30 cm thick.

MORAINE: Any type of constructional topographic form con-consisting of till and resulting from glacial deposition.

MULCH: A loose surface horizon that forms naturally or may be produced by cultivation and consists of either inorganic or organic materials.

PENEPLAIN: A large flat or gently undulating area. Its formation is attributed to progressive erosion by rivers and rain, which con-tinues until almost all the elevated portions of the land surface are worn down. When a peneplain is elevated, it may become a plateau which then forms the initial stage in the development of a second peneplain.

PERCOLATION (soil water): The downward or lateral movement of water through soil.

PERMEABILITY: The ease with which air, water, or plant roots penetrate into or pass through a specific horizon.

pH SOIL: The negative logarithm of the hydrogen-ion activity of a soil. The degree of acidity (or alkalinity) of a soil expressed in terms of the pH scale, from 2 to 10.

PHYSICAL WEATHERING: The comminution of rocks into smaller fragments by physical forces such as by frost action and exfoliation.

PLAGIOCLIMAX: A plant community which is maintained by con-tinuous human activity of a specific nature.

PLEISTOCENE PERIOD: The period following the Pliocene period, extending from 10,000 – 2,000,000 years BP. In Europe and North America, there is evidence of four or five periods of intense cold, when large areas of the land surface were covered by ice – glacial periods. During the interglacial periods the climate ameliorated and the glaciers retreated.

PLUVIAL PERIOD: A period of hundreds or thousands of years of heavy rainfall.

POLDER: A term used in Holland for an area of land reclaimed from the sea or a lake. A dyke is constructed around the area which is then drained by pumping the water out. Polders form valuable agricultural land or pasture land for cattle.

QUATERNARY ERA: The period of geological time following the Tertiary Era, it includes the Pleistocene and Holocene periods and extends from 0 – 2,000,000 B.P.

RAISED BEACH: A beach raised by earth movement thus forming a narrow Coastal Plain. There may be raised beaches at different levels resulting from repeated earth movement.

REGOLITH: The unconsolidated mantle of weathered rock, soil and superficial deposits overlying solid rock.

ROCHES MOUTONNÉES: Small hills of rock smoothed and striated by glacier on the upstream side and roughened on the downstream side, from which fragments have been plucked during ice movement.

SALINISATION: The process of accumulation of salts in soil.

SELF-MULCHING SOIL: A soil with a naturally formed well aggregated surface which does not crust and seal under the impact of raindrops.

SOIL: The natural space-time continuum occurring at the surface of the earth and supporting plant life.

SOIL HORIZON: See Horizon

SOIL PROFILE: A section of *two dimensions* extending vertically from the earth's surface so as to expose all the soil horizons and a part of the relatively unaltered underlying material.

SOLIFLUCTION: Slow flowage of material on sloping ground, characteristic, though not confined to regions subjected to alternate periods of freezing and thawing.

SOLUM: That part of the soil above the relatively unaltered material.

STRIP CROPPING: The practice of growing crops in strips along the contour in an attempt to reduce run off, thereby preventing erosion or conserving moisture.

TERRACE: A broad-surface running along the contour. It can be a natural phenomenon or specially constructed to intercept run off, thereby preventing erosion or conserving moisture. Sometimes they are built to provide adequate rooting depth for plants.

TERTIARY PERIOD: The period of time extending from 2,000,000 – 75,000,000 B.P.

THERMOPHILIC BACTERIA: Bacteria which have optimum activity between about 45° and 55°C.

THORN FOREST: A deciduous forest of small, thorny trees, developed in a tropical semi-arid climate.

TILE DRAIN: Concrete or pottery pipe placed in the soil or sub-soil to remove excess moisture.

TILL: An intrastratified or crudely stratified glacial deposit consisting of a stiff matrix of fine rock fragments and old soil containing subangular stones of various sizes and composition many of which may be striated (scratched). It forms a mantle from less than 1 m to over 100 m in thickness covering areas which carried an ice-sheet or glaciers during the Pleistocene and Holocene periods.

TILL PLAIN: A level or undulating land surface covered by glacial till.

TOPOSEQUENCE: A sequence of soils whose properties are determined by their particular topographic situation.

TRANSLOCATION: Migration of material in solution or suspension from one horizon to another.

TRIASSIC: A period of geological time extending from 150,000,000 to 190,000,000 years B.P.

TROPICAL RAIN FOREST: see Equatorial forest

VARNISH (Desert): A dark shiny coating on stones in deserts, probably composed of compounds of iron and manganese.

VENTIFACT: A facetted or moulded pebble caused by wind action, usually forms in polar and desert areas. The flat facets meet at sharp angles.

VOLCANIC ASH (VOLCANIC DUST): Fine particles of lava ejected during a volcanic eruption. Sometimes the particles are shot high into the atmosphere and carried long distances by the wind.

WATERLOGGED: Saturated with water.

WATER TABLE: The upper limit in the soil or underlying material permanently saturated with water.

WATER-TABLE (Perched): The upper limit in the soil of a zone of permanent saturation; overlying an unsaturated zone. Usually due to impeded drainage caused by an impermeable layer.

REFERENCES

AFANASEVA, E. A.	1927	In Vilensky, D. G., 1957. *Soil Science*, pp. 488. Moscow.
AFANASIEV, J. N.	1947	The classification problem in Russian soil science. *Russ. Pedol.*, **5**, pp. 51.
ANDERSON, B.	1957	*A Survey of Soils in the Kongwa and Nachingwea districts of Tanganyika.* Univ. of Reading in conjunction with Tanganyika Agricultural Corporation, pp. 120.
AUBERT, G. and DUCHAUFOUR, Ph.	1956	Projet de classification des sols. *Trans. 6th Int. Cong. Soil Sci.*, Paris, Vol. E, 597–604.
AVERY, B. W., STEPHEN, I., BROWN, G., and YAALON, D. H.	1959	The origin and development of Brown Earths on clay-with-flints and Coombe deposits. *J. Soil Sci.*, **10**, 177–195.
BALDWIN, M., KELLOGG, C. E., and THORP, J.	1938	*Soil classification* in *Soils and Men*, (Yearbook of Agriculture, 1938). Washington, U.S.D.A., pp. 1232.
BARLEY, K. P.	1959	Earthworms and soil fertility. *Aust. J. Agric. Res.*, **10**, 171–185.
BASINSKI, J. J.	1959	The Russian approach to soil classification and its recent development. *J. Soil Sci.*, **10**, 14–26.
BIDWELL, O. W., and HOLE, F. D.	1964(a)	Numerical taxonomy and soil classification. *Soil Sci.*, **97**, 58–62.
BIDWELL, O. W., and HOLE, F. D.	1964(b)	An experiment in the numerical classification of some Kansas soils. *Soil Sci. Soc. Amer. Proc.*, **28**, 263–268.
BLACK, R. F.	1964	Periglacial studies in the United States 1959–1963, *Biuletyn Peryglacjalny*, **14**, 5–29.
BLOOMFIELD, C.	1964	*Mobilization and immobilization phenomena in soils. Palaeoclimatology*, Ed. A. E. M. Nairn. Interscience. New York.
BONNEVIE-SVENDSEN, C. and GJEMS, O.	1957	Amount and chemical composition of the litter from larch, beech, Norway spruce and Scots pine stands and its effect on the soil. *Medd. Norske Skogsforsoksv*, **48**, 111–174.
BOWSER, W. E., PETERS, T. W. and KJEARSGAARD, A. A.,	1963	*Soil Survey of the eastern portion of St. Mary and Milk Rivers Development Irrigation Project.* Alberta Soil Survey Report No. 22, pp. 49.
BRAMMER, H.	1962	*Soils.* Chapter 6, Agriculture and Land Use in Ghana. Oxford Univ. Press, 88–126.
BREWER, R.	1960	Cutans: their definition, recognition and interpretation. *J. Soil Sci.*, **11**, 280–292.
BREWER, R.	1964	*Fabric and Mineral Analysis of Soils.* John Wiley & Sons Inc., pp. 470.
BREWER, R. and HALDANE, A. D.	1956	Preliminary experiments in the development of clay orientation in soils. *Soil Sci.*, **84**, 301.
BROWN, G.	1961	*The X-Ray identification and crystal structures of clay minerals.* Editor, Mineralogical Society (Clay Minerals Group), London.
BUTLER, B. E.	1959	*Periodic phenomena in Landscape as a basis for soil studies.* CSIRO., Australia Soil Pub. No. 14, pp. 20.
CHESHIRE, M. V., CRANWELL, P. A., FRANSHAW, C. P., FLOYD, A. J. and HOWARTH, R. D.	1967	Humic acid – II, Structure of humic acid. *Tetrahedron*, **23**, 1668–1682.
CHRISTIAN, C. S. and STEWART, G. A.	1953	*General report on a survey of the Katherine Darwin region*, 1946, CSIRO, Aust. Land Res. Ser. No. 1.
CLARK, F. E.	1967	Bacteria in Soil, in *Soil Biology*, Ed. A. Burges and F. Raw. Academic Press, 15–19.
CLARK, F. W.	1924	*The data of geochemistry* U.S. Geol. Survey Bull., 770.
CLEMENTS, F. E. and WEAVER, J. E.	1924	*Experimental Vegetation*, Carnegie Institution, 192.
CLINE, A. J., WHITNEY, R. S. and RETZER, J. L.	1960	Section IV: Colorado. Guide to Tour III. 7th Inter. Cong. Soil Sci. Madison 25–39.

COFFEY, G. N.   1912   *A Study of the Soils of the United States*, U.S. Dep. Agr. Bureau of Soils Bull., **85**, 114.

CRITCHFIELD, H. J.   1966   *General climatology*. 2nd ed. Prentice-Hall, Englewood Cliffs, New Jersey.

CROWTHER, E. M.   1953   The sceptical soil chemist. *J. Soil Sci.*, **4**, 107–122.

DANCKELMANN, B.   1887   Streuertragstafel für Buchen- und Fichtenhochwaldunger *Z. Forst- u. Jagdw*, **19**, 577–587.

DAVIS, W. M.   1954   *Geographical essays*. Dover Publ., pp. 777.

DE CONINCK, F. and LARUELLE, J.   1964   Soil development in sandy materials of the Belgian Campine. *Soil Micromorphology*, Ed. A. Jongerius, 169–188.

DEER, W. A., HOWIE, R. A. and ZUSSMAN, J.   1966   *An introduction to the rockforming minerals*. Longmans, pp. 528.

DEL VILLAR, E.   1937   *Los suelos de la Peninsula Luso-iberica*. English translation by G. W. Robinson. Madrid, 1937, Murby, London, 1937.

DESPANDE, T. L., GREENLAND, D. J. and QUIRK, J. P.   1964   Role of iron oxide in the bonding of soil particles. *Nature*, London, **201**, 107–108.

DOKUCHAEV, V. V.   1883   *Russian chernozem*, St. Petersburg.

DOUGLAS, L. A. and TEDROW, J. C. F.   1960   Tundra soils of Arctic Alaska. *Trans. 7th Inter. Cong. Soil Sci.*, IV, 291–304.

DUDAL, R.   1965   *Dark Clay soils of Tropical and Subtropical Regions*. F.A.O., Rome. pp. 161.

DURY, G.   1959   *The Face of the Earth*. Pelican Books, pp. 223.

EMERSON, W. W.   1959   The structure of soil crumbs. *J. Soil Sci.*, **10**, 235–244.

ESWARAN, H.   1967   Micromorphology study of a 'cat-clay' soil. *Pedologie*, **17**, 259–265.

EVERETT, D. H.   1961   The thermodynamics of frost damage to porous solids. *Trans. Faraday Soc.*, **57**, 1541–1551.

EYRE S. R.   1963   *Vegetation and Soils, A World Picture*. Edward Arnold pp. 324.

FARNHAM, R. S. and FINNEY, H. R.   1965   Classification and properties of organic soils. *Adv. Agron.*, **17**, 115–162.

F.A.O.   1964   Meeting on the classification and correlation of soils from volcanic ash. Report 14, pp. 169.

FIELDS, M. and SWINDALE, L. D.   1954   Chemical weathering of silicates in soil formation. *N.Z. J. Sci. Tech.*, 36B, 140–154.

FITZPATRICK, E. A.   1956   An indurated soil horizon formed by permafrost. *J. Soil Sci.*, **7**, 248–254.

FITZPATRICK, E. A.   1958   An introduction to the periglacial geomorphology of Scotland. *Scott Geog. Mag.*, **74**, 28–36.

FITZPATRICK, E. A.   1967   Soil nomenclature and classification. *Geoderma*, **1**, 91–105.

FITZPATRICK, E. A.   1969   Some aspects of soil evolution in north-east Scotland. *Soil Sci.*, **107**, 403–408.

FLAIG, W.,   1960   Comparative chemical investigations of natural humic compounds and their model substances. *Sci. Proc. Roy. Dublin Soc.*, **1**, 149–162.

FRASER, G. K.   1933   Studies of certain Scottish moorlands in relation to tree growth. *For. Comm. Bull.*, **15**, pp. 112, HMSO.

FRYE, J. C., SCHAFFER, P. R., WILLMAN, H. B. and EKBLAW, G. E.   1960a   Accretion-gley and the gumbotil dilemma. *Amer. J. Sci.*, **258**, 185–190.

FRYE, J. C., WILLMAN, H. B. and GLASS, H. D.   1960b   *Gumbotil, accretion-gley, and the weathering profile*. Ill. State Geol. Survey, Circular 295, pp. 39.

FRYE, J. C., WILLMANS, H. B. and BLACK, R. F.   1965   *Outline of Glacial Geology of Illinois and Wisconsin. The Quaternary of the United States*. Edited by H. E. Wright and D. G. Frey.

GAULD, J. H.,   1968   A study of four characteristic Malaysian soils. Unpublished thesis for Ph.D. degree. University of Aberdeen.

GEDROIZ, K. K.   1929   Der adsorbierende Bodenkomplex und die adsorbierten Bodenkationen als Grundlage der genetischen Boden-Klassification Kolloidchem. Beihafte, 1–112.

GIBBONS, F. R. and DOWNES, R. G   1964   *A Study of the Land in South-Western Victoria*. Soil Conservation authority, Victoria, pp. 289.

GLENTWORTH, R. and MUIR, J. W.   1963   The Soils of the Country round Aberdeen, Inverurie and Fraserburgh. *Mem. Soil Surv. Great Britain, Scotland*, pp. 371, H.M.S.O.

GLINKA, K. D.   1914   *Die Typen der Bodenbildung*, Berlin.

GRIM, R. E.   1969   *Clay Mineralogy*, 2nd Edition, New York, McGraw-Hill. pp. 596.

GUPPY, E. M., and SABINE, P. A.,   1956   Chemical analysis of igneous rocks, metamorphic rocks and minerals. *Mem. Geol. Surv. Great Britain*, pp. 78, H.M.S.O.

HALLSWORTH, E. G., ROBERTSON, G. K., and GIBBONS, F. R.   1955   Studies in pedogenesis in New South Wales. VII. The Gilgai Soils. *J. Soil Sci.*, **6**, 1–31.

HANDLEY, W. R. C.   1954   Mull and Mor formation in relation to forest soils. *For. Comm. Bull.*, **23**, pp. 115, H.M.S.O.

HARDON, H. F.   1936   *Factoren, die het organische staf-en het stikstofgehalts van tropische gronden beheerschen, Medeleelingen alg.* Proefst Landbouw, 18, Buitenzorg.

HARRISON, Sir J. B.   1933   *The katamorphism of igneous rocks under humid tropical conditions.* Imp. Bureau of Soil Science, Rothamsted Experimental Station, Harpenden, pp. 79.

HILGARD, E. W.   1914   *Soils.* MacMillan Company, New York.

HOLE, F. D.   1953   Suggested terminology for describing soils as three dimensional bodies. *Soil Sci. Soc. Amer. Proc.*, **17**, 131–135.

D'HOORE, J.   1963   *La carte des sols d'Afrique au 1/5,000,000.* Colloque CCTA/FAO, Lovanium (Congo-Leopoldville).

JACKSON, E. A.   1962   *Soil Studies in central Australia: Alice Springs – Hermannsburg – Rodinga Areas.* CSIRO, Australia, Soil Pub., No. 19, pp. 81.

JACKSON, M. L.,   1958   *Soil chemical analysis.* Prentice-Hall Inc., Englewood Cliffs, N.J., pp. 498.

JACKSON, M. L.   1964   *Chemical composition of soils,* in *Chemistry of the soil.* Ed. F. E. Bear, p. 71–141.

JACKSON, M. L. and SHERMAN, G. D.   1953   Chemical weathering of minerals in soils. *Advan. Agron.*, **5**, 219–318.

JENNY, H.   1941   *Factors of Soil Formation.* McGraw-Hill Book Co. Inc., pp. 281.

JONES, T. A.   1959   Soil classification – a destructive criticism. *J. Soil Sci.*, **10**, 196–200.

KAY, G. F.   1916   Gumbotil, a new term in Pleistocene geology. *Science*, **44**, 637–638.

KENDRICK, W. B. and BURGES, A.   1962   Biological aspects of the decay of *Pinus sylvestris* leaf litter. *Nova Hedwigia*, **4**, 313–342.

KING, L. C.   1967   *Morphology of the earth.* Oliver and Boyd, Edinburgh, pp. 726.

KONONOVA, M. M.   1961   *Soil Organic Matter.* 2nd Ed. Pergamon Press, pp. 544.

KOVDA, V. A.,   1946   *Origin and Conditions of Salinified Soils,* 1–2, in Russian. Publ. Akad. SSSR, M–L.

KOVDA, V. A., LOBOVA. YE. V., and ROZANOV, V. V.   1967   Classification of the world's soils. *Soviet Soil Sci.* 851–863.

KUBIËNA, W. L.   1938   *Micropedology*, Collegiate Press Inc., Ames, Iowa.

KUBIËNA, W. L.   1953   *The Soils of Europe.* Murby, London, pp. 317.

KUBIËNA, W. L.   1958   The classification of soils. *J. Soil Sci.*, **9**, 9–19.

LANG, R.   1915   *Versuch einer exakten klassifikaten der böden in klimatischen und geologischen hinsicht. Int. Mitt. Bodenk.*, **5**, 312–346.

LEEPER, G. W.   1956   The classification of soils. *J. Soil Sci.*, **7**, 59–64.

McCALEB, S. B.   1959   The genesis of red yellow podzolic soils. *Soil Sci. Soc. Amer. Proc.*, **23**, 164–168.

MACKNEY, D. and BURNHAM, C. P.   The Soils of the Church Stretton district of Shropshire. *Mem. Soil Surv. Great Britain. England and Wales.* Rothamsted Experimental Station, Harpenden, pp. 247.

MAIGNIEN, R.   1966   *Review of research on laterites.* UNESCO, Paris, pp. 148.

MARBUT, C. F.   1928   A scheme for soil classification. *Proc. 1st. Int. Congr. Soil Sci*, **4**, 1–31.

MATTSON, S. and LÖNNEMARK, H.   1939   The pedology of hydrologic podsol series. *I. Ann. Agr. Coll. Sweden*, **7**, 185–227.

MEYER, A.   1926   Uber einige Zusammenhänge Zwichen Klima und Böden in Europe. *Chem. der Erde*, 290–347.

MILITARY ENGINEERING EXPERIMENTAL ESTABLISHMENT   1965   Description of a land system Report 955.

MILJKOVIC, N.   1965   General review of the salt-affected ('Slatina') Soils of Yugoslavia and their classification. *Agrokemia es Talajtan.* 14 Supplement, 235–242.

MILNE, G.   1935   Some suggested units of classification particularly for E. African soils. *Soil Res.*, **4**, 183–198.

MILNER, H. B.   1962   *Sedimentary petrography.* Vol. I. *Methods in sedimentary petrography*, pp. 643. Vol. II *Principles and application.* pp. 715. Alen and Unwin.

MINISTRY OF AGRICULTURE OF THE USSR.   1964   Short guide to soil excursion, Moscow to Kherson. 8th Inter. Cong. Soil Sci. pp. Kolos, Moscow.

MOORMANN, F. R.   1963   Acid sulfate soils (Cat-clays) of the tropics. *Soil Sci.*, **95**, 271–275.

MÜCKENHAUSEN, E.   1962   The soil classification system of the Federal Republic of Germany. *Trans. Int. Soil Conf., N.Z.*, 377–387.

MUIR, A.   1934   The soils of Teindland state forest. *Forestry*, **8**, 25–55.

MUIR, A,   1961   The podzol and podzolic soils. *Adv. Agron.*, **13**, 1–56.

MULCAHY, M. J. and   1961   *The development and distribution of the soil of the York–Quairading area, Western Australia,*
HINGSTON, F. J.   *in relation to landscape evolution.* CSIRO, Aust. Soil Pub., **17**, pp. 43.

MÜLLER, P. E.   1879   Studies over Skovjord. *Tidsskr. Skovbrug.*, **3**, 1–124.

MÜLLER, P. E.   1884   Studies over Skovjord. *Tidsskr. Skovbrug.*, **7**, 1–232.

MÜLLER, P. E.   1887   *Studien üder die natürlichen Humus formen.* Springer, Berlin. pp. 324.

Munsell Soil Colour Charts.   1954   Munsell Color Co., U.S.A.

NEUSTREUEV, S. S.   1926   *Bull. Geogr. Inst.* Russian. In J. N. Afanasiev, 1927. The Classification problem in Soil
                                 Science'. *Russ. Pedol.*, **5**, pp. 51.

NEILSON, C. O.   1967   Nematoda. In *Soil Biology*. Ed. A. Burges and F. Raw. Academic Press, 197–211.

NORTHCOTE, K. H.   1960   *Factual Key for the recognition of Australian soils.* CSIRO. Aust. Soils Div. Rep.,
                                 pp. 60.

NORTHCOTE, K. H.   1962   The factual classifications of soils and its use in soil research. *Trans. Int. Soil Conf. N.Z.*,
                                 291–297.

OERTEL, A. C.   1961   Pedogenesis of some red brown soils based on trace element profiles. *J. Soil Sci.*, **12**,
                                 242–258.

PELTIER, L.   1950   The geographic cycle in periglacial regions as it is related to climatic geomorphology.
                                 Ann. Assoc. Amer. Geog., **40**, 214–236.

PENCK, W.   1953   *Morphological Analysis of Land-forms.* Translated by H. Czech and K. C. Boswell.
                                 Macmillan & Co. pp. 429.

PENMAN, H. L.   1956   Evaporation: an introductory survey. *Neth. J. Agric. Sci.*, **4**, 9–29.

PÉWÉ, J. L.   1959   Sand-wedge polygons (Tesselations) in the McMurdo Sound Region, Antartica – a
                                 progress report. *Amer. J. Sci.*, **257**, 545–552.

PIPER, C. S.   1947   *Soil and plant analysis.* Adelaide Univ. Press.

RACZ, Z.   1964   The heath and bracken soil of Kordun, *J. Sci. Agric. Res.*, **17**, 5–26. Yugoslavia.

RACZ, Z., SILJAK, M. and 1967   Višeslojni profili na području kontinentalnog krša hrvatske i pitanje porijekla
MALEZ, M   pojedinih horizonta. *Zemljište i biljka*, **16**, 581–590.

RAMANN, E.,   1911   *Bodenkunde.* Berlin.

REEDER, S. W. and ODYNSKY, WM   1965   *Reconnaissance soil survey of the Cherry Point and Hines Creek area.* Alberta Soil Survey
                                 Report No. 23, pp. 102.

REUTEUR, G. VON,   1958   Anwendung genetisch-morphologischer Horizontbezeichnungen. *Dtsch. Akad. Landw.*
                                 Berlin.

RICHTOFEN, F. VON   1886   Führer fur Forschungsreisende. In Glinka, 1914, pp. 23–24.

RILEY, D. and YOUNG, A.   1966   *World Vegetation.* Cambridge University Press. pp. 96.

RIQUIER, J.   1960   Les Phytolithes de Certain Sols Tropicaux et des Podzols. *Trans. 7th Inter. Cong. Soil
                                 Sci.*, IV, 425–431.

RUHE, R. V., DANIELS, R. B.   1967   *Landscape evolution and soil formation in south western Iowa.* USDA, Tech. Bull., 1349,
and CADY, J. G.   pp. 242.

SAITÔ, T.   1957   Chemical changes in beech litter under microbiological decomposition. *Ecol. Rev.*
                                 Japan, **14**, 209–216.

SAITÔ, T.   1965   Microbiological decomposition of beech litter. *Ecol. Rev. Japan*, **14**, 141–147.

SAUSSURE, THÉODORE DE.   1804   *Recherches chimiques sur la végétation*, Paris.

SHARPE, C. F. S.   1938   *Landslides and Related Phenomena, a study of mass-movement of soil and rock.* Columbia
                                 University Press, New York. pp. 137.

SIBIRTSEV, N. M.   in Glinka, 1914.

DE' SIGMOND, A. A. J.   1933   Principles and scheme of a general soil system. *Soil Res.*, **3**, 103–126.

SIMONSON, R. W.   1949   Genesis and classification of red-yellow podzolic soils. *Soil Sci. Soc. Amer. Proc.*, **14**,
                                 316–319.

SIMONSON, R. W. and   1960   Concept and functions of the pedon. *Trans. 7th Int. Congr. Soil Sci.*, **4**, 127–131.
GARDINER, D. R.

SIMONSON, R. W.   1968   Concept of soil. *Adv. Agron.*, **20**, 1–47.

SINGH, S.   1956   The formation of dark-coloured clay-organic complexes in black soils. *J. Soil Sci.*
                                 7, 43–58.

SMYTH, A. J. and   1962   *Soils and land use in central western Nigeria.* Ibadan pp. 265.
MONTGOMERY, R. F.

SOIL SURVEY STAFF.   1960   *Soil classification, a comprehensive system. 7th Approximation.* U.S. Government Printing
                                 Office, Washington D.C., pp. 265.

SOIL SURVEY STAFF. 1960 *Guide to Tour III, 7th Int. Cong. Soil Sci.*, Madison, U.S.A.

SOILEAU, J. M. and McCRACKEN, R. J. 1967 Free iron and coloration in certain well-drained coastal plain soils in relation to their other properties and classification. *Soil Sci. Soc. Amer. Proc.*, **31**, 248–255.

STACE, H. T. C., HUBBLE, G. D., BREWER, R., NORTHCOTE, K. H., SLEEMAN, J. R., MULCAHY, M. J. and HALLSWORTH, E. G. 1968 *A manual of Australian soils.* Rellim Tech. Press, Adelaide.

STEBUTT, A. 1930 *Lehrbuch der allgemeinen Bodenkunde*, Berlin. pp. 293.

STEPHEN, I. 1952 A study of rock weathering with reference to the soils of the Malvern Hills I: Weathering of biotite and granite. *J. Soil Sci.*, **3**, 20–33.

STEWART, V. I. and ADAMS, W. A. 1968 The quantitative description of soil moisture states in natural habitats with special reference to moist soils. In *The measurement of environmental factors in terrestrial ecology*, edited by R. M. Wadsworth. Blackwell Scientific Publications, Oxford and Edinburgh, 161–173.

SWEDERSKI, W. 1931 Untersuchungen über die Gebirgsböden in den Ostkarpaten, I, *Mémoires de L'Institut National Polonais d'Economie Rurale a Pulawy*, **12**, 115–154.

TAVERNIER, R. and SMITH, G. D. 1957 The concept of braunerde (brown forest soil) in Europe and the United States. *Adv. Agron.*, **9**, 217–289.

TAVERNIER, R. and MARECHAL, R. 1962 Soil Survey and soil classification in Belgium. *Trans. Int. Soil Conf. N.Z.*, 298–307.

THORP, J. 1931 The effects of vegetation and climate upon soil profiles in northern and northwestern Wyoming. *Soil Sci.* **32**, 283–301.

THORP, J., CADY, J. G. and GAMBLE, E. E. 1959 Genesis of Miami silt loam. *Soil Sci. Soc. Amer. Proc.*, **23**, 153–161.

THORP, J., JOHNSON, W. M. and REED, E. C. 1951 Some post-Pliocene buried soils of central United States. *J. Soil Sci.*, **2**, 1–19.

TRANSEAU, E. N. 1905 Forest centers of eastern America. *Am. Naturalist*, **39**, 875–889.

U.S. DEPARTMENT OF AGRICULTURE. 1951 *Soil Survey Manual.*

VAN DER SPEK, J. 1950 *Katteklei. Verslag. Landbouwk, Onderzoek*, 56.2: 1–40.

VILENSKY, D. G. 1927 Concerning the principles of a genetic soil classification Contributions to the study of the soils of Ukrania. *Ist. Inter. Cong. Soil Sci.*, **6**, 129–151.

WAKSMAN, S. A. and IYER, K. R. N. 1932 Contribution to our knowledge of the chemical nature and origin of humus. *Soil Sci.*, **33**, 43–69.

WALKER, G. F. 1949 The Decomposition of Biotite in the Soil. *Mineral Mag.*, **28**, 693–703.

WALSCHER, H. L., ALEXANDER, J. D., RAY, B. W., BEAVERS, A. H. and ODELL, R. T. 1960 *Characteristics of soils associated with glacial tills in north eastern Illinois.* Bull. 665, Univ. of Ill. pp. 155.

WARCUP, J. H. 1967 Fungi in Soil, in *Soil Biology.* E. A. Burges and H. Raw, Academic Press, 51–100.

WEBSTER, R. 1965 A catena of soils on the northern Rhodesia plateau. *J. Soil Sci.*, **16**, 31–43.

WEBSTER, R. 1968 Fundamental objections to the 7th Approximation. *J. Soil Sci.*, **19**, 354–366.

WEST, R. G. 1961 Late and postglacial vegetational history in Wisconsin, particularly changes associated with the Valders readvance. *Amer. J. Sci.*, **259**, 766–783.

WHITESIDE, E. P. 1953 Some relationships between the classification of rocks by geologists and the classification of soils by soil scientists. *Soil Sci. Soc. Amer. Proc.*, **17**, 138–143.

WITKAMP, M. and VAN DER DRIFT, J. 1961 Breakdown of forest litter in relation to environmental factors. *Plant and Soil*, **15**, 295–311.

YAALON, D. H. 1959 Weathering reactions. *J. Chem. Education.*, **36**, 73–76.

YAALON, D. H. 1960 Some implication of fundamental concepts of pedology in soil classification. *Trans. 7th, Inter. Cong. Soil Sci.*, IV, 119–123.

ZAKHAROV, S. A. 1931 *A course in Pedology.* Ed. 2 Moscow.

ZAKHAROV, S. A. 1946 Dokuchaev as the founder and organiser of the new science of genetic soil science. *Pedology*, **6**, 361–365.

INDEX